ELECTRICIAN'S
Technical Reference

Theory and Calculations

Online Services

Delmar Online

To access a wide variety of Delmar products and services on the World Wide Web, point your browser to:

http://www.delmar.com/delmar.html

or email: info@delmar.com

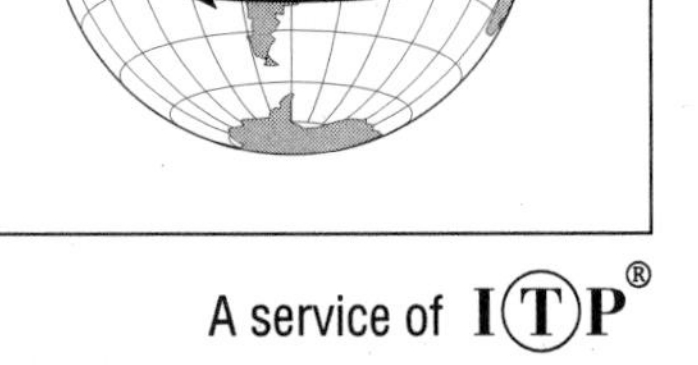

thomson.com

To access International Thomson Publishing's home site for information on more than 34 publishers and 20,000 products, point your browser to:

http://www.thomson.com

or email: findit@kiosk.thomson.com

A service of I(T)P®

ELECTRICIAN'S
Technical Reference

Theory and Calculations

Stephen L. Herman

an International Thomson Publishing company

Albany • Bonn • Boston • Cincinnati • Detroit • London • Madrid
Melbourne • Mexico City • New York • Pacific Grove • Paris • San Francisco
Singapore • Tokyo • Toronto • Washington

Notice to the Reader

Publisher does not warrant or guarantee any of the products described herein or perform any independent analysis in connection with any of the product information contained herein. Publisher does not assume, and expressly disclaims, any obligation to obtain and include information other than that provided to it by the manufacturer.

The reader is expressly warned to consider and adopt all safety precautions that might be indicated by the activities herein and to avoid all potential hazards. By following the instructions contained herein, the reader willingly assumes all risks in connection with such instructions.

The publisher makes no representation or warranties of any kind, including but not limited to, the warranties of fitness for particular purpose or merchantability, nor are any such representations implied with respect to the material set forth herein, and the publisher takes no responsibility with respect to such material. The publisher shall not be liable for any special, consequential, or exemplary damages resulting, in whole or part, from the readers' use of, or reliance upon, this material.

Delmar Staff
Acquisitions Editor: Mark Huth
Developmental Editor: Jeanne Mesick
Production Manager: Larry Main
Art Director: Nicole Reamer
Editorial Assistant: Dawn Daugherty
Cover Design: Nicole Reamer

Printed in the United States of America

For more information, contact:

Delmar Publishers
3 Columbia Circle, Box 15015
Albany, New York 12212-5015

International Thomson Publishing Europe
Berkshire House
168-173 High Holborn
London, WC1V 7AA
England

Thomas Nelson Australia
102 Dodds Street
South Melbourne, 3205
Victoria, Australia

Nelson Canada
1120 Birchmount Road
Scarborough, Ontario
Canada, M1K 5G4

International Thomson Editores
Campos Eliseos 385, Piso 7
Col Polanco
11560 Mexico D F Mexico

International Thomson Publishing GmbH
Konigswinterer Strasse 418
53227 Bonn
Germany

International Thomson Publishing Asia
60 Albert St.
#15-01 Albert Complex
Singapore 189969

International Thomson Publishing—Japan
Hirakawacho Kyowa Building, 3F
2-2-1 Hirakawacho
Chiyoda-ku, Tokyo 102
Japan

1 2 3 4 5 6 7 8 9 10 XXX 04 03 02 01 00 99

Library of Congress Cataloging-in-Publication Data

Herman, Stephen L.
Electrician's technical reference theory and calculations/ by
Stephen L. Herman
p. cm.
Includes index.
ISBN 0-8273-7885-8 (alk. paper)
1. Electric enginerring--Mathematics. I. Title. 97-16485
TK153.H43 1999 CIP
621.3--dc21

Contents

Preface

The Electrician's Technical Reference—Theory and Calculations opens with a section of electrical terms and definitions. This section gives quick explanation to many of the technical terms used throughout the electrical field. Formulas that deal with measurements are shown with the terms.

The remainder of the book gives detailed explanation and examples of how to solve circuit problems dealing with common electrical connections such as series, parallel, and combination circuits. Ohm's law, Kirchoff's law, Thevenin's theorem, and Norton's theorem, and superposition theorem are also discussed. Mathematical calculations are presented in an easy to follow, step-by-step procedure.

Alternating current circuits that contain elements of resistance, inductance, and capacitance are also discussed. Detailed explanation of phase angle, power factor, and power factor correction is given for both single-phase and three-phase power systems. The *Q* of components, circuit resonance, and bandwidth is explained.

The Electrician's Technical Reference—Theory and Calculations is a must-have book for electricians who desire to know the hows and whys of their business.

Safety

Any study of electricity should begin with thoughts of safety. Electricity is a form of pure energy and has the potential to cause great harm or even death to individuals who come in contact with it. Many organizations, such as the National Fire Protection Association, publishers of the National Electrical Code, and OSHA, a government agency concerned with safety in all types of fields, are concerned with the safe handling of electrical equipment, and almost every large factory has strict safety procedures for the handling of electrical equipment. In many of these plants, failure to follow established safety procedures will result in the loss of your job. Safety is a serious matter. It is not to be taken lightly or disregarded.

Regardless of the safety policies or regulations, safety is ultimately **your** responsibility. No one can guard your life and well-being as well as you can. Five general rules concerning safety should be followed at all times.

General Safety Rules

Think

Of all the rules concerning safety, this one is probably the most important. No amount of safe-guarding or "idiot-proofing" a piece of equipment can protect a person as well as the person taking time to think before acting. Many electricians have been killed by supposedly "dead" circuits. Do not depend on circuit breakers, fuses, or someone else to open a circuit. Test it for yourself before you touch it. If you are working on high-voltage equipment, use insulated gloves and meter probes designed to be used on the voltage being tested. When electrical equipment is being serviced, most plants require the use of lock-out tags and/or padlocks. When possible, remove the fuses and take them with you when working on a piece of equipment. Your life is your own, so *think* before you touch something that can take it away from you.

Avoid Horseplay

Jokes and horseplay have a time and place, but the time or place is not when someone is working on a live electrical circuit or a piece of moving machinery. Another extremely dangerous practice is charging a capacitor and then handing it to someone. Capacitors have the ability to supply an almost infinite amount of current for a short period of time. For example, if a capacitor is charged to 1000 volts and then handed to someone, when the person touches across the capacitor terminals there is no difference in touching a 1000-volt line with infinite current capacity for a fraction of a second. If this current should flow from hand to hand through the heart, there is real danger that it can cause a person's heart to go into fibrillation. After all, the operating principle of a defibrillator is to discharge a charged capacitor through the heart. Do not be the cause of someone being injured or killed, and do not let someone else be the cause of you being injured or killed.

Do Not Work Alone

This is especially true when working in a hazardous location or on a live circuit. Have someone with you to turn off the power or give first aid. One of the effects of electrocution is that it causes breathing difficulty and often causes the heart to go into fibrillation.

Work with One Hand When Possible

One of the worst conditions for electrical shock is when the current path is from one hand to the other. This causes the current path to be through the heart. A person may survive a severe shock between the hand and one foot that would otherwise cause death if the current path was from one hand to the other. Another dangerous situation is when a current path exists through the brain. This can be caused by someone accidentally bumping into an overhead energized circuit. Electricians should always wear both a safety helmet made of a good insulating material and shoes with insulating soles. Many plants require the use of steel-toed shoes and boots to protect against pinch hazards to the foot. These are always a good safety precaution, but the electrician should make sure that the shoes also have insulated soles. Avoid shoes that have any kind of metal tacks or nails in the soles or heels.

Learn First Aid

Anyone working in the electrical field should make an effort to learn first aid and CPR. This is especially true for electricians who work in an industrial environment and are exposed to high voltages and currents. Often, it is necessary to troubleshoot live circuits or work in areas of extreme cold or heat. In these instances there is not only the danger of electrocution, but also of frostbite or heat exhaustion or heatstroke. Know the area you work in and know the hazards that exist, but most of all know the first-aid procedures for the problems you may encounter.

Occupational Safety and Health Administration

Under the Occupational Safety and Health Act of 1970, the Occupational Safety and Health Administration (OSHA) was created to encourage employers and employees to reduce workplace hazards and implement new or improved safety and health programs; establish separate but dependent responsibilities and rights for employers and employees to achieve better safety and health conditions; maintain a reporting and recordkeeping system to monitor job-related injuries and illnesses; develop mandatory job safety and health standards and enforce them; and provide for the development, analysis, evaluation, and approval of state occupational safety and health programs. The act provides six distinct provisions for protecting the safety and health of federal workers on the job. OSHA also encourages a broad range of voluntary workplace improvement efforts, including consultation programs, training and education efforts, grants to establish safety and health competence, and a variety of similar programs.

OSHA established that there were an average of 12,976 lost workdays and 86 fatalities of electric power generation, transmission, and distribution employees annually. It was also estimated that 1,634 lost workdays and 61 deaths could be prevented annually. To help prevent these lost workdays and deaths, OSHA initiated regulations concerning the training of employees. All of these regulations are too numerous to mention, but some are as follows:

> Beginning August 6, 1991, the training practices of the employer for qualified and unqualified employees shall be evaluated to assess whether the training provided is appropriate to the tasks being performed or to be performed.
>
> 1. All employees who face a risk of electric shock, burns or other related injuries, not reduced to a safe level by the installation safety requirements of Subpart S, must be trained in safety-related work practices required by 29 CFR 1910.331–.335.
> 2. In addition to being training in and familiar with safety-related work practices, unqualified employees must be trained in the inherent hazards of electricity, such as high voltages, electric current, arcing, grounding, and lack of guarding. Any electrically related safety practices not specifically addressed by Sections 1910.331 through 1910.335 but necessary for safety in specific workplace conditions shall be included.
> 3. The training of qualified employees must include at the minimum the following:
>   a. The ability to distinguish exposed live parts from other parts of electric equipment.
>   b. The ability to determine the nominal voltage of live parts.
>   c. The knowledge of clearance and/or approach distances specified in 1910.333 (c).
> 4. During walkaround inspections, compliance officers shall evaluate any electrical-related work being performed to ascertain conformance with the employer's written procedures as required by 1910.333(b)(2)(i) and all safety-related work practices in Sections 1910.333 through 910.335.
> 5. Any violations found must be documented adequately, including the actual voltage level.

Interpretive Guidance

The following guidance is provided relative to specific provisions of the standard for Electrical Safety-Related Work Practices:

1. Definitions: Qualified/Unqualified Persons.
   a. The standard defines a qualified person as one familiar with the construction and operation of the equipment and the hazards involved. "Qualified

Persons" are intended to be only those who are well acquainted with and thoroughly conversant in the electric equipment and electrical hazards involved with the work being performed.

(1) Whether an employee is considered to be a "qualified person" will depend on various circumstances in the workplace. It is possible and, in fact, likely for an individual to be considered "qualified" with regard to certain equipment in the workplace, but "unqualified" as to other equipment. (See 29 CFR 1910.332(b)(3) for training OSHA Instruction STD 1-16.7 JUL 1, 1991 Directorate of Compliance Programs requirements that specifically apply to qualified persons.) Only qualified persons may place and remove locks and tags.

(2) An employee who is undergoing on-the-job training, who, in the course of such training, has demonstrated an ability to perform duties safely at his or her level of training, and who is under the direct supervision of a qualified person is considered to be a qualified person for the performance of those duties.

b. Where the term "may not" is used in these standards, the term bears the same meaning as "shall not."

c. Training requirements apply to all employees in occupations that carry a risk of injury due to electrical hazards that are not sufficiently controlled under 29 CFR 1910.303 through 1910.308.

2. Scope/Coverage of the Standard

   a. The provisions of the standard cover all employees working on, near or with premises wiring, wiring for connection to supply, other wiring, such as outside conductors on the premises and optical fiber cable, where the fiber cable installations are made along with electric conductors and the optical fiber cable types are those that contain noncurrent-carrying conductive members such as metallic strength members and metallic vapor barriers.

OSHA regulations cover hundreds of elements related to the electrical field and are by far too numerous to discuss in detail. The following is an example of OSHA requirements concerning locking and tagging equipment.

1. If a tag is used without a lock, it shall be supplemented by at least one additional safety measure that provides a level of safety equivalent to that obtained by the use of a lock. Examples of additional safety measures include the removal of an isolating circuit element, blocking of a controlling switch, or opening of an extra disconnecting device.

   Each lock and tag shall be removed by the employee who applied it. However, if the employee is absent from the workplace, then the lock or tag may be removed by a qualified person designated to perform this task, provided that the employer ensures:

   a. That the employee who applied the lock or tag is not available at the workplace, and
   b. That the employee is informed that the lock or tag has been removed before he or she resumes work at the workplace, and
   c. That there is to be a visual determination that all employees are clear of the circuits and equipment prior to lock and tag removal.

Although OSHA was established to help ensure safe working conditions for people in all areas of employment, it is the responsibility of each individual to safeguard his or her own well-being.

Effects of Electric Current

The effect electric current has on the body depends mainly on the amount of current that flows. Some people have the very dangerous misconception that direct current (DC) won't hurt anyone. This misconception originates from the fact that when people touch the terminals of an automobile battery they don't feel any sensation of electrical shock. The reason that they don't feel anything is that 12 volts will generally not push enough *current* through the resistance of a person's body to cause a sensation of shock. The truth is that anytime a current path is provided across a potential, some amount of current will flow. The following list shows different amounts of current flow and the effect they will have on the average person (A = amperes).

0.002–0.003 A	A slight tingling sensation
0.009–0.010 A	Moderate tingling sensation
0.010–0.020 A	Very painful
0.020–0.030 A	Unable to let go of the circuit
0.030–0.040 A	Muscular paralysis
0.040–0.060 A	Breathing difficulty
0.060–0.100 A	Extreme difficulty in breathing
0.100–0.200 A	**DEATH**—This range applied across the heart generally causes fibrillation. When the heart is in this condition, it vibrates at a fast rate like a "quiver" and ceases to pump blood to the rest of the body. Currents above 200 ma. (0.200 A) will often cause the heart muscle to squeeze shut, and normal operation will continue only when the current is removed. This is the principle of operation behind a defribillator.

CHAPTER 1

Atomic Structure

Atoms

Atoms are the basic building blocks of all matter. An atom is the smallest part of an element. There are 103 known elements: 92 are natural and 11 are artificial or manmade. Atoms are composed of three principal particles—the *electron,* the *proton,* and the *neutron.* Protons and neutrons form the nucleus. All atoms, with the exception of hydrogen, contain both protons and neutrons in the nucleus. A hydrogen atom contains one proton and one electron (see Figure 1–1). Electrons orbit around the nucleus.

Electrons

Although electrons are physically larger than protons or neutrons, they are much less massive. The approximate diameter of an electron is 10^{10-15} meters, or 10 fm (femtometers) (femto is an engineering unit equal to 10^{-15}). To better understand the size of an electron as compared to an object with a diameter of 1 meter, compare a time of 10 seconds to approximately 320,000 years.

The mass of an electron is 9.1×10^{-31} kg, and it has a negative electrical charge of -1.6×10^{-19} C (coulomb) (a coulomb is a quantity measure of electrons equal to 6.25×10^{18} electrons).

Protons and Neutrons

Protons and neutrons are extremely massive particles and account for practically all the mass and weight of an atom. Protons have a mass of approximately 1.673×10^{-27} kg and a positive electrical charge of $+1.6 \times 10^{-19}$ C. The proton is about one-third the physical size of the electron, with an approximate diameter of 3 fm, but its combined mass and weight is approximately 1836 times greater than that of the electron.

The neutron has a slightly greater mass than the proton (1.675×10^{-27} kg), but it has no electrical charge. Although all atoms, with the exception of hydrogen, contain both protons and neutrons in the nucleus, they might or might not contain the same number. An example of a beryllium atom is shown in Figure 1–2. The atom contains four electrons and

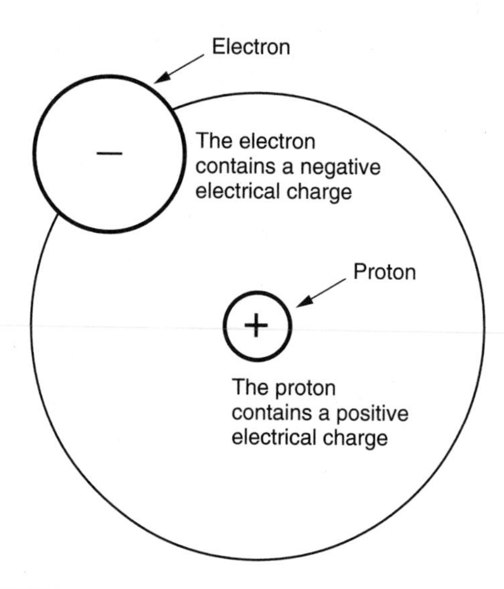

Figure 1–1 The hydrogen atom contains one proton and one electron.

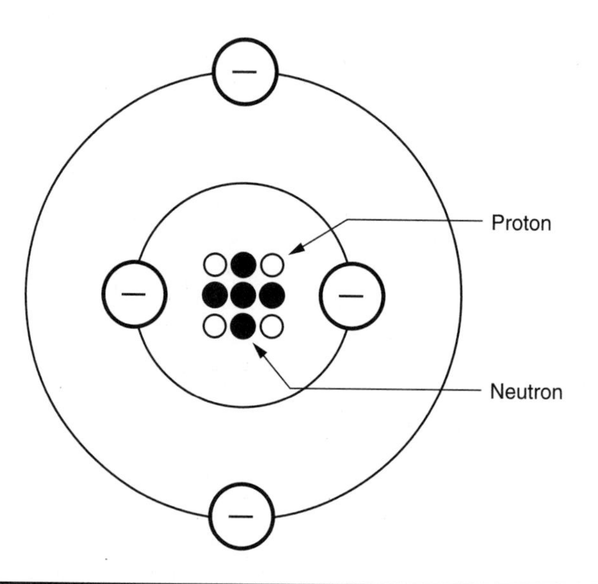

Figure 1–2 The beryllium atom does not contain the same number of protons and neutrons in the nucleus.

the nucleus contains four protons and five neutrons. Large, heavy atoms such as uranium may contain about 1.5 times more neutrons in the nucleus than protons. Although the nucleus of an atom accounts for almost all its mass and weight, the nucleus is only about 1/10,000 the diameter of the average atom. Atoms contain a great amount of empty space.

Electron Orbits

The number of electrons that can be contained in an orbit or energy level of an atom is limited. The formula $2N^2$ can be used to determine the number of electrons that can be contained in a particular orbit or shell. The letter N represents the number of the orbit or shell. The first orbit, or energy level, of an atom can hold a maximum of 2 electrons.

$$2 \times 1^2$$

$$2 \times 1 = 2$$

The second orbit or energy level can contain a maximum of 8 electrons.

$$2 \times 2^2$$

$$2 \times 4 = 8$$

The third orbit can contain a maximum of 18 electrons.

$$2 \times 3^2$$

$$2 \times 9 = 18$$

The maximum number of electrons that can be contained in any orbit or energy level is 32, with the exception of the outer, or valence, shell. The outermost orbit or shell of an atom is called the *valence shell,* and any electrons contained in this shell are called *valence electrons.* The maximum number of valence electrons that an atom may contain is 8. The number of valence electrons an atom has greatly determines its electrical and chemical properties.

Atomic Numbers and Weights

The *atomic number* of an atom is determined by the number of protons in the nucleus. Since atoms are electrically neutral, they will have the same number of electrons as protons. The negative charges of the electrons and the positive charges of the protons cancel each other. The *atomic weight* of an atom is determined by the number of protons and neutrons in the nucleus. If an atom has an atomic weight approximately twice that of the atomic number, the atom contains an equal number of protons and neutrons in the nucleus. If the atomic weight is much greater than twice the value of the atomic number, the atom contains more neutrons than protons in the nucleus. Figure 1–3 presents the atomic number, atomic weight, symbol, and number of valence electrons for different elements.

Atomic Number	Name	Symbol	Valence Electrons	Atomic Weight	Atomic Number	Name	Symbol	Valence Electrons	Atomic Weight	Atomic Number	Name	Symbol	Valence Electrons	Atomic Weight
1	Hydrogen	H	1	1.00797	37	Rubidium	Rb	1	85.47	73	Tantalum	Ta	2	180.948
2	Helium	He	2	4.0026	38	Strontium	Sr	2	87.62	74	Tungsten	W	2	183.85
3	Lithium	Li	1	6.939	39	Yttrium	Y	2	88.905	75	Rhenium	Re	2	186.22
4	Beryllium	Be	2	9.0122	40	Zirconium	Zr	2	91.22	76	Osmium	Os	2	190.2
5	Boron	B	3	10.811	41	Niobium	Nb	1	92.906	77	Iridum	Ir	2	192.2
6	Carbon	C	4	12.01115	42	Molybdenum	Mo	1	95.94	78	Platinum	Pt	1	195.09
7	Nitrogen	N	5	14.0067	43	Technetium	Tc	2	99	79	Gold	Au	1	196.967
8	Oxygen	O	6	15.9994	44	Ruthenium	Ru	1	101.07	80	Mercury	Hg	2	200.59
9	Fluoride	F	7	18.9984	45	Rhodium	Rh	1	102.91	81	Thallium	Tl	3	204.37
10	Neon	Ne	8	20.183	46	Palladium	Pd	–	106.4	82	Lead	Pb	4	207.19
11	Sodium	Na	1	22.9898	47	Silver	Ag	1	107.87	83	Bismuth	Bi	5	208.98
12	Magnesium	Ma	2	24.312	48	Cadmium	Cd	2	112.4	84	Polonium	Po	6	210
13	Aluminum	Al	3	26.9815	49	Indium	In	3	114.82	85	Astatine	At	7	210
14	Silicon	Si	4	28.086	50	Tin	Sn	4	118.69	86	Radon	Rd	8	222
15	Phosphorus	P	5	30.9738	51	Antimony	Sb	5	121.75	87	Francium	Fr	1	223
16	Sulfur	S	6	32.064	52	Tellurium	Te	6	127.6	88	Radium	Ra	2	226
17	Chlorine	Cl	7	35.453	53	Iodine	I	7	126.9044	89	Actinium	Ac	2	227
18	Argon	A	8	39.948	54	Xenon	Xe	8	131.3	90	Thorium	Th	2	232.038
19	Potassium	K	1	39.102	55	Cesium	Cs	1	132.905	91	Protactinium	Pa	2	231
20	Calcium	Ca	2	40.08	56	Barium	Ba	2	137.34	92	Uranium	U	2	238.03
21	Scandium	Sc	2	44.956	57	Lanthanum	La	2	138.91					
22	Titanium	Ti	2	47.9	58	Cerium	Ce	2	140.12					
23	Vanadium	V	2	50.942	59	Praseodymium	Pr	2	140.907	Artificial Elements				
24	Chromium	Cr	1	51.996	60	Neodymium	Nd	2	144.24	93	Neptunium	Np	2	237
25	Manganese	Mn	2	54.938	61	Promethium	Pm	2	145	94	Plutonium	Pu	2	244
26	Iron	Fe	2	55.847	62	Samarium	Sm	2	150.35	95	Americium	Am	2	243
27	Cobalt	Co	2	58.9332	63	Europium	Eu	2	151.96	96	Curium	Cm	2	247
28	Nickel	Ni	2	58.71	64	Gadolinium	Gd	2	157.25	97	Berkelium	Bk	2	247
29	Copper	Cu	1	63.546	65	Tebrium	Tb	2	158.924	98	Californium	Cf	2	251
30	Zinc	Zn	2	65.37	66	Dysprosium	Dy	2	162.5	99	Einsteinium	E	2	254
31	Gallium	Ga	3	69.72	67	Holmium	Ho	2	164.93	100	Fermium	Fm	2	253
32	Germanium	Ge	4	72.59	68	Erbium	Er	2	167.26	101	Mendelevium	Mv	2	256
33	Arsenic	As	5	74.9216	69	Thulium	Tm	2	168.934	102	Nobelium	No	2	254
34	Selenium	Se	6	78.96	70	Ytterbium	Yb	2	173.04	103	Lawrencium	Lw	2	257
35	Bromine	Br	7	79.909	71	Lutetium	Lu	2	174.97					
36	Krypton	Kr	8	83.8	72	Hafnium	Hf	2	178.49					

Figure 1–3 This table of elements shows the atomic number, atomic weight, symbol, and valence electrons for each element.

CHAPTER 2

Electrical Measurements and Terms

This chapter is an alphabetical listing of terms used in measuring electricity. Knowledge of the descriptions and equations in this chapter will be useful in following chapters.

Admittance (Y or y)

Admittance is the opposite of impedance. It can be computed by taking the reciprocal of the impedance value ($Y = \frac{I}{Z}$). The admittance value incorporates all values of conductance and susceptance in an alternating-current circuit. Admittance is measured in mhos (*ohms* spelled backward) or siemens.

Alternating Current (AC)

Alternating current reverses its direction of flow at regular intervals. Alternating-current waveforms may vary, such as square wave or triangle wave, but the most common of all the alternating-current waveforms is the sine wave (Figure 2–1.) The sine wave is produced by all rotating armatures, such as the one used in an alternator. One complete sine wave contains 360 degrees. The maximum positive voltage is reached at 90 degrees, the waveform returns to zero volt at 180 degrees, maximum negative voltage occurs at 270 degrees, and the voltage returns to zero at 360 degrees. Each complete 360-degree wave is called a cycle. The number of complete cycles per second is called the frequency. Frequency is measured in hertz (Hz). The standard frequency throughout the United States and Canada is 60 Hz.

Ampere (A) or Intensity (I)

The ampere is a flow rate of electric current defined as 1 coulomb per second. The ampere, or amperage, describes the actual amount of electrons flowing through a complete circuit. The Chicago International Electrical Congress of 1893 defined the ampere as

> that unvarying current, which, when passed through a solution of nitrate of silver in water in accordance with standard specifications, deposits silver at the rate of one thousand one hundred and eighteen millionths (0.001118) of a gram per second.

This definition was legalized by an act of U.S. Congress in 1894.

In 1948 the Conférence Générale des Poids et Measures adopted the following as the definition of the ampere for the standard SI (Système International, or International System of Units) metric unit.

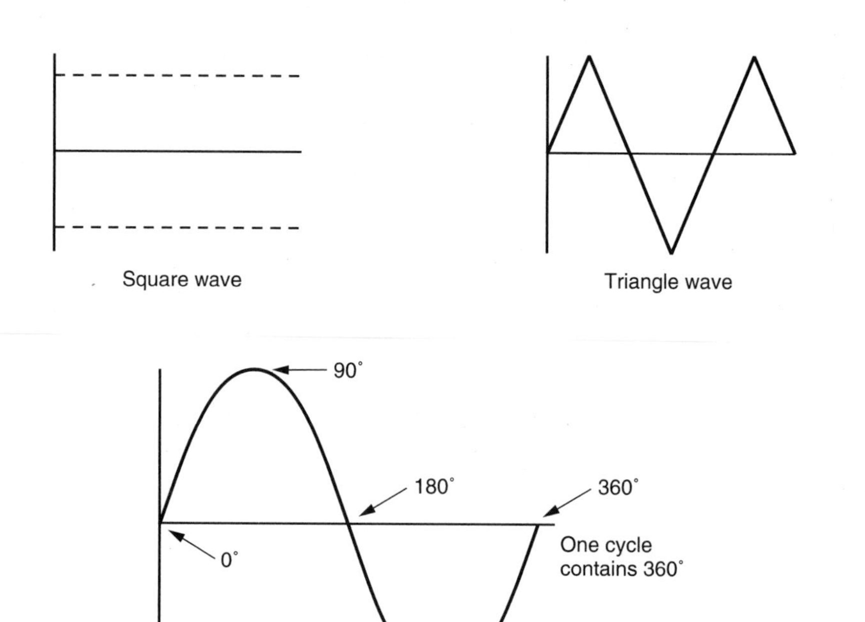

Figure 2–1 Alternating current waveforms.

> The ampere is that constant current which, if maintained in two straight parallel conductors of infinite length, of negligible circular cross section, and placed one meter apart in vacuum, would produce between these conductors a force equal to 2×10^{-7} Newton per meter of length.

The letter I is generally used in Ohm's law formulas to represent current. The letter I stands for *intensity* of current. Amperage can be computed using the following formulas:

$$I = \frac{E}{R} \text{ or } I = \frac{P}{E} \text{ or } I = \sqrt{\frac{P}{R}}$$

where E = voltage; R = resistance; and P = watts. These terms are further described in this chapter.

Angle Theta (∅)

Angle theta expresses the amount of phase angle shift between voltage and current in an alternating-current circuit containing inductive or capacitive components. The decimal power factor (PF) is the cosine of angle theta.

$$PF = \cos \theta$$

Example: An AC circuit has a power factor of 84 percent. How many degrees are the current and voltage out of phase with each other?

$$84\% = 0.84$$

0.84 is the cosine of angle theta. Angle theta is 32.86 degrees.

To change	to	Multiply by
Peak	RMS	0.707
Peak	Average	0.637
Peak	Peak-to-Peak	2
RMS	Peak	1.414
RMS	Average	0.9
Average	Peak	1.567
Average	RMS	1.111

Figure 2–2 Conversions for alternating current values.

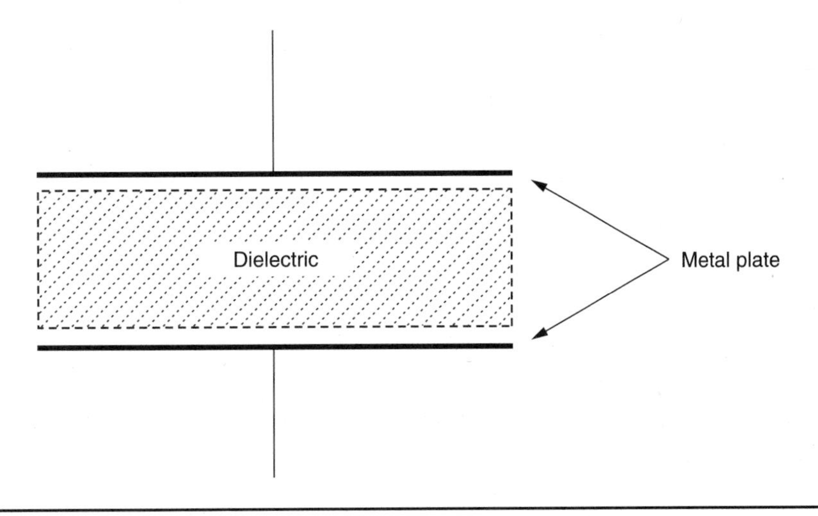

Figure 2–3 A capacitor is two metal plates separated by an insulating material.

Average Values

The average values of voltage and current are actually direct-current values. When alternating current is converted into direct current with a rectifier, the DC voltage or current produced is called the average value. Refer to the chart in Figure 2–2 for the conversion factors.

Capacitance (C)

Capacitance is the ability of an object to store electrical energy in the form of an electrostatic field. The basic construction of a capacitor is two conductive elements, such as metal plates, separated by an insulating material called the dielectric, Figure 2–3. The basic unit of capacitance is the farad (F). A capacitor has a capacitance of 1 farad when a voltage change of 1 volt across its plates results in a change of 1 coulomb. The farad is an extremely large amount of capacitance and is seldom used for measuring the size of capacitors or the amount of capacitance in a circuit. Other units such as the microfarad (μF), nanofarad (nF), or picofarad (pF) are commonly used.

Capacitive Reactance (X_C)

Capacitive reactance is the unit of opposition to current flow in a circuit containing pure capacitance. It is measured in ohms. Capacitive reactance is a counter charge that opposes the applied voltage and is proportional to the capacitance of the capacitor and the frequency of the applied voltage. When the capacitance and frequency of the circuit are known, capacitive reactance can be computed using the formula:

$$X_C = \frac{1}{2\pi FC}$$

where F is frequency in hertz and C is capacitance in farads.

Capacitive reactance does not oppose the flow of current by converting electrical energy into some other form, but stores the energy in the form of an electrostatic–static field. The stored energy is returned to the circuit each half cycle. Since capacitive reactance is the current limiting element of a pure capacitive circuit, capacitive reactance can also be computed using Ohm's law in similar formulas as those for computing resistance.

$$X_C = \frac{E}{I} \text{ or } X_C = \frac{E^2}{VARs} \text{ or } X_C = \frac{VARs}{I^2}$$

where E = voltage; I = intensity; and $VARs$ = volt-amps reactive.

Capacitors

Capacitors are devices that store electrical energy in the form of an electrostatic field. The basic capacitor is constructed by separating two metal plates by an insulating material called the dielectric (see Figure 2–3). Three factors determine the amount of capacitance a capacitor will have:

1. The surface area of the plates
2. The distance between the plates
3. The type of dielectric employed

A capacitor can be charged by connecting it to a source of direct current (see Figure 2–4.) The positive side of the power source attracts electrons from the plate it is connected to, causing a deficiency of electrons at that plate. The negative side of the power source supplies electrons to the plate it is connected to, creating an excess of electrons on that plate. This flow of electrons continues until a voltage is developed across the plates of the capacitor that is equal to the voltage of the power source. Current can flow only during the time that the capacitor is either charging or discharging. Once the voltage across the plates of the capacitor are equal to the voltage of the power source, no more current can flow.

Once a capacitor has been charged, it can be disconnected from the power source and remain in a charged state. In theory, the capacitor should retain its charge forever, but in practice it will not. No dielectric material is a perfect insulator; it will eventually permit the electrons on the negative plate to leak across to the positive plate. The rate of leakage is determined by two factors:

1. The insulating quality of the dielectric
2. The potential between the two plates

The basic unit of capacitance measure is the farad. A capacitor has a capacitance of 1 farad when a voltage change of 1 volt across its plates results in a change in charge of 1 coulomb.

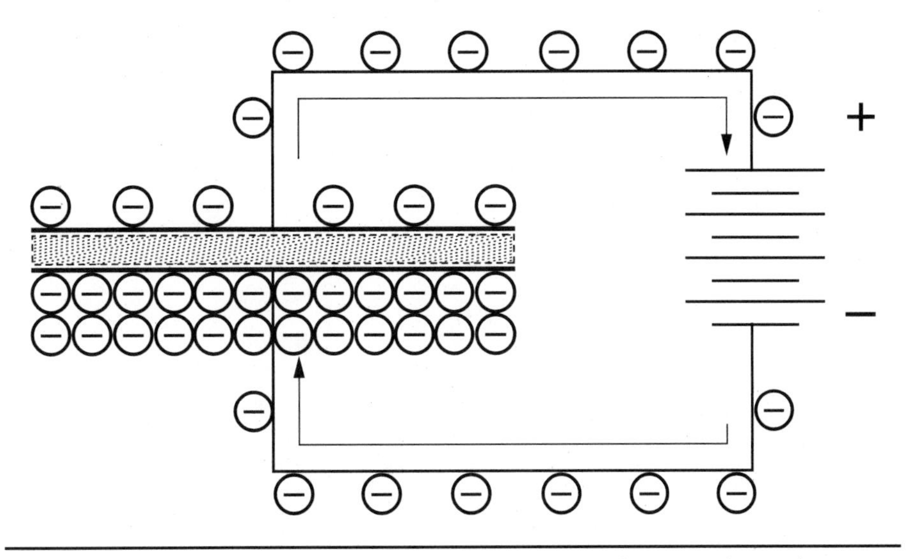

Figure 2–4 A capacitor is charged by removing electrons from one plate and depositing them on the other.

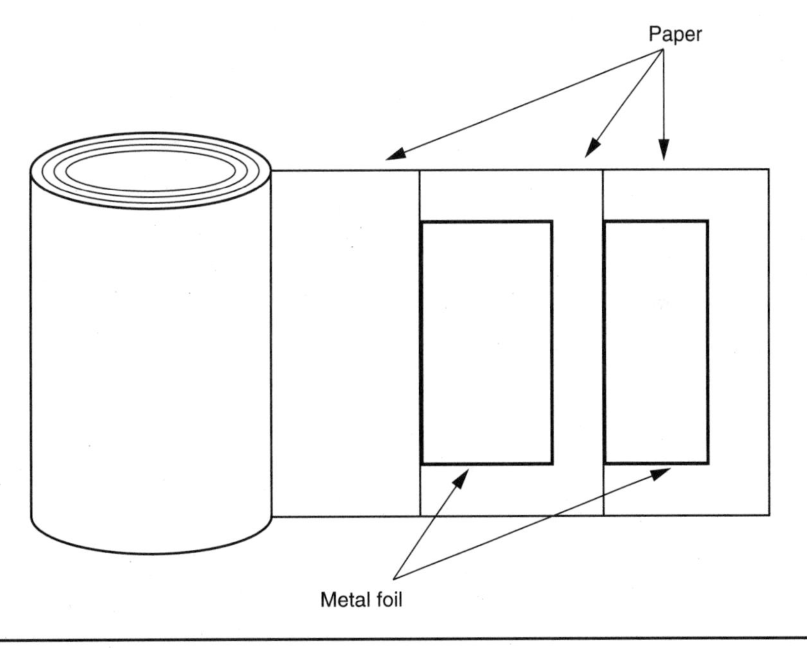

Figure 2–5 In an oil-filled paper capacitor, metal plates are separated by oil-filled paper, which acts as an insulator.

Capacitors can be divided into two basic types: nonpolarized and polarized. Nonpolarized capacitors are often referred to as *AC capacitors* because they can be connected to either direct or alternating current without harming the capacitor. Most nonpolarized capacitors are constructed by separating two metal plates by some type of insulating material. The construction of a typical oil-filled paper capacitor is shown in Figure 2–5.

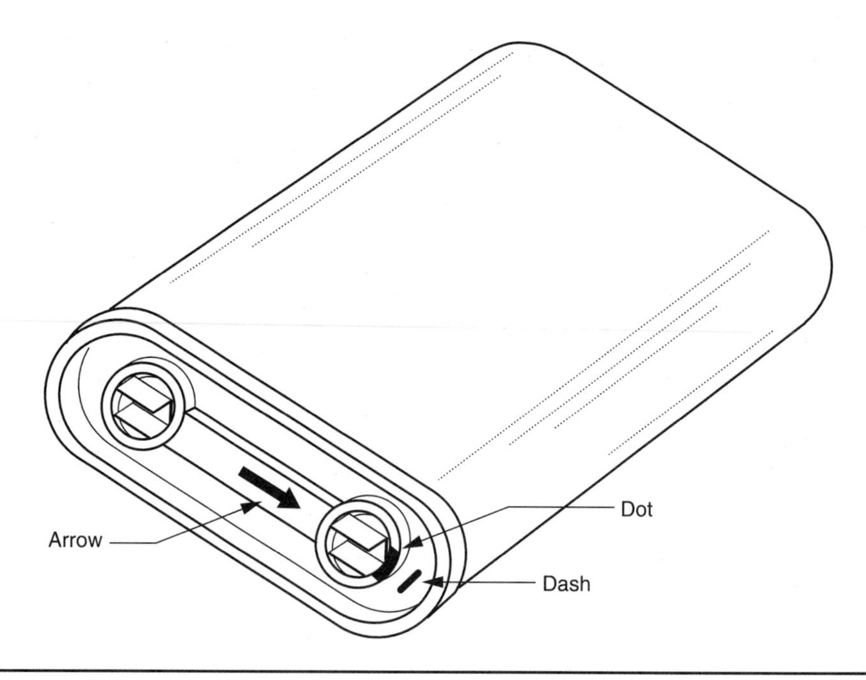

Figure 2–6 Marking indicate plate nearest capacitor case.

Capacitors of this type often contain a terminal that is marked with an arrow, a dash stamped into the metal, or a painted dot, shown in Figure 2–6. This marked terminal is connected to the plate located nearer to the case of the capacitor.

Polarized capacitors are often called *electrolytic capacitors.* These capacitors are used in direct current circuits only because the polarity must be maintained to the proper terminal. Electrolytic capacitors will identify one terminal as positive or negative. The advantage of electrolytic capacitors is that they can contain a large amount of capacitance in a small case size. There are two basic types of electrolytic capacitors: the wet type and the dry type. Wet-type electrolytic capacitors have a positive plate of aluminum and the negative plate is actually an electrolyte of borax solution (Figure 2–7). The positive terminal is connected to the piece of aluminum used as the positive plate and the negative terminal is connected to the case of the capacitor that is in contact with the borax solution. When the capacitor is connected to the proper polarity, the borax solution forms a thin layer of oxide film on the positive plate. This oxide film, which is only a few molecules thick, forms the dielectric of the capacitor. It is the small distance between the plates that permits the electrolytic capacitor to have a high capacitance in a small case size.

A special wet-type electrolytic is known as the *AC electrolytic capacitor.* This capacitor is constructed by connecting two wet-type electrolytic capacitors back-to-back in a single case. During each half cycle one of the capacitors reverses polarity and becomes a short circuit while the other reforms to become a working capacitor. The capacitors are often used in low-power applications such as electronic circuits or ceiling fans, or as the starting capacitor for many single-phase motors.

Dry-type electrolytic capacitors are similar to the wet type except that the borax solution is suspended in a gauze material. This has the advantage of making the capacitor leakproof, but it does have a disadvantage. If the incorrect polarity is connected to the termi-

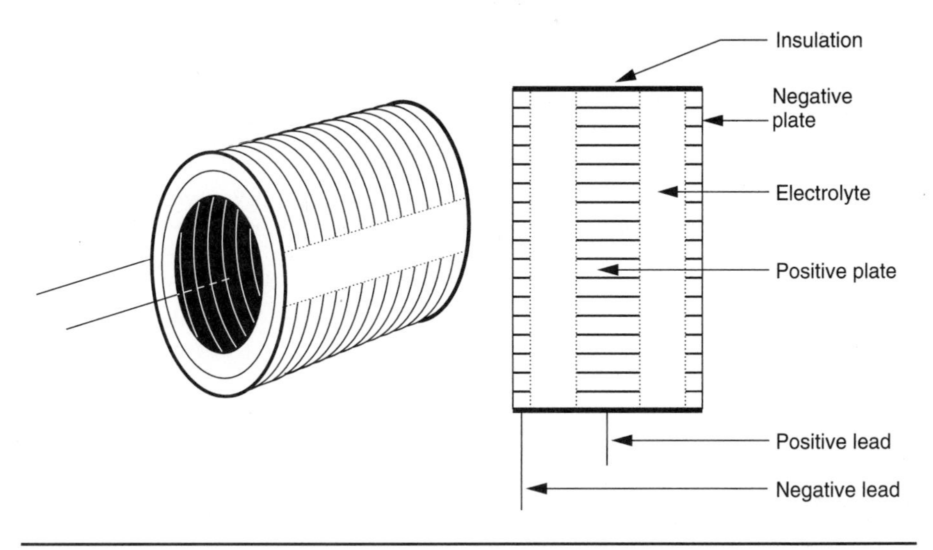

Figure 2–7 Wet-type electrolytic capacitor.

nals, the capacitor will be destroyed and it cannot be reformed by reconnecting it to the proper polarity.

Conductance (*G* or *g*)

Conductance is the opposite of resistance. The conductance value can be computed by taking the reciprocal of the resistance, *R*.

$$G = \frac{1}{R}$$

Conductance is measured in mhos (*ohms* spelled backward), or siemens. Some of the basic formulas using conductance are shown here, where I = intensity of current.

$$I = EG \qquad E = \frac{I}{G} \qquad G = \frac{I}{E}$$

Coulomb (C)

The coulomb is a quantity measurement of electrons equal to 6.25×10^{18} (6,250,000,000,000,000,000), or 6 billion, 250 million, million electrons. Since it is a quantity measurement, it is similar to a gallon, quart, or liter. The coulomb as an individual unit of measure is almost never used in practical application, but is used extensively in a flow rate where 1 coulomb per second is equal to an ampere.

Direct Current (DC)

An electric current that flows in only one direction is direct current. In accord with the electron theory of current flow, the direction of current flow will be from a more negative source to a more positive source.

Electromotive Force (EMF)

Electromotive force is generally described as the force that drives electrons through a circuit. There are several ways in which EMF may be produced, such as electromagnetic induction (generators and alternators), chemical action (batteries), thermal action (thermocouples), and vibrating or heating crystals (piezoelectric).

Horsepower (hp)

Horsepower is a rate of doing work. James Watt originally established the unit to rate the output power of a steam engine. Watt determined that an average horse could lift a weight of 550 pounds 1 foot in 1 second. The basic unit of horsepower is established as 1 hp = 550 ft.-lb./sec. This is also equal to 33,000 ft.-lb./min. For electrical purposes, it has been established that 1 hp = 746 watts. Since electric motors are rated in horsepower, it is sometimes necessary to compute their output power when the torque and speed are known. The formula shown can be used to determine the output horsepower of a motor.

$$\text{hp} = \frac{2\pi \times RPM \times L \times P}{33{,}000}$$

Where:

$$\begin{aligned} 2\pi &= 6.283 \\ RPM &= \text{speed in revolutions per minute} \\ L &= \text{length in feet} \\ P &= \text{pounds} \\ 33{,}000 \text{ (constant) } 1 \text{ hp} &= 33{,}000 \text{ ft.-lb./min.} \end{aligned}$$

Example: A motor is operating at a speed of 1725 RPM and producing a torque of 28 in.-lb. What is the output horsepower of the motor?

$$\text{hp} = \frac{6.283 \times 1725 \times 0.08333(1/12 \text{ ft.}) \times 28}{33{,}000}$$

$$\text{hp} = 0.766$$

Impedance (*Z*)

Impedance is the unit of opposition to current flow in any alternating current circuit. It may be composed of resistance, inductive reactance, capacitive reactance, or a combination of all three. Impedance is measured in ohms. The formula used to calculate the value of impedance is determined by the type of circuit and the current-limiting elements contained in the circuit. In a circuit containing pure resistance, the impedance value would be the same as the resistance value. In a circuit containing pure capacitance, the impedance value would be the same as the capacitive reactance. Different types of circuits and the formulas for calculating impedance are shown in Figure 2–8.

Inductance (*L*)

Inductance is the ability of a component to store electrical energy in the form of an electromagnetic field. It is also characterized by the component's ability to induce a voltage into itself or another conductor. Inductors are commonly referred to as *chokes* or *reactors.* Inductance is measured in units called henrys (H). An inductor has an inductance of 1 henry when a current change of 1 ampere per second results in an induced voltage of 1 volt. When the inductive reactance at a particular frequency of a component is known, the inductance can be determined using the formula:

$$L = \frac{X_L}{2\pi F}$$

where X_L is inductive reactance and F is frequency in hertz.

Inductive Reactance (X_L)

Inductive reactance is the unit of opposition to the flow of current in a circuit containing inductance. Inductive reactance is measured in ohms. It does not oppose the flow of current by converting electrical energy into heat, as does resistance. Inductive reactance is a counter voltage produced by an inductor. This counter voltage opposes the applied volt-

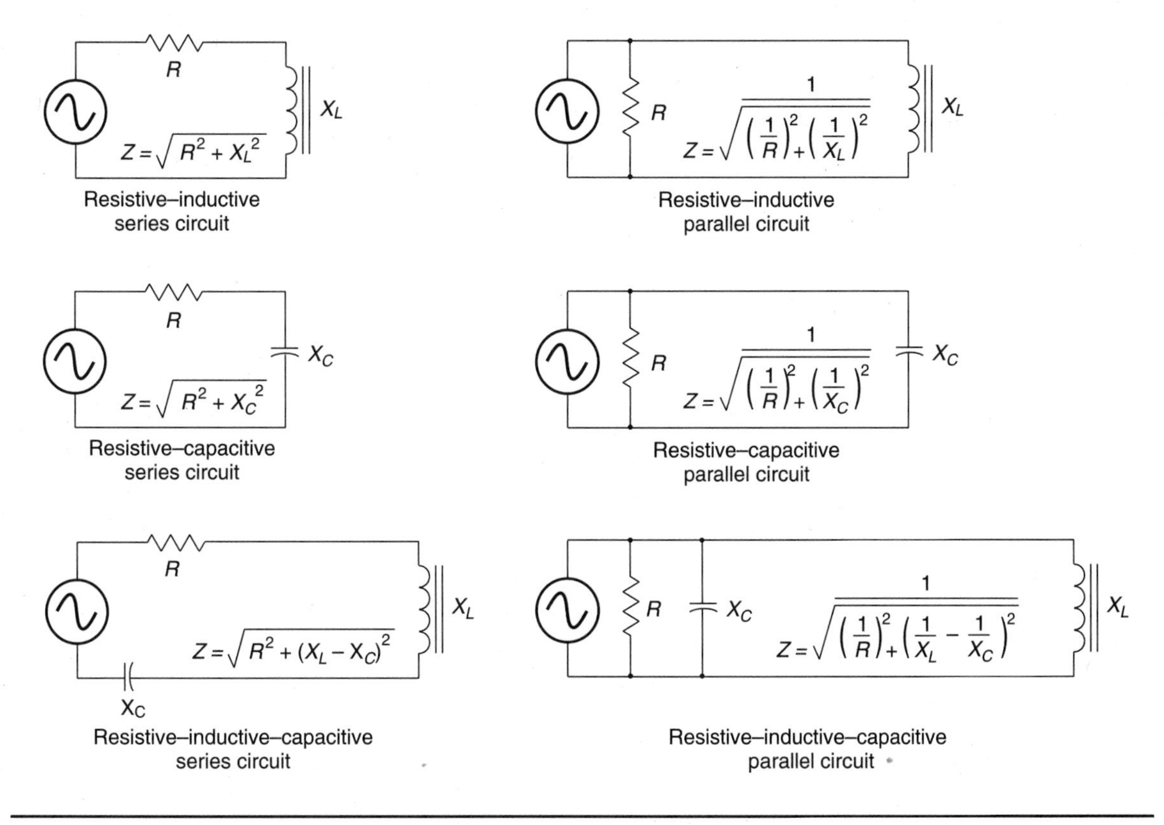

Figure 2–8 Impedance formulas.

age, thus limiting the flow of current. Inductive reactance is proportional to the inductance of the inductor measured in henrys and the frequency of the circuit. If inductance and frequency are known, inductive reactance can be computed using the formula:

$$X_L = 2\pi FL$$

where F is frequency in hertz and L is inductance in henrys.

Inductive reactance can also be computed using the same basic formulas as resistance if pure inductive values are used. In a theoretical circuit containing pure inductance (Figure 2–9), an alternating-current source is connected to a pure inductor with negligible resistance. In this circuit the only current-limiting factor is inductive reactance. Therefore, inductive reactance can be computed using the formulas:

$$X_L = \frac{E}{I} \text{ or } X_L = \frac{E^2}{VARs} \text{ or } X_L = \frac{VARs}{I^2}$$

where E = voltage; I = intensity; and $VARs$ = volt-amps reactive.

Inductors

Inductors are devices that have the ability to store electrical energy in the form of an electromagnetic field. Inductors are actually coils of wire wound on some type of core material. If some type of nonmagnetic core material is employed, such as wood or plastic, the inductor is known as an *air-core* inductor. If the core is made of a magnetic material, such

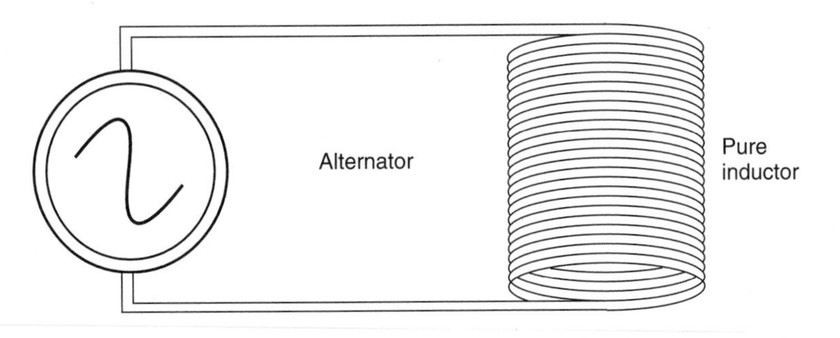

Figure 2–9 Inductor connected to an AC source.

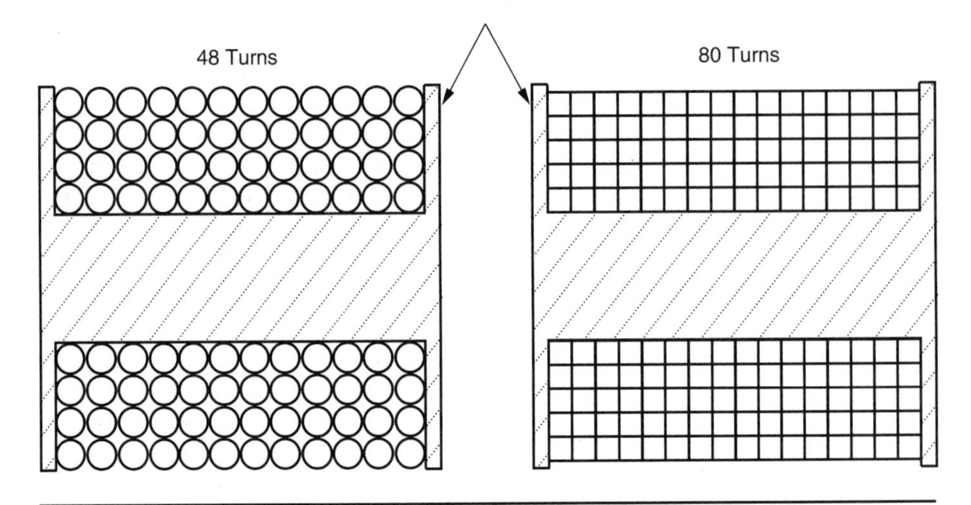

Figure 2–10 Square wire permits more turns in the same area than round wire.

as soft iron or silicon steel, the inductor is known as an *iron-core* inductor. Inductors are often referred to as chokes or coils. The amount of inductance an inductor has is determined by several factors, such as the type and shape of core material used, the type of wire used to wind the inductor, and the closeness with which the wires can be wound. Winding turns of wire closer together improves magnetic coupling and increases inductance for a particular size and shape of a coil. Large inductors and electromagnets are often wound with square or rectangular wire to permit them to have more turns of wire in a smaller space (see Figure 2–10).

The unit of inductance is the henry. An inductor has an inductance of 1 henry when a current change of 1 ampere per second results in an induced voltage of 1 volt.

Joule (*J*)

The joule is the metric unit in the SI (Système Internatíonal or International System of Units) system equivalent to the wattsecond. It is described as the force of 1 newton working through a distance of 1 meter (N•m). One joule is equal to 1 wattsecond or 10^7 ergs. One kilowatt-hour is equal to 3.6×10^6 joules.

Ohm (Ω)

The *ohm* is the practical unit of electrical resistance or reactance and is symbolized by the Greek letter omega (Ω). The ohm is the unit used to describe the opposition to electron flow; it may or may not be associated with resistance. Units of inductive reactance, capacitive reactance, resistance, and impedance are all measured in ohms. Although ohms are not necessarily a measurement of resistance, the international ohm unit is defined as

> the resistance offered to an unvarying electric current by a column of mercury, at a temperature of 0° C, which has a mass of 14.4521 grams, and a constant cross section, and a length of 106.3 centimeters.

One ohm will oppose the flow of current to the extent that when 1 volt is applied across 1 ohm, the flow of current will be limited to 1 ampere.

Peak Value

The peak value of voltage and current is the maximum amount of voltage or current produced by an alternating-current waveform. In a sine wave, the positive peak value occurs at 90 degrees and the negative peak occurs at 270 degrees. Refer to the sine wave shown in Figure 2–1. To convert peak values to RMS or average, refer to the conversion factors listed in Figure 2–2.

Power Factor (*PF*)

The power factor is the ratio of true power or watts as compared to the apparent power or volt-amperes.

$$PF = \frac{\text{Watts}}{\text{Volt} - \text{Amperes}}$$

The power factor is expressed as a percentage. The decimal number obtained from the formula is generally changed to a percent.

Example: An electric motor is connected to a 240-volt single-phase AC line and has a current of 14 amperes. A wattmeter indicates that the true power of the circuit is 2250 watts. What is the power factor?

$$VA = 240 \times 14 \quad VA = 3360$$

$$PF = \frac{2250}{3360}$$

$$PF = 0.6696 \text{ or } 66.96\%$$

Resistance (*R*)

Resistance is the name given to a phenomenon exhibited by certain materials that impede the flow of electric current through them. The unit measure of resistance is the ohm (Ω). Resistance is characterized by the conversion of electrical energy into heat in accord with the formulas:

$$P = I^2 \times R \text{ or } P = \frac{E^2}{R}$$

where P = watts; I = intensity; R = resistance; and E = voltage.

All natural materials have some amount of resistance. Metals generally have very little resistance and are generally used as conductors. Materials that have a high resistance are used as insulators. Resistance can be computed using the following formulas:

$$R = \frac{E}{I} \text{ or } R = \frac{E^2}{P} \text{ or } R = \frac{P}{I^2}$$

Root Mean Square (RMS)

RMS is the value of alternating current that will produce the same amount of power as an equivalent value of direct current. There are RMS equivalent values for voltage, current, and power. The RMS value is indicated by almost all voltmeters, ammeters, and wattmeters. When alternating-current values for voltage and current are specified, they are RMS values unless specifically stated as peak, average, or peak-to-peak. A chart for converting peak, RMS, and average values is shown in Figure 2–2.

Sine Waves

Sine waves are the most common of all alternating-current waveforms. They are produced by rotating armatures, as in an alternator. Sine waves are so named because the voltage at any point on the waveform is equal to the maximum or peak voltage times the sine of the angle of rotation.

Example: A sine wave has a peak value of 100 volts. What is the voltage at 45 degrees?

$$E_{\text{INST}} = E_{\text{MAX}} \times \sin\angle \qquad E_{\text{INST}} = 100 \times 0.707 \qquad E_{\text{INST}} = 70.7V$$

The formula to find E_{MAX} when instantaneous voltage and angle are known, or the formula to find the angle when the maximum and instantaneous voltages are known are:

$$E_{\text{MAX}} = \frac{E_{\text{INST}}}{\sin\angle} \qquad \sin\angle = \frac{E_{\text{INST}}}{E_{\text{MAX}}}$$

Susceptance (*B* or *b*)

Susceptance is the opposite of reactance. It can be computed by taking the reciprocal of the reactance value ($B = \frac{I}{X_L}$) or ($B = \frac{I}{X_C}$). Susceptance is measured in mhos or siemens.

Voltage (*E* or *V*)

Voltage is the unit of electromotive force often described as potential difference or electrical pressure. By international agreement, 1 volt is the amount of electromotive force necessary to cause a current of 1 ampere to flow through a resistance of 1 ohm.

Volt–Amperes (*VA*)

Volt–amperes is referred to as the *apparent* power because it is the product of the volts and amperes for any circuit. Although one of the formulas for determining the true power or watts for an electric circuit is to multiply the voltage and amperes, the product of the voltage and current may or may not be the true power. In a direct-current circuit, the product of volts and amperes will equal watts. The product of volts and amperes in an alternating-current circuit containing only resistive loads will equal watts. Multiplying the volts and amperes in an alternating-current circuit containing inductive or capacitive loads will not equal watts.

$$\text{Single Phase: } VA = E \times I \text{ or } VA = I^2 \times Z \text{ or } VA = \frac{E^2}{Z}$$

$$\text{Three Phase: } VA = E_{\text{Line}} \times I_{\text{Line}} \times \sqrt{3}$$

where E = voltage; I = intensity; Z = impedance.

Volt-Amps Reactive (VARs)

Often referred to as wattless or reactive power, VARs are the reactive counterpart of watts in a resistive circuit. In a pure reactive circuit, whether inductive or capacitive, electric power is stored during one half cycle and then returned to the circuit in the next half cycle. It is not converted into some other form. VARs can be computed in the same manner as watts, except that reactive values of voltage and current are used. The power rating of inductors or chokes and capacitors is given in VARs. The formulas for computing VARs are similar to those for computing watts except that the current and voltage values that apply to pure reactive components are used in the calculation.

$$VARs = \frac{E^2}{X_L \text{ or } X_C} \text{ or } VARs = E \times I \text{ or } VARs = I^2 \times (X_L \text{ or } X_C),$$

where: E = voltage; I = intensity; X_L = inductive reactance; X_C = capacitive reactance

Watt (*P* or *W*)

The watt is the unit of true power in an electric circuit. Watts are a measure of the amount of electrical energy that is converted into some other form—heat energy in the case of resistance, for example, or mechanical energy in the case of a motor. Watts can be computed using the formulas

$$P = \frac{E^2}{R} \text{ or } P = E \times I \text{ or } P = I^2 \times R$$

where E = voltage; I = intensity; and R = resistance.

Since the watt is a measurement of power, it can be converted into other basic units that are used to measure power or energy. Some of these conversions are:

1 horsepower =	745.6 watts
1 horsepower =	550 ft.-lb./second
1 watt =	0.00134 horsepower
1 watt =	3.412 BTU/hour
1 watt/second =	1 joule
1 BTU/second =	1.055 watts
1 calorie/second =	4.19 watts
1 ft.-lb./second =	1.36 watts
1 BTU =	1050 joules
1 joule =	0.2389 calorie
1 calorie =	4.186 joules

Wattsecond

The wattsecond is a unit of power equal to the force necessary to move 1 coulomb between two points with a potential of 1 volt. A wattsecond can also be described as the energy stored in a capacitor equal to:

$$\text{Watt second} = \frac{CV^2}{2}$$

where C is capacitance in farads and V is volts.

CHAPTER 3

Magnetic Measurement

A magnet is any body or device with the ability to attract iron or steel. Magnets have two poles identified as *north* and *south,* (Figure 3–1). Magnetic lines of force connect the north and south poles and are commonly referred to as *flux* or *lines of flux.* Although magnetic flux lines do not flow, it is assumed that they are in a direction of north to south. The magnetic lines of flux are strongest or more concentrated at the poles and decrease in strength as the distance from the pole increases. *Magnetism* is the force by which a magnet can attract an object.

The following terms are listed in alphabetical order and describe aspects of magnetic measurement. Knowledge of these terms will be important in understanding the following chapters.

Ampere-Turns

The flux density of an electromagnet is determined by the ampere-turns. These are the number of turns of wire and the amount of current flowing through the coil of an electromagnet.

Coercive Force (*H*)

The amount of demagnetizing force necessary to remove residual magnetism from a magnetic material is called coercive force.

Flux Density

Flux density is an English-system measurement. It is used to measure the strength of a magnetic field expressed in lines of flux per square inch. The Greek letter phi (Φ) is generally used to represent lines of flux. The letter B is used to represent flux density.

$$B \text{ (flux density)} = \frac{\Phi \text{ (flux lines)}}{A \text{ (area)}}$$

Lodestones

Lodestones are natural magnets. Centuries ago it was discovered that if certain stones were suspended from a string or placed on a piece of wood in a dish of water, the stone would align itself north and south, creating a simple compass. Since these stones had the ability to find north, they became known as *leading* stones.

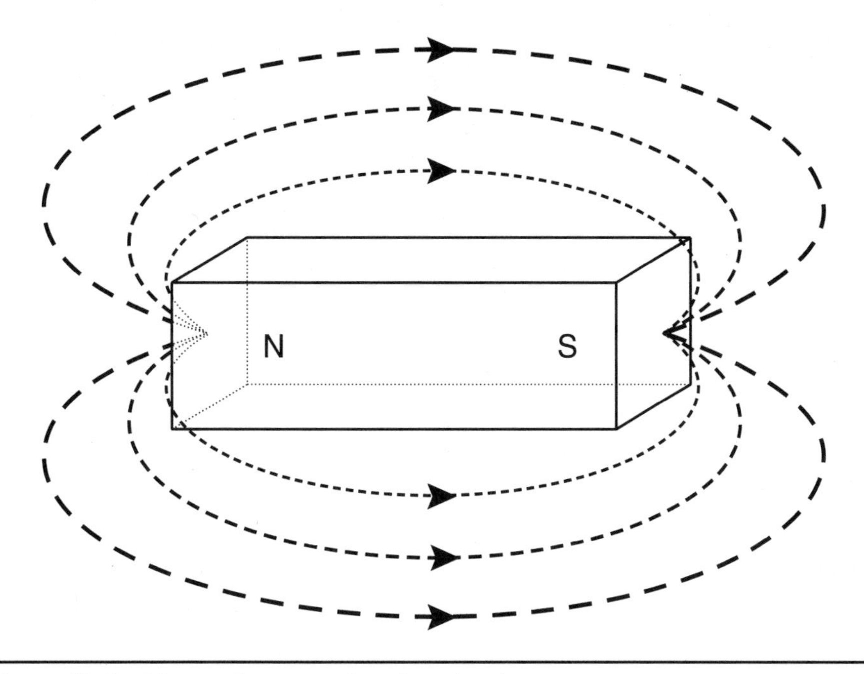

Figure 3–1 Magnets have a north and south pole.

Magnetic Flux

A magnetic field is often depicted as individual lines of force that exist between the north and south poles. Iron filings sprinkled on a piece of paper that has been placed over a magnet will produce a definite pattern of lines. The lines are called *flux lines.*

Magneto-Motive Force (mmf)

Magnetomotive force is an English-system measurement used to describe the total force producing a magnetic field. The mmf is equal to the product of the total flux lines and the reluctance.

$$\text{mmf} = \Phi \times \text{rel (reluctance)}$$

Permanent Magnet

A magnetic material that retains its magnetism after the magnetizing force has been removed is permanent. Natural permanent magnets are known as lodestones. Manmade permanent magnets are much stronger and can retain their magnetism longer than natural permanent magnets.

Permeability (μ)

Permeability is a measure of a material's ability to conduct magnetic lines of flux as compared to air. Air is assigned a permeability of 1. If the flux density of an air-core electromagnet increases 50 times after a metal core in inserted, the material has a permeability of 50.

Reluctance

Reluctance, resistance to magnetism, is equal to the magnetomotive force divided by the flux.

$$\text{rel} = \frac{\text{mmf}}{\Phi}$$

Residual Magnetism

The amount of magnetism left in an object after the magnetizing force has been removed is called residual magnetism.

Retentivity

Retentivity is the ability of a magnetic material to retain magnetism after the magnetizing force has been removed. Retentivity is determined by measuring the amount of demagnetizing force that is necessary to remove the residual magnetism.

Saturation

The point at which a magnetic material can hold no more lines of force is saturation. It occurs when all the magnetic molecules in a material are aligned.

CHAPTER 4

Resistors and Ohm's Law

Resistors

Resistors are one of the most common components found in electrical circuits. The unit of measure for resistance is the *ohm,* which was named for German scientist Georg S. Ohm. The ohm is represented by the uppercase Greek letter omega (Ω). Resistors come in various sizes, types, and ratings to accommodate the needs of almost any circuit applications.

Resistors are commonly used to perform two functions in a circuit. One is to limit the flow of current through the circuit. In Figure 4–1 a 30-ohm resistor is connected to a 15-volt battery. The current in this circuit is limited to a value of 0.5 amp.

$$I = \frac{E}{R}$$

$$I = \frac{15}{30}$$

$$\mathrm{I} = 0.5\ \mathrm{amp}$$

If this resistor were not present, the circuit current would be limited only by the resistance of the conductor—which would be very low, causing a large amount of current to flow. Assume, for example, that the wire has a resistance of 0.0001 ohm. When the wire is connected across the 15-volt power source, a current of 150,000 amps would try to flow through the circuit (15/0.0001 = 150,000). This is commonly known as a *short circuit.*

The second principal function of resistors is to produce a voltage divider as shown in Figure 4–2. These three resistors are connected in series with a 17.5-volt battery. If the leads of a voltmeter were connected between different points in the circuit, it would indicate the following voltages:

A–B	1.5 volts
A–C	7.5 volts
A–D	17.5 volts
B–C	6 volts
B–D	16 volts
C–D	10 volts

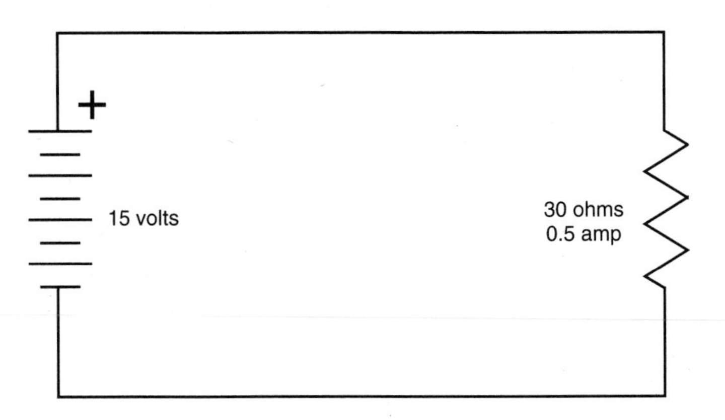

Figure 4–1 Resistors can be used to limit the flow of current.

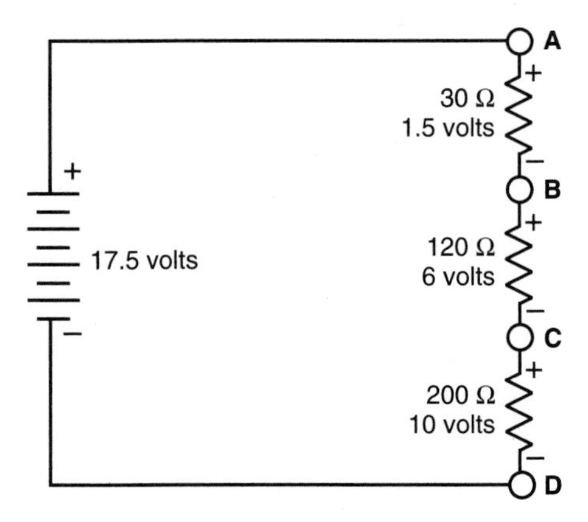

Figure 4–2 Resistors can be used as a voltage divider.

By connecting resistors of the proper value, almost any voltage can be obtained. Voltage dividers were used to a large extent in vacuum-tube circuits many years ago. Voltage-divider circuits are still used today in applications involving field effect transistors (FETs) and in multirange voltmeter circuits.

Fixed Resistors

Fixed resistors have only one ohmic value, which cannot be changed or adjusted. There are several different types of fixed resistors. One of the most common types of fixed resistors is the composition carbon resistor. Carbon resistors are made from a compound of carbon graphite and a resin bonding material. The proportions of carbon and resin material determine the value of resistance. This compound is enclosed in a case of nonconductive material with connecting leads (see Figure 4–3).

Carbon resistors are popular for most applications, because they are inexpensive and readily available. They are made in standard values ranging from about 1 ohm to about 22

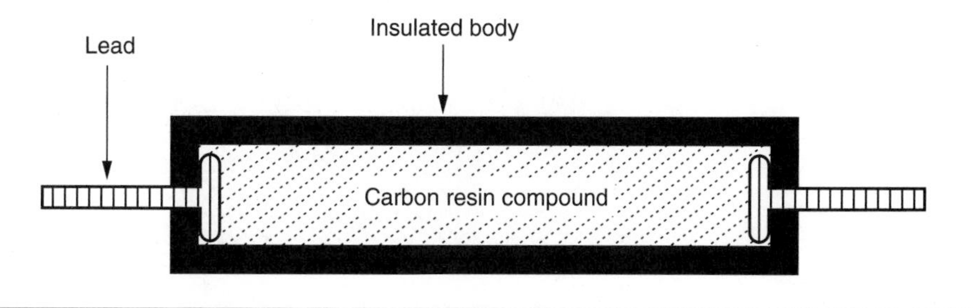

Figure 4–3 Composition carbon resistor.

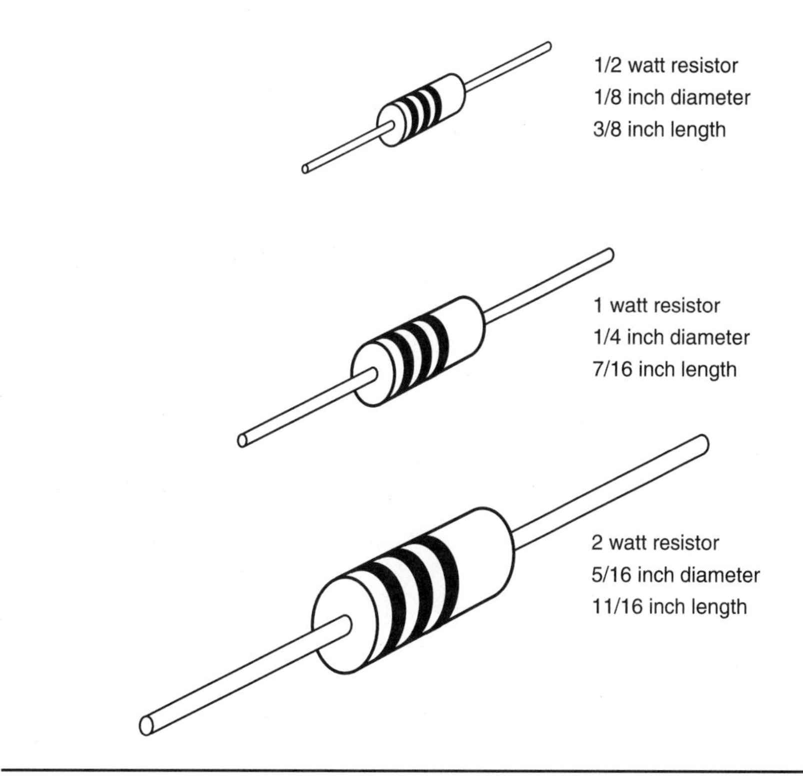

Figure 4–4 Power rating is indicated by size.

million ohms, and they can be obtained in power rating of 1/8, 1/4, 1/2, 1, and 2 watts. The power rating of the resistor is indicated by its size. A 1/2-watt resistor is approximately 3/8 inch in length and 1/8 inch in diameter. A 2-watt resistor is approximately 11/16 inch long and approximately 5/16 inch in diameter, illustrated in Figure 4–4. The 2-watt resistor is larger than the 1/2-watt or 1-watt resistors because it must have a larger surface area to be able to dissipate more heat. Although carbon resistors have a lot of desirable characteristics, one of their characteristics is not desirable. Carbon resistors will change their value with age or if they are overheated. Carbon resistors generally increase instead of decrease in value.

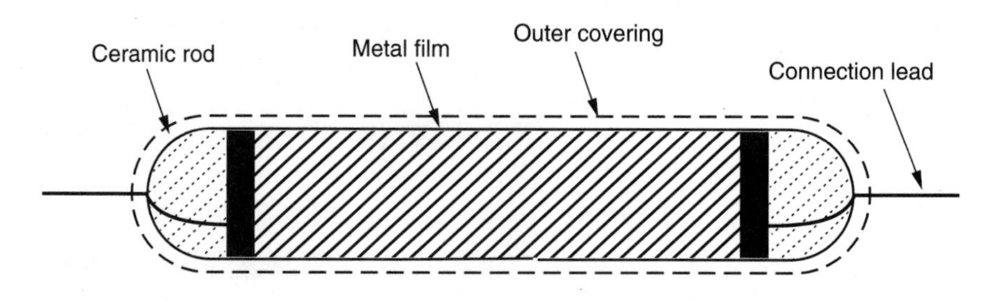

Figure 4–5 In a metal-film resistor, a film of metal is applied to a ceramic rod.

Metal-Film Resistors

Another type of fixed resistor is the metal film resistor. Metal-film resistors are constructed by applying a film of metal to a ceramic rod in a vacuum, as shown in Figure 4–5. The resistance is determined by the type metal used to form the film and the thickness of the film. Typical thickness for the film is from 0.00001 to 0.00000001 inch. Leads are then attached to the film coating, and the entire assembly is covered with a coating. These resistors are superior to carbon resistors in several respects. Metal-film resistors do not change their value with age, and their tolerance is generally better than that of carbon resistors. Carbon resistors commonly have a tolerance range of 20 percent, 10 percent, or 5 percent. Metal-film resistors generally range in tolerance from 2 percent to 0.1 percent. The disadvantage of the metal film resistor is that it is higher in cost.

Carbon-Film Resistors

The carbon-film resistor is constructed by coating a ceramic rod with a film of carbon instead of metal. Carbon-film resistors are less expensive to manufacture than metal-film resistors and can have a higher tolerance rating than composition carbon resistors.

Metal-Glaze Resistors

The metal-glaze resistor is similar to the metal-film resistor. This resistor is made by combining metal with glass. The compound is then applied to a ceramic base as a thick film. The resistance is determined by the amount of metal used in the compound. Tolerance rating of 2 percent and 1 percent are common.

Wire-Wound Resistors

Wire-wound resistors are made by winding a piece of resistive wire around a ceramic core (see Figure 4–6). The resistance of a wire-wound resistor is determined by three factors:

- The type of material used to make the resistive wire
- The diameter of the wire
- The length of the wire

Wire-wound resistors can be found in various case styles and sizes. These resistors are generally used when a high power rating is needed. Wire-wound resistors can operate at higher temperatures than any other type of resistor. This type of resistor should be mounted vertically, not horizontally. The center of the resistor is hollow for a good reason.

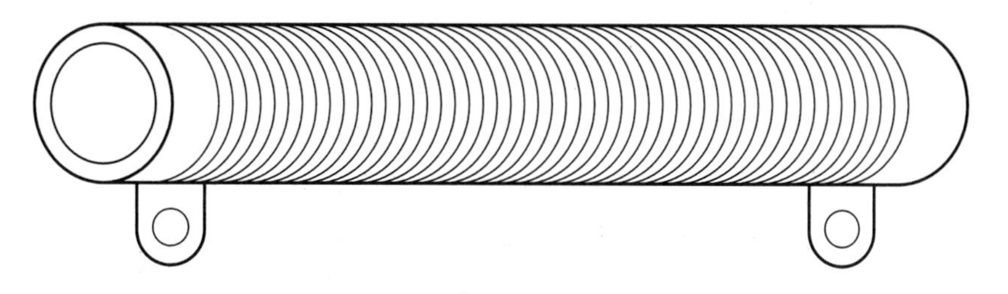

Figure 4–6 In a wire-wound resistor, resistive wire is wound around a ceramic core.

When the resistor is mounted vertically, the heat from the resistor produces a chimney effect and causes air to circulate through the center. This increase of air flow dissipates heat at a faster rate to help keep the resistor from overheating. The disadvantage of wire-wound resistors is they are expensive and generally require a large amount of space for mounting. They can also exhibit an amount of inductance in circuits that operate at high frequencies. This added inductance can cause problems to the rest of the circuit. Inductance will be covered in later chapters.

Resistor Color Code

The values of a resistor can often be determined by the color code. Many resistors use bands of color to determine the resistance value, tolerance, and, in some cases, reliability. The color bands represent numbers. Each color represents a different numerical value. The chart shown in Figure 4–7 lists the color and the number value assigned to that color. The resistors shown in the color code chart illustrate how to determine the value of a resistor.

Resistor Values Determined from Color Bands

Resistors can have from three to five bands of color. Resistors with a tolerance of ±20 percent have only three color bands. Most resistors contain four bands of color. For resistors with tolerances that range from ±10 percent to ±2 percent, the first two color bands represent number values. The third color band is called the multiplier. This means to multiply the first two numbers by 10 the number of times indicated by the value of the third band. The fourth band indicates the tolerance. For example, assume a resistor has color bands of brown, green, red, and silver (see Figure 4–8). The first two bands represent the numbers 1 and 5, (brown is 1 and green is 5). The third band is red, which has a number value of 2. The number 15 should be multiplied by 10 two times. The value of the resistor is 1,500 ohms. Another method, which is simpler to understand, is to add the number of zeros to the first two numbers indicated by the multiplier band. The multiplier band in this example is red, which has a numeric value of 2. Add two zeros to the first two numbers. The number 15 becomes 1,500.

The fourth band is the tolerance band. The tolerance band in this example is silver, which means ±10 percent. This resistor should be 1,500 ohms plus or minus 10 percent. To determine the value limits of this resistor, find 10 percent of 1,500.

$$1{,}500 \times 0.10 = 150$$

The value can range from 1,650 ohms (1,500 + 10 percent, or 1,500 + 150) to 1,350 ohms (1,500 – 10 percent or 1,500 – 150).

Resistor Color Codes

Color	Value	Tolerance	
Black	0	No band	20%
Brown	1	Silver	10%
Red	2	Gold	5%
Orange	3	Red	2%
Yellow	4	Brown	1%
Green	5		
Blue	6		
Violet	7		
Gray	8		
White	9		

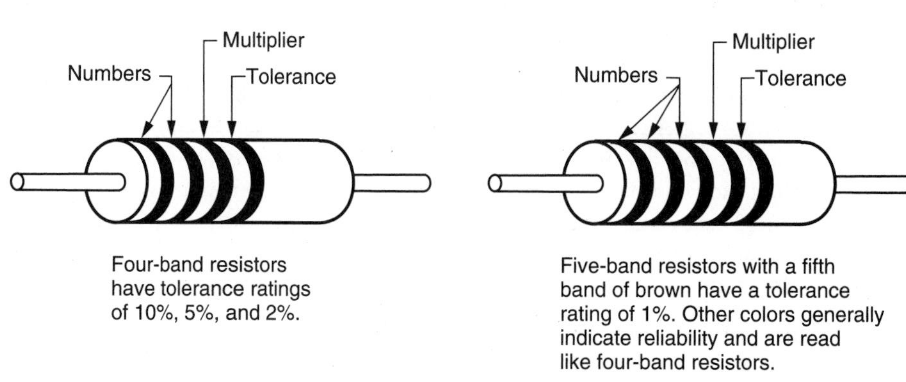

Figure 4–7 Resistor color code chart lists the colors and their number values.

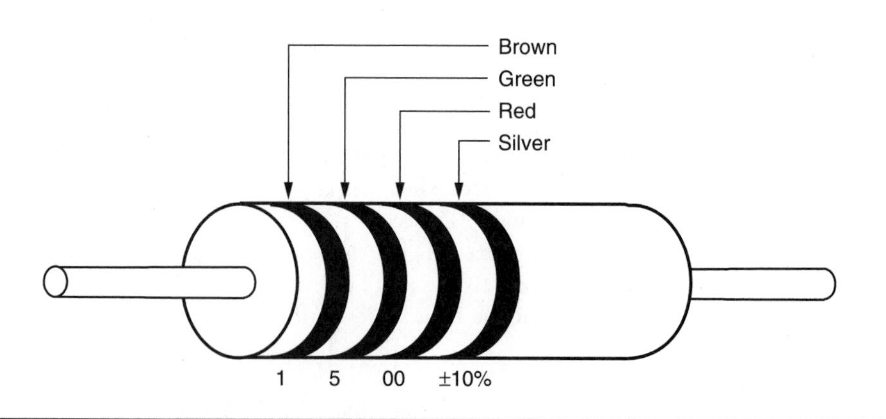

Figure 4–8 Determining resistor values using the color code.

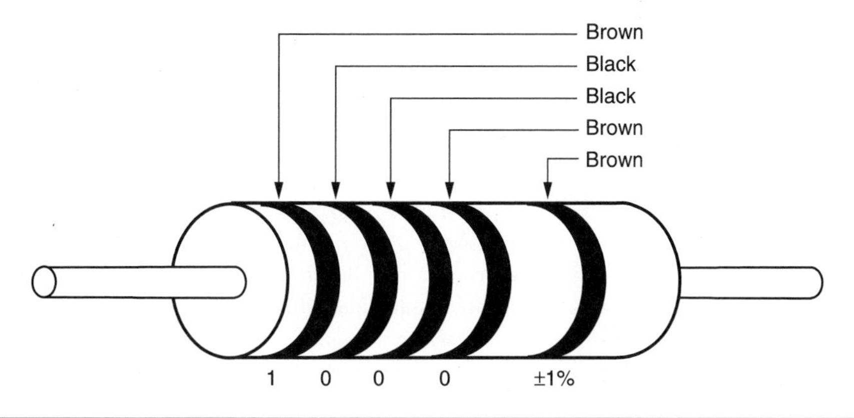

Figure 4–9 Determining the value of a ±1% resistor.

Resistors with a tolerance of ±1 percent—and some military resistors—contain five bands of color.

Example #1: The resistor shown in Figure 4–9 contains the following bands of color:

First band = brown
Second band = black
Third band = black
Fourth band = brown
Fifth band = brown

The brown fifth band indicates that this resistor has a tolerance of ±1 percent. To determine the value of a 1 percent resistor, the first three bands are numbers and the fourth band is the multiplier. In this example, the first band is brown, which has a number value of 1. The next two bands are black, which represent a number value of 0. The fourth band is brown, which means add one 0 to the first three numbers. The value of this resistor is 1,000 Ω ±1 percent.

Example #2: A five-band resistor has the following color bands:

First band = red
Second band = orange
Third band = violet
Fourth band = red
Fifth band = brown

The first three bands represent number values. Red is 2, orange is 3, and violet is 7. The fourth band is the multiplier; in this case, red represents 2. Add two 0s to the number 237. The value of the resistor is 23,700 ohms. The fifth band is brown, which indicates a tolerance of ±1 percent.

Military resistors often have five bands of color, also. These resistors are read in the same manner as a resistor with four bands of color. The fifth band can represent different things. A fifth band of orange or yellow is used to indicate reliability. Resistors with a fifth band of orange have a reliability good enough to be used in missile systems, and a resistor with a fifth band of yellow can be used in space flight equipment. A military resistor with a fifth band of white indicates the resistor has solderable leads.

Resistors with tolerance ratings ranging from 0.5 percent to 0.1 percent will generally have their values printed directly on the resistor.

Gold and Silver as Multipliers

The colors gold and silver are generally found in the fourth band of a resistor, but they can be used in the multiplier band, also. When the color gold is used as the multiplier band it means to divide the first two numbers by 10. If silver is used as the multiplier band it means to divide the first two numbers by 100. For example, assume a resistor has color bands of orange, white, gold, gold. The value of this resistor is 3.9 ohms with a tolerance of ±5 percent (orange = 3, white = 9, gold means to divide 39 by 10 which = 3.9, and gold in the fourth band means ±5 percent tolerance).

Standard Resistance Values

Fixed resistors are generally produced in standard values. The higher the tolerance value, the fewer resistance values available. Standard resistor values are listed in the chart shown in Figure 4–10. In the column under 10 percent, there are only twelve values of resistors listed. These standard values, however, can be multiplied by factors of 10. Notice that one of the standard values listed is 33 ohms. There are also standard values in 10 percent resistors of 0.33, 3.3, 330, 3,300, 33,000, 330,000, and 3,300,000 ohms. The 5 percent column shows there are 24 resistor values, and the 1 percent column list 96 values. All of the values listed in the chart can be multiplied by factors of 10 to obtain other resistance values.

Power Ratings

Resistors also have a power rating in watts that should not be exceeded or damage will occur to the resistor. The amount of heat that must be dissipated by the resistor can be determined by the use of one of the following formulas.

$$P = I^2 \times R$$

$$P = \frac{E^2}{R}$$

$$P = EI$$

Example: Assume a resistor has a value of 100 ohms and a power rating of 1/2 watt. If the resistor is connected to a 10-volt power supply, will it be damaged?

Solution: Using the formula $P = \frac{E}{R}$, determine the amount of heat that will be dissipated by the resistor.

$$P = \frac{10 \times 10}{100}$$

$$P = \frac{100}{100}$$

$$P = 1 \text{ watt}$$

Since the resistor has a power rating of 1/2 watt and the amount of heat that will be dissipated is 1 watt, the resistor will be damaged.

Variable Resistors

A variable resistor is a resistor whose values can be changed or varied over a range. Variable resistors can be obtained in different case styles and power ratings. Figure 4–11 illustrates how a variable resistor is constructed. In this example, a resistive wire is wound in a circular pattern, and a sliding tap makes contact with the wire. The value of resistance can be

STANDARD RESISTANCE VALUES

.1%, .25% .5%	1%	.1%, .25% .5%	1%	.1%, .25% .5%	1%	.1%, .25% .5%	1%	.1%, .25% .5%	1%
10.0	10.0	17.2	—	29.4	29.4	50.5	—	86.6	86.6
10.1	—	17.4	17.4	29.8	—	51.1	51.1	87.6	—
10.2	10.2	17.6	—	30.1	30.1	51.7	—	88.7	88.7
10.4	—	17.8	17.8	30.5	—	52.3	52.3	89.8	—
10.5	10.5	18.0	—	30.9	30.9	53.0	—	90.9	90.9
10.6	—	18.2	18.2	31.2	—	53.6	53.6	92.0	—
10.7	10.7	18.4	—	31.6	31.6	54.2	—	93.1	93.1
10.9	—	18.7	18.7	32.0	—	54.9	54.9	94.2	—
11.0	11.0	18.9	—	32.4	32.4	55.6	—	95.3	95.3
11.1	—	19.1	19.1	32.8	—	56.2	56.2	96.5	—
11.3	11.3	19.3	—	33.2	33.2	56.9	—	97.6	97.6
11.4	—	19.6	19.6	33.6	—	57.6	57.6	98.8	—
11.5	11.5	19.8	—	34.0	34.0	58.3	—		
11.7	—	20.0	20.0	34.4	—	59.0	59.0		
11.8	11.8	20.3	—	34.8	34.8	59.7	—		
12.0	—	20.5	20.5	35.2	—	60.4	60.4		
12.1	12.1	20.8	—	35.7	35.7	61.2	—		
12.3	—	21.0	21.0	36.1	—	61.9	61.9		
12.4	12.4	21.3	—	36.5	36.5	62.6	—		
12.6	—	21.5	21.5	37.0	—	63.4	63.4		
12.7	12.7	21.8	—	37.4	37.4	64.2	—	**2%, 5%**	**10%**
12.9	—	22.1	22.1	37.9	—	64.9	64.9	10	10
13.0	13.0	22.3	—	38.3	38.3	65.7	—	11	—
13.2	—	22.6	22.6	38.8	—	66.5	66.5	12	12
13.3	13.3	22.9	—	39.2	39.2	67.3	—	13	—
13.5	—	23.2	23.2	39.7	—	68.1	68.1	15	15
13.7	13.7	23.4	—	40.2	40.2	69.0	—	16	—
13.8	—	23.7	23.7	40.7	—	69.8	69.8	18	18
14.0	14.0	24.0	—	41.2	41.2	70.6	—	20	—
14.2	—	24.3	24.3	41.7	—	71.5	71.5	22	22
14.3	14.3	24.6	—	42.2	42.2	72.3	—	24	—
14.5	—	24.9	24.9	42.7	—	73.2	73.2	27	27
14.7	14.7	25.2	—	43.2	43.2	74.1	—	30	—
14.9	—	25.5	25.5	43.7	—	75.0	75.0	33	33
15.0	15.0	25.8	—	44.2	44.2	75.9	—	36	—
15.2	—	26.1	26.1	44.8	—	76.8	76.8	39	39
15.4	15.4	26.4	—	45.3	45.3	77.7	—	43	—
15.6	—	26.7	26.7	45.9	—	78.7	78.7	47	47
15.8	15.8	27.1	—	46.4	46.4	79.6	—	51	—
16.0	—	27.4	27.4	47.0	—	80.6	80.6	56	56
16.2	16.2	27.7	—	47.5	47.5	81.6	—	62	—
16.4	—	28.0	28.0	48.1	—	82.5	82.5	68	68
16.5	16.5	28.4	—	48.7	48.7	83.5	—	75	—
16.7	—	28.7	28.7	49.3	—	84.5	84.5	82	82
16.9	16.9	29.1	—	49.9	49.9	85.6	—	91	—

Figure 4–10 Chart lists standard resistance values.

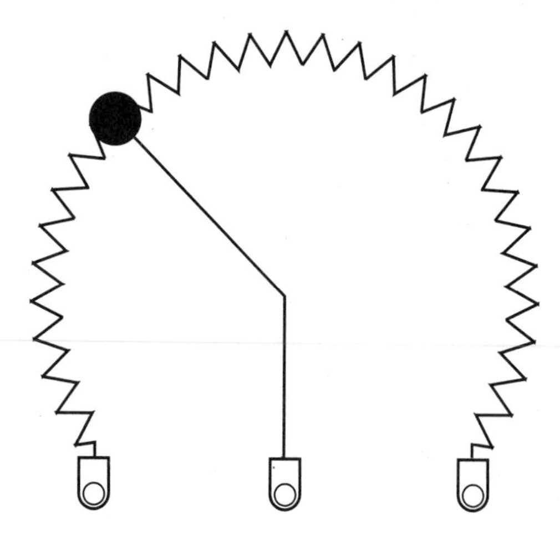

Figure 4–11 In a variable resistor, a resistive wire is wound in a circular pattern, and a sliding tap makes contact with the wire.

adjusted between one end of the resistive wire and the sliding tap. If the resistive wire has a total value of 100 watts, the resistor can be set between the values of 0 and 100 ohms.

This type of resistor has a wiper arm inside the case that makes contact with the resistive element. The full resistance value is between the two outside terminals, and the wiper arm is connected to the center terminal. The resistance between the center terminal and either of the two outside terminals can be adjusted by turning the shaft and changing the position of the wiper arm. Wire-wound variable resistors of this type can be obtained, also. The advantage of the wire-wound type is a higher power rating.

Variable Resistor Terminology

Variable resistors are known by several common names. The most popular name is *pot,* which is shortened from the word *potentiometer.* Another common name is *rheostat.* A rheostat is actually a variable resistor that has only two terminals instead of three, but three-terminal variable resistors are often referred to as rheostats, as well. A potentiometer describes how a variable resistor is used rather than some specific type of resistor. The word *potentiometer* comes from the word *potential,* or voltage. A potentiometer is a variable resistor used to provide a variable voltage as shown in Figure 4–12. In this example, one end of a variable resistor is connected to +12 volts, and the other end is connected to ground. The middle terminal or wiper is connected to the positive terminal of a voltmeter and the negative lead is connected to ground. If the wiper is moved to the upper end of the resistor, the voltmeter will indicate a potential of 12 volts. If the wiper is moved to the bottom, the voltmeter will indicate a value of 0 volts. The wiper can be adjusted to provide any value of voltage between 12 and 0 volts.

Schematic Symbols for Resistors

Electrical schematics use symbols to represent the use of a resistor. Unfortunately, the symbol used to represent a resistor is not standard. Figure 4–13 illustrates several schematic symbols used to represent both fixed and variable resistors.

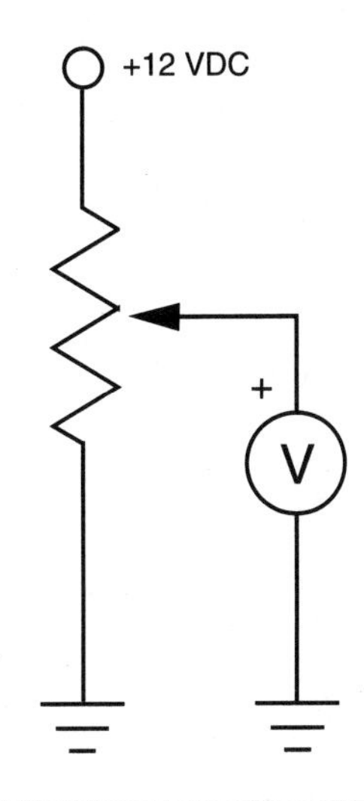

Figure 4–12 A variable resistor used as a potentiometer can provide variable voltage.

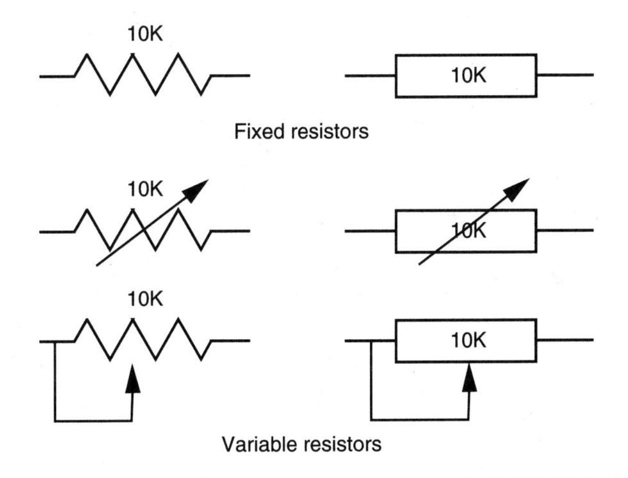

Figure 4–13 These schematic symbols are used to represent resistors.

Ohm's Law

Georg S. Ohm discovered that the voltage across any element of a direct-current circuit was proportional to the product of current flowing through that element in amperes and the resistance or impedance of the element in ohms. Three basic Ohm's law formulas are:

$$E = I \times R \quad I = \frac{E}{R} \quad R = \frac{E}{I}$$

or

$$E = I \times Z \quad I = \frac{E}{Z} \quad Z = \frac{E}{I}$$

Where:

E = EMF or voltage
I = amperes (intensity of current)
R = resistance in ohms
Z = impedance in ohms

A chart showing formulas for values of voltage, current, resistance, and watts (P for power) is shown in Figure 4–14.

Figure 4–14 Chart of Ohm's law formulas.

CHAPTER 5

Series Circuits

Series circuits are characterized by the fact that they contain only one path for current flow. A series circuit may contain a single resistor or multiple resistors, shown in Figure 5–1. Regardless of the number of resistors or other electrical devices contained in the circuit, the primary characteristic of a series circuit is that it contains only one path for the flow of current.

Series circuits have several characteristics.

1. Since there is only one path for the current to flow in a series circuit, the current must be the same at all points in the circuit.
2. The total impedance of a series circuit is the sum of all elements that impede the flow of current. These elements may be resistive, inductive, capacitive, or a combination of these elements.
3. The sum of the voltage drops across all the elements in a series circuit must equal the applied voltage.

Determining Total Impedance for a Series Circuit

The formula used to determine the total impedance of a series circuit is dictated by the elements contained within the circuit. The total impedance of a series circuit that contains only one type of load, such as a circuit containing only resistance, or a circuit containing only inductors or only capacitors, can be determined by adding the current limiting elements together, shown in Figure 5–2.

Determining Series Impedance for Circuits With Different Loads

When different types of loads are connected in series, vector addition must be used to determine total impedance. In the circuit shown in Figure 5–3, a resistor and inductor are connected in series. The resistor has a resistance of 300 Ω and the inductor has an inductive reactance of 275 Ω. Although this circuit is connected in series, the total impedance of the circuit is not the algebraic sum of the two quantities. The total impedance can be determined using the formula:

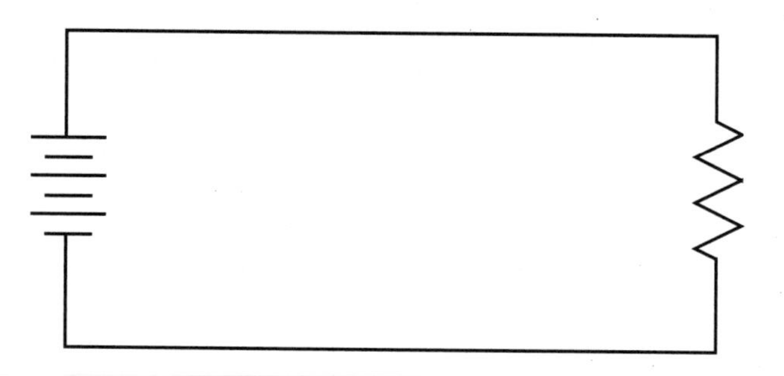

Series circuit containing one resistor

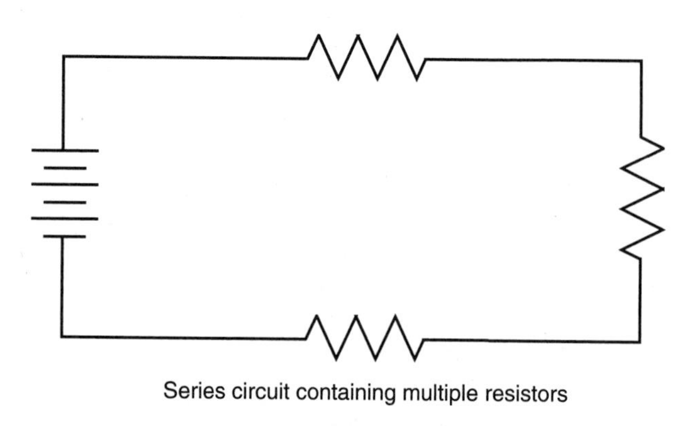

Series circuit containing multiple resistors

Figure 5–1 Series circuits have only one path for current flow.

$$Z = \sqrt{R^2 + X_L{}^2}$$

$$Z = \sqrt{300^2 + 275^2}$$

$$Z = 406.97\,\Omega$$

Similar formulas can be used to determine the total impedance of series circuits containing resistance and capacitance, or resistance, inductance, and capacitance.

For a circuit containing resistance and capacitance, use the formula:

$$Z = \sqrt{R^2 + X_C{}^2}$$

For a circuit containing elements of resistance, inductance, and capacitance, use the formula:

$$Z = \sqrt{R^2 + (X_L - X_C)^2}$$

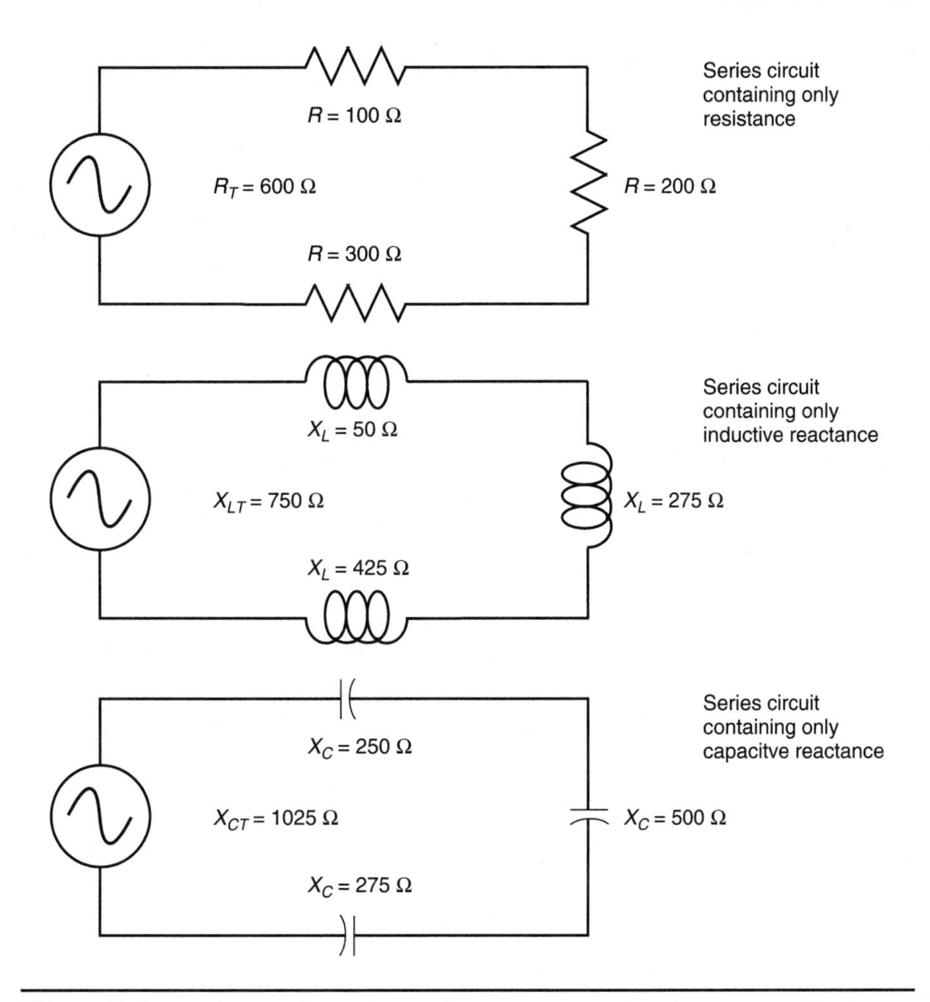

Figure 5–2 Finding total impedance of series circuits with only one type of load.

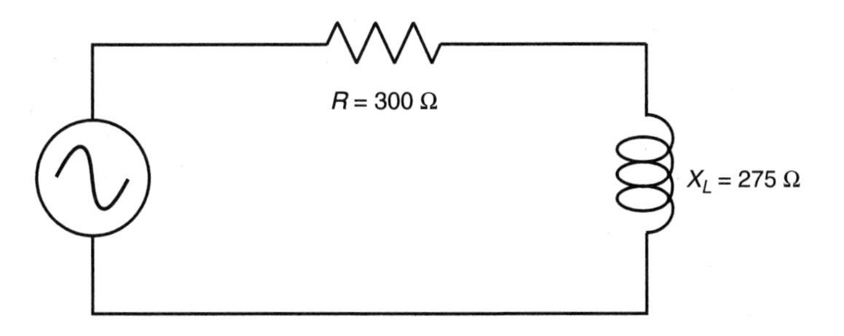

Figure 5–3 A resistor and inductor connected in series.

CHAPTER 6

Parallel Circuits

Parallel circuits are defined by the fact that they contain more than one path for current flow. The circuit shown in Figure 6–1 contains three separate paths that current can follow. Three characteristics are true for any parallel circuit:

1. The voltage is the same across any branch of a parallel circuit.
2. The total current is the sum of the currents flowing through each circuit branch.
3. The total impedance is less than the impedance of any single branch.

Determining Total Impedance for Parallel Circuits

In a parallel circuit the total circuit impedance is always less than any individual branch because each additional branch provides another current path that increases the total current flow. Any additional current paths reduce the total impedance to the power source. Three basic formulas are used to determine the impedance of a parallel circuit. One of these formulas can be used only when all parallel resistive or reactive values are the same, such as three parallel resistors with a value of 75 Ω each, or five inductors with an inductive reactance of 200 Ω each. To determine the total resistance of three resistors with a value of 75 Ω each, use the formula:

$$R_T = \frac{R}{N}$$

In this formula, *R* stands for the resistance of one single branch, and *N* stands for the number of branches.

$$R_T = \frac{75}{3}$$

$$R_T = 25\,\Omega$$

Another formula for determining parallel impedance is called the product-over-sum formula:

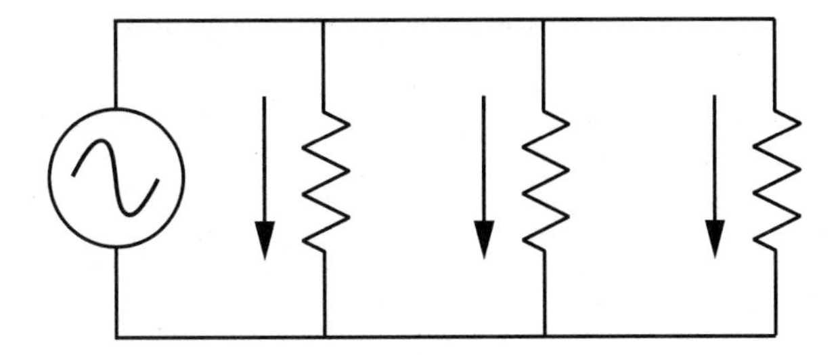

Figure 6–1 A parallel circuit has more than one path for current flow.

$$R_T = \frac{R_1 \times R_2}{R_1 + R_2}$$

This formula can be used to determine the total resistance of two resistors at a time. Assume a circuit contains three parallel resistors with resistance values of 40 Ω, 60 Ω, and 120 Ω. To use the product-over-sum formula, find the total resistance of two resistors at a time until all values have been used.

$$R_T = \frac{40 \times 60}{40 + 60}$$

$$R_T = \frac{2400}{100}$$

$$R_T = 24\ \Omega$$

This procedure is repeated using the total value for the first two resistors as R_1 and the value of the next resistor as R_2.

$$R_T = \frac{24 \times 120}{24 + 120}$$

$$R_T = \frac{2880}{144}$$

$$R_T = 20\ \Omega$$

The third formula for determining total resistance of resistors connected in parallel is the reciprocal formula. This formula states that the reciprocal of the total resistance is equal to the sum of the reciprocals for each branch.

$$\frac{1}{R_T} = \frac{1}{R_1} + \frac{1}{R_2} + \frac{1}{R_3} + \frac{1}{R_N}$$

or

$$R_T = \frac{1}{\frac{1}{R_1} + \frac{1}{R_2} + \frac{1}{R_3} + \frac{1}{R_N}}$$

The total resistance of the three resistors will be computed using the reciprocal formula.

$$R_T = \frac{1}{\frac{1}{4}+\frac{1}{60}+\frac{1}{120}}$$

$$R_T = \frac{1}{0.025+0.0167+0.00833}$$

$$R_T = \frac{1}{0.05}$$

$$R_T = 20\ \Omega$$

When reactive components such as inductive reactance or capacitive reactance are connected in parallel, the total impedance must be computed to include the phase angle shift between voltage and current. In the circuit shown in Figure 6–2, a resistor and inductor are connected in parallel. The resistor has a resistance of 50 Ω and the inductor has an inductive reactance of 80 Ω. The total impedance of the circuit can be found using the formula:

$$Z = \frac{1}{\sqrt{\left(\frac{1}{R}\right)^2+\left(\frac{1}{X_L}\right)^2}}$$

$$Z = \frac{1}{\left(\frac{1}{50}\right)^2+\left(\frac{1}{80}\right)^2}$$

$$Z = \frac{1}{\sqrt{0.0004+0.00015625}}$$

$$Z = 42.399\ \Omega$$

The formula for determining the impedance of a resistor and capacitor connected in parallel is similar to the formula for determining the impedance of a resistor and inductor connected in parallel.

$$Z = \frac{1}{\sqrt{\left(\frac{1}{R}\right)^2+\left(\frac{1}{X_C}\right)^2}}$$

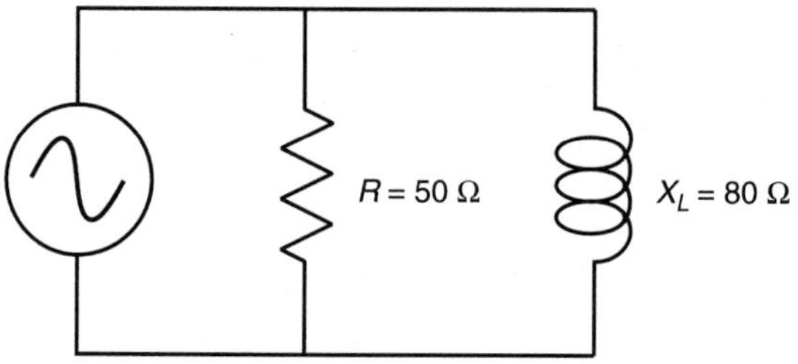

Figure 6–2 A resistor and inductor connected in parallel.

When a circuit contains elements of resistance, inductance, and capacitance connected in parallel, as shown in Figure 6–3, the following formula can be used to determine impedance:

$$Z = \frac{1}{\sqrt{\left(\frac{1}{R}\right)^2 + \left(\frac{1}{X_L} - \frac{1}{X_C}\right)^2}}$$

In the circuit shown in Figure 6–4, the resistor has a resistance of 40 Ω, the inductor has an inductive reactance of 80 Ω, and the capacitor has a capacitive reactance of 50 Ω.

$$Z = \frac{1}{\sqrt{\frac{1}{40} + \left(\frac{1}{80} - \frac{1}{50}\right)^2}}$$

$$Z = \frac{1}{\sqrt{0.025^2 + (0.0125 - 0.02)^2}}$$

$$Z = \frac{1}{\sqrt{0.000625 + 0.00005625}}$$

$$Z = 38.313\ \Omega$$

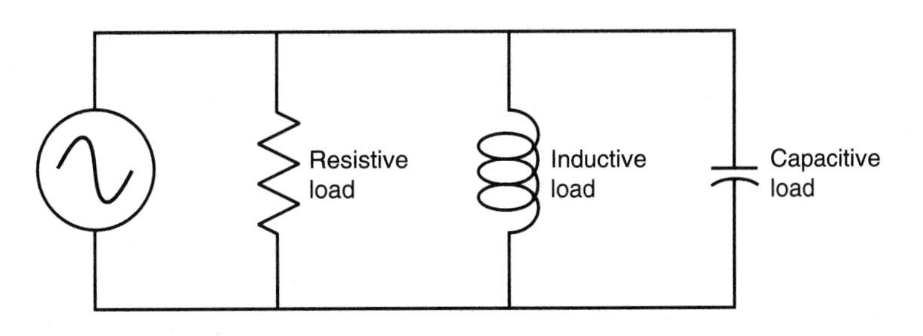

Figure 6–3 A parallel circuit containing resistive, inductive, and capacitive branches.

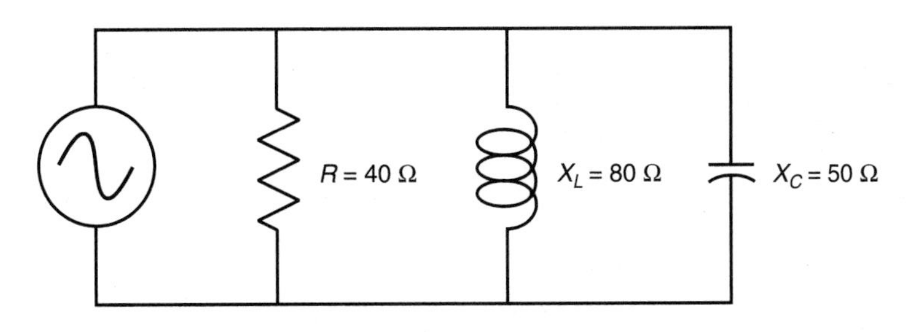

Figure 6–4 Resistance, inductance, and capacitance connected in parallel.

Current Relationships in a Parallel Circuit

In a parallel circuit the total current is equal to the sum of the currents through each branch. As long as all the branches are the same—such as all resistive or all inductive or all capacitive—the total current can be determined by adding the currents through all branches.

$$I_{\text{Total(Resistive)}} = I_{1\,\text{(Resistive)}} + I_{2\,\text{(Resistive)}} + I_{3\,\text{(Resistive)}}$$

$$I_{\text{Total(Inductive)}} = I_{1\,\text{(Inductive)}} + I_{2\,\text{(Inductive)}} + I_{3\,\text{(Inductive)}}$$

$$I_{\text{Total(Capacitive)}} = I_{1\,\text{(Capacitive)}} + I_{2\,\text{(Capacitive)}} + I_{3\,\text{(Capacitive)}}$$

When a parallel circuit contains different types of loads, where one branch is resistive and another inductive, vector addition must be used to determine the total.

Example: Assume the resistive branch to have a current flow of 1.2 amperes and the inductive branch to have a current flow of 1.6 amperes. To find the total current, use the formula:

$$I_{\text{Total}} = \sqrt{I_R^{\,2} + I_L^{\,2}}$$

$$I_{\text{Total}} = \sqrt{1.2^2 + 1.6^2}$$

$$I_{\text{Total}} = 2 \text{ amperes}$$

A similar formula can be used to determine the total current of a parallel circuit containing resistance and capacitance.

$$I_{\text{Total}} = \sqrt{I_R^{\,2} + I_C^{\,2}}$$

When a parallel circuit contains branches of all three elements (see Figure 6–3), this formula can be used to determine total current:

$$I_{\text{Total}} = \sqrt{I_R^{\,2} + (I_L - I_C)^2}$$

CHAPTER 7

Combination Circuits

Combination circuits are circuits that contain a combination of both series and parallel elements. A simple combination circuit is shown in Figure 7–1. To determine which components are in parallel and which are in series, trace the flow of current through the circuit. Series elements can be identified by the fact that all the circuit current must pass through them, and parallel elements can be identified by the fact that the circuit current will divide, and part of the current will flow through each element.

In Figure 7–1, it will be assumed that the current will flow from point A to point B. All the current in the circuit must flow through resistor R_1. Resistor R_1 is, therefore, in series with the rest of the circuit. When the current reaches the junction point of resistors R_2 and R_3, however, it splits. Part of the current flows through resistor R_2 and part flows through resistor R_3. These two resistors are in parallel with each other.

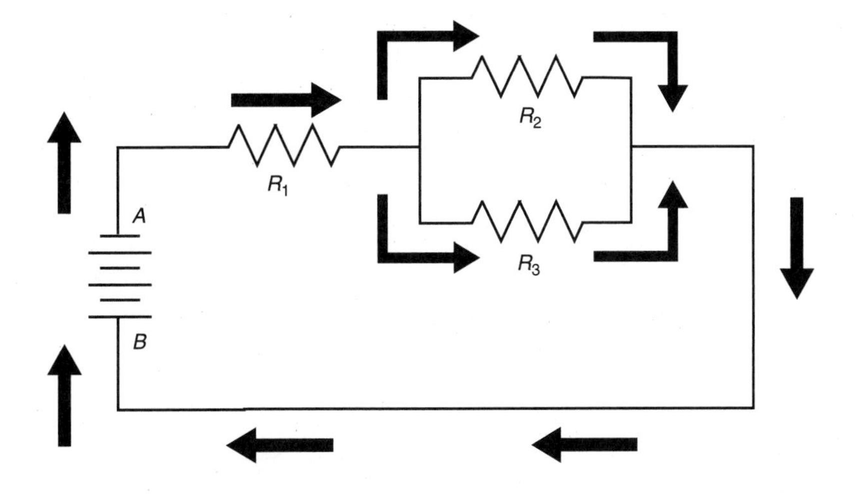

Figure 7–1 Simple combination circuit.

Example Circuit #1

The circuit shown in Figure 7–2 contains four resistors with values of 325 ohms, 275 ohms, 150 ohms, and 250 ohms. The circuit has a total current flow of 1 ampere. When the path of current flow is traced, it can be seen that there are two separate paths by which current can flow from the negative terminal to the positive terminal. One path is through resistors R_1 and R_2, and the other path is through resistors R_3 and R_4. These two paths are, therefore, in parallel with each other. However, the same current must flow through resistors R_1 and R_2. These two resistors are in series with each other. The same is true for resistors R_3 and R_4. Resistors R_3 and R_4 are in series with each other.

The circuit shown in Figure 7–2 can be reduced or simplified to a simple parallel circuit as shown in Figure 7–3. Since resistors R_1 and R_2 are connected in series, their values

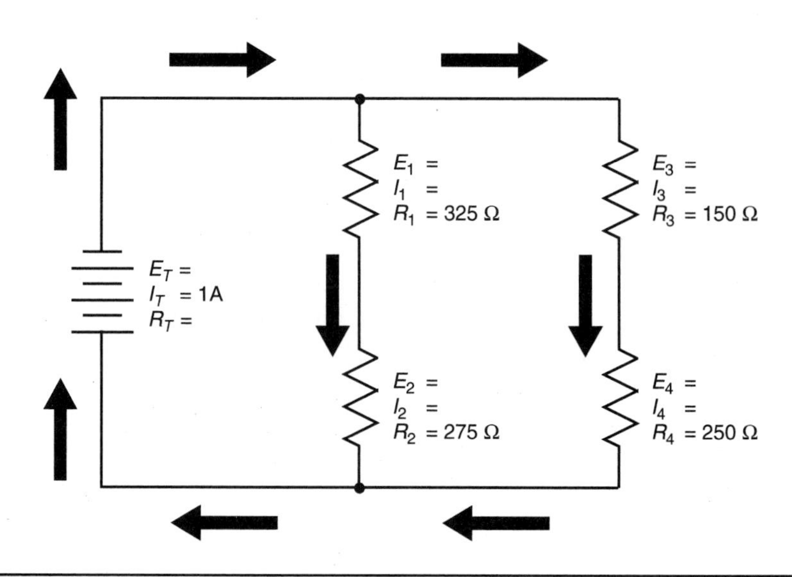

Figure 7–2 Tracing the current paths through combination circuit #1.

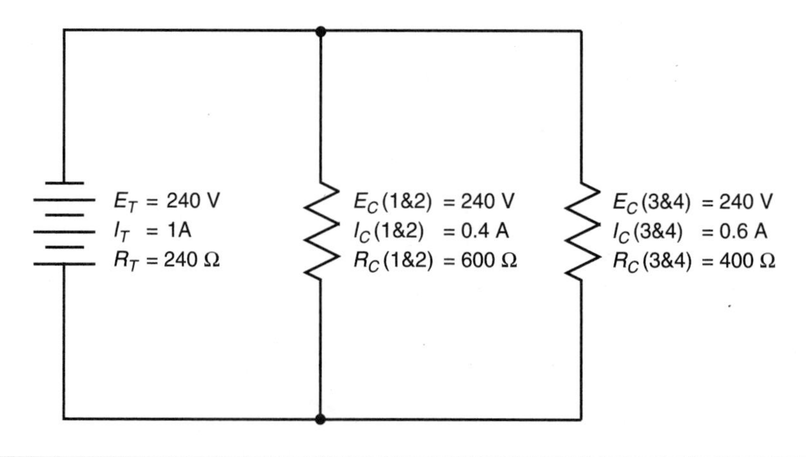

Figure 7–3 Simplifying combination circuit #1.

can be added to form one equivalent resistor, [$R_{C(1\&2)}$], which stands for a combination of resistors 1 and 2. The same is true for resistors R_3 and R_4. Their values are added to form resistor $R_{C(3\&4)}$. Now that the circuit has been reduced to a simple parallel circuit, the total resistance can be found.

$$R_T = \frac{1}{\frac{1}{R_{C(1\&2)}} + \frac{1}{R_{C(3\&4)}}}$$

$$R_T = \frac{1}{\frac{1}{600} + \frac{1}{400}}$$

$$R_T = \frac{1}{0.0016667 + 0.0025}$$

$$R_T = \frac{1}{0.0041667}$$

$$R_T = 240 \text{ ohms}$$

Now that the total resistance has been determined, the other circuit values can be computed. The applied voltage can be found using Ohm's law.

$$E_T = I_T \times R_T$$

$$E_T = 1 \times 240$$

$$E_T = 240 \text{ volts}$$

Branches connected in parallel must have the same voltage. Therefore, the voltage drop across resistors $R_{C(1\&2)}$ and $R_{C(3\&4)}$ is the same. Since the voltage drop is known and the resistance is known, Ohm's law can be used to find the current flow through each branch.

$$I_{C(1\&2)} = \frac{E_{C(1\&2)}}{R_{C(1\&2)}}$$

$$I_{C(1\&2)} = \frac{240}{600}$$

$$I_{C(1\&2)} = 0.4 \text{ amp}$$

$$I_{C(3\&4)} = \frac{E_{C(3\&4)}}{R_{C(3\&4)}}$$

$$I_{C(3\&4)} = \frac{240}{400}$$

$$I_{C(3\&4)} = 0.6 \text{ amp}$$

These values can now be used to solve the missing values in the original circuit. Resistor $R_{C(1\&2)}$ is actually a combination of resistors R_1 and R_2. The values of voltage and current that apply to $R_{C(1\&2)}$ therefore apply to resistors R_1 and R_2. Since resistors R_1 and R_2 are connected in series, 0.4 amp of current flows through each resistor. The voltage drop across each can be computed using Ohm's law.

$$E_1 = I_1 \times R_1$$
$$E_1 = 0.4 \times 325$$
$$E_1 = 130 \text{ volts}$$

$$E_2 = I_2 \times R_2$$
$$E_2 = 0.4 \times 275$$
$$E_2 = 110 \text{ volts}$$

The values of voltage and current for resistor $R_{C(3\&4)}$ apply to resistors R_3 and R_4. The same amount of current that flows through resistor $R_{C(3\&4)}$ flows through resistors R_3 and R_4. The voltage across these two resistors can now be computed using Ohm's law.

$$E_3 = I_3 \times R_3$$
$$E_3 = 0.6 \times 150$$
$$E_3 = 90 \text{ volts}$$

$$E_4 = I_4 \times R_4$$
$$E_4 = 0.6 \times 250$$
$$E_4 = 150 \text{ volts}$$

These values of voltage and current can now be added to the circuit in Figure 7–2.

Example Circuit #2

The second example circuit is shown in Figure 7–4. The first step in finding the missing values is to trace the current path through the circuit to determine which resistors are connected in series and which are connected in parallel. All the current must flow through resistor R_1. Resistor R_1 is, therefore, in series with the rest of the circuit. When the current reaches the junction of resistors R_2 and R_3, it divides, and part of the current flows through

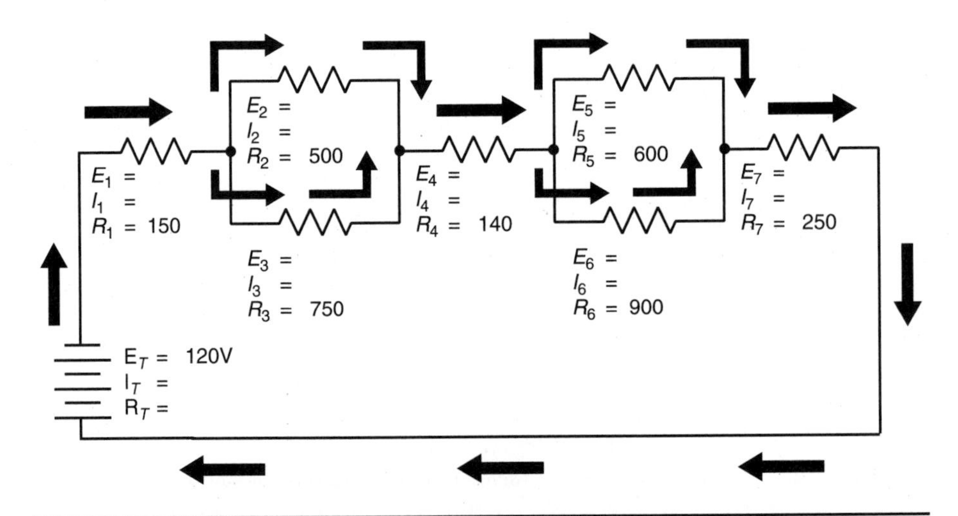

Figure 7–4 Tracing the flow of current through combination circuit #2.

each resistor. Resistors R_2 and R_3 are in parallel with each other. All the current must then flow through resistor R_4 to the junction of resistors R_5 and R_6. The current path is divided between these two resistors. Resistors R_5 and R_6 are connected in parallel with each other. All the circuit current must then flow through resistor R_7.

The next step in solving this circuit is to reduce it to a simpler circuit. If the total resistance of the first parallel block formed by resistors R_2 and R_3 is found, this block can be replaced by a single resistor.

$$R_{C(2\&3)} = \frac{1}{\frac{1}{R_2} + \frac{1}{R_3}}$$

$$R_{C(2\&3)} = \frac{1}{0.002 + 0.0013333}$$

$$R_{C(2\&3)} = \frac{1}{0.00333333}$$

$$R_{C(2\&3)} = 300\ \Omega$$

The equivalent resistance of the second parallel block can be computed in the same way.

$$R_{C(5\&6)} = \frac{1}{\frac{1}{R_5} + \frac{1}{R_6}}$$

$$R_{C(5\&6)} = \frac{1}{\frac{1}{600} + \frac{1}{900}}$$

$$R_{C(5\&6)} = \frac{1}{0.0016667 + 0.00011111}$$

$$R_{C(5\&6)} = \frac{1}{0.00277778}$$

$$R_{C(5\&6)} = 360\ \Omega$$

Now that the total resistance of the second parallel block is known, the circuit can be redrawn as a simple series circuit as shown in Figure 7–5. The first parallel block has been

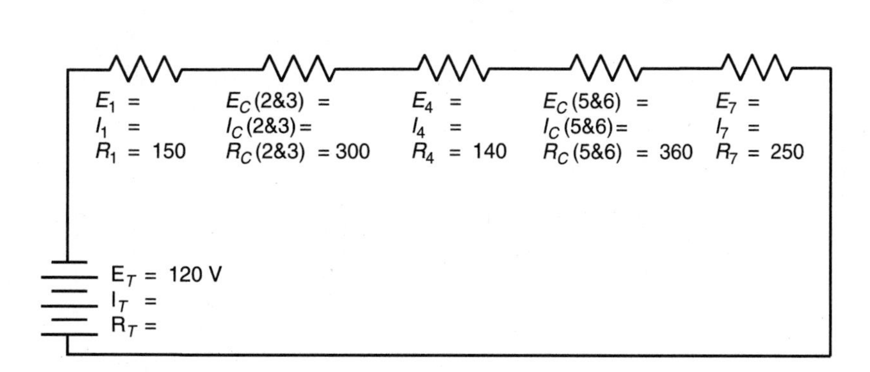

Figure 7–5 Simplifying combination circuit #2.

replaced with a single resistor of 300 ohms labeled $R_{C(2\&3)}$, and the second parallel block has been replaced with a single 360-ohm resistor labeled $R_{C(5\&6)}$. Ohm's law can be used to find the missing values in this series circuit.

R_T (total resistance of the circuit) can be computed by adding the resistance of all resistors together.

$$R_T = R_1 + R_{C(2\&3)} + R_4 + R_{C(5\&6)} + R_1$$

$$R_T = 500 + 300 + 140 + 360 + 250$$

$$R_T = 1200 \text{ ohms}$$

Since the total voltage and total resistance is known, the total current flow through the circuit can be computed.

$$I_T = \frac{E_T}{R_T}$$

$$I_T = \frac{120}{1200}$$

$$I_T = 0.1 \text{ amp}$$

The current flow through each resistor is 0.1 amp. The voltage drop across each resistor can now be computed using Ohm's law.

$$E_1 = I_1 \times R_1$$

$$E_1 = 0.1 \times 150$$

$$E_1 = 15 \text{ volts}$$

$$E_{C(2\&3)} = I_{C(2\&3)} \times R_{C(2\&3)}$$

$$E_{C(2\&3)} = 0.1 \times 300$$

$$E_{C(2\&3)} = 30 \text{ volts}$$

$$E_4 = I_4 \times R_4$$

$$E_4 = 0.1 \times 140$$

$$E_4 = 14 \text{ volts}$$

$$E_{C(5\&6)} = I_{C(5\&6)} \times R_{C(5\&6)}$$

$$E_{C(5\&6)} = 0.1 \times 360$$

$$E_{C(5\&6)} = 36 \text{ volts}$$

$$E_7 = I_7 \times R_7$$

$$E_7 = 0.1 \times 250$$

$$E_7 = 25 \text{ volts}$$

These values can be used to solve missing parts in the original circuit.

Resistor $R_{C(2\&3)}$ is actually the parallel block containing resistors R_2 and R_3. Since 30 volts is dropped across resistor $R_{C(2\&3)}$, the same 30 volts is dropped across resistors R_2 and R_3. The current flow through these resistors can be computed using Ohm's law.

$$I_2 = \frac{E_2}{R_2}$$

$$I_2 = \frac{30}{500}$$

$$I_2 = 0.06 \text{ amp}$$

$$I_3 = \frac{E_3}{R_3}$$

$$I_3 = \frac{30}{750}$$

$$I_3 = 0.04 \text{ amp}$$

The values of resistor $R_{C(5\&6)}$ can be applied to the parallel block composed of resistors R_5 and R_6. $E_{C(5\&6)}$ is 36 volts. This is the voltage dropped across resistors R_5 and R_6. The current flow through these two resistors can be computed using Ohm's law.

$$I_5 = \frac{E_5}{R_5}$$

$$I_5 = \frac{36}{600}$$

$$I_5 = 0.06 \text{ amp}$$

$$I_6 = \frac{E_6}{R_6}$$

$$I_6 = \frac{36}{900}$$

$$I_6 = 0.04 \text{ amp}$$

Example Circuit #3

Both of the preceding circuits were solved by first determining which parts of the circuit were in series and which were in parallel. The circuits were then reduced to a simple series or parallel circuit. This same procedure can be used for any combination circuit. The circuit shown in Figure 7–6 will be reduced to a simpler circuit first. Once the values of the simple circuit are found, they can be placed back in the original circuit to find other values.

The first step will be to reduce the top part of the circuit to a single resistor. This part consists of resistors R_3 and R_4. Since these two resistors are connected in series with each

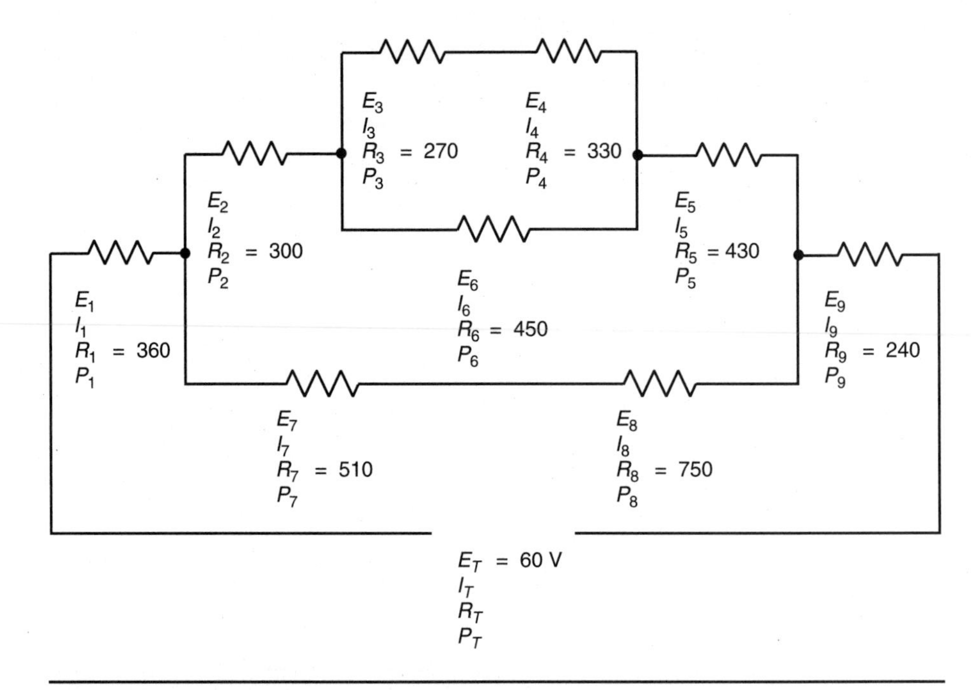

Figure 7–6 Circuit #3.

other, their values can be added to form one single resistor. This combination will form R_{C1}, shown in Figure 7–7.

$$R_{C1} = R_3 + R_4$$

$$R_{C1} = 270 + 330$$

$$R_{C1} = 600 \text{ ohms}$$

The top part of the circuit is now formed by resistors R_{C1} and R_6. These two resistors are in parallel with each other. If their total resistance is computed, they can be changed into one single resistor with a value of 257.143 ohms. This combination will become resistor R_{C2}, shown in Figure 7–8.

$$R_{C2} = \frac{1}{\frac{1}{R_{C1}} + \frac{1}{R_6}}$$

$$R_{C2} = \frac{1}{\frac{1}{600} + \frac{1}{450}}$$

$$R_T = 257.143 \text{ ohms}$$

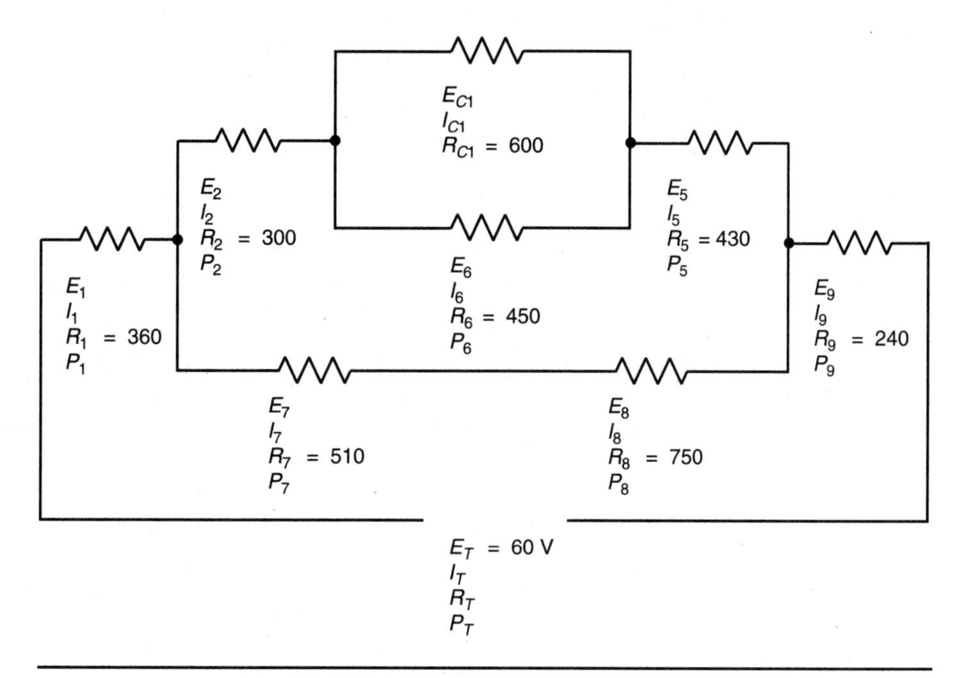

Figure 7–7 Resistors R_1 and R_2 are combined to form R_{C1}.

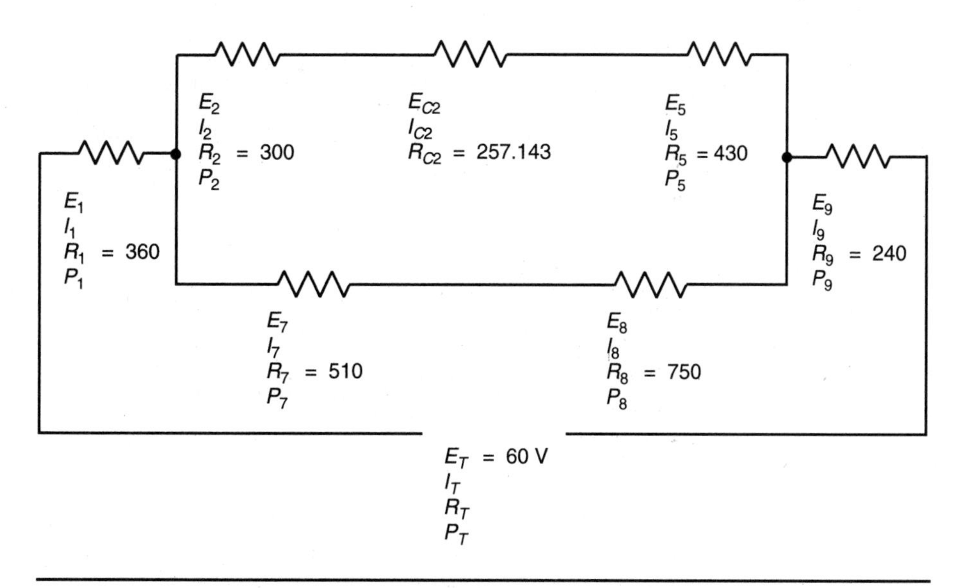

Figure 7–8 Resistors R_{C1} and R_6 are combined to form R_{C2}.

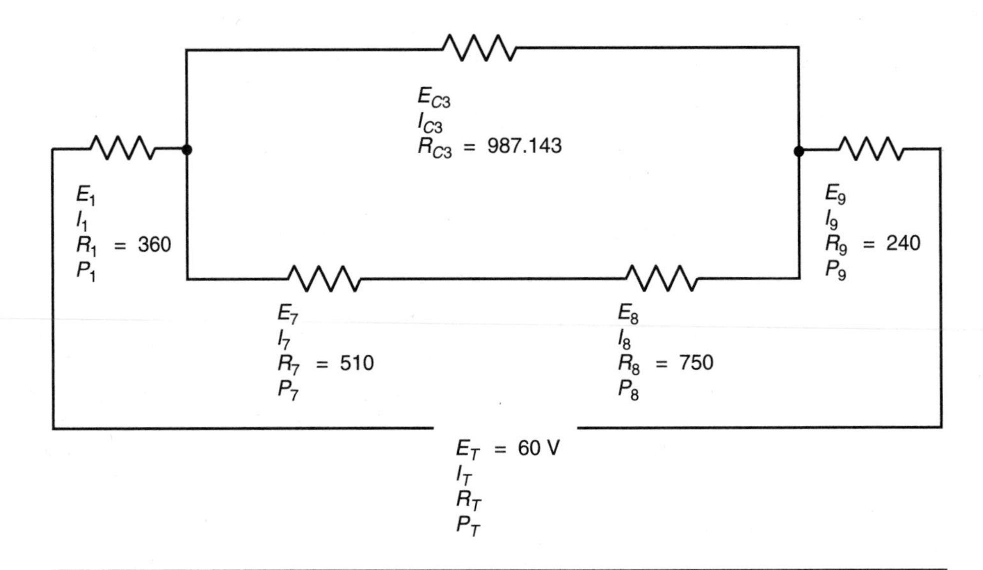

Figure 7–9 Resistors R_2, R_{C2}, and R_5 are combined to form resistor R_{C3}.

The top of the circuit now consists of resistors R_2, R_{C2}, and R_5. These three resistors are connected in series with each other. They can be combined to form resistor R_{C3} by adding their resistance's together, shown in Figure 7–9.

$$R_{C3} = R_2 + R_{C2} + R_5$$

$$R_{C3} = 300 + 257.143 + 430$$

$$R_{C3} = 987.143 \text{ ohms}$$

Resistors R_7 and R_8 are connected in series with each other, also. These two resistors will be added to form resistor R_{C4}, shown in Figure 7–10.

$$R_{C4} = R_7 + R_8$$

$$R_{C4} = 510 + 750$$

$$R_{C4} = 1260 \text{ ohms}$$

Resistors R_{C3} and R_{C4} are connected in parallel with each other. There total resistance can be computed to form resistor R_{C5}, shown in Figure 7–11.

$$R_{C5} = \frac{1}{\frac{1}{R_{C3}} + \frac{1}{R_{C4}}}$$

$$R_{C5} = \frac{1}{\frac{1}{987.143} + \frac{1}{1260}}$$

$$R_{C5} = 553.503 \text{ ohms}$$

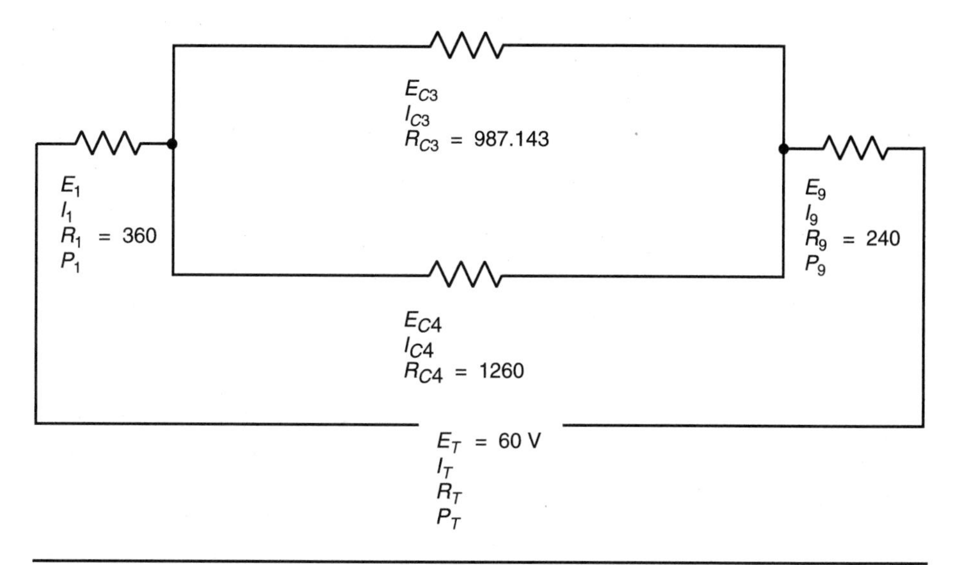

Figure 7–10 Resistors R_7 and R_8 are combined to form R_{C4}.

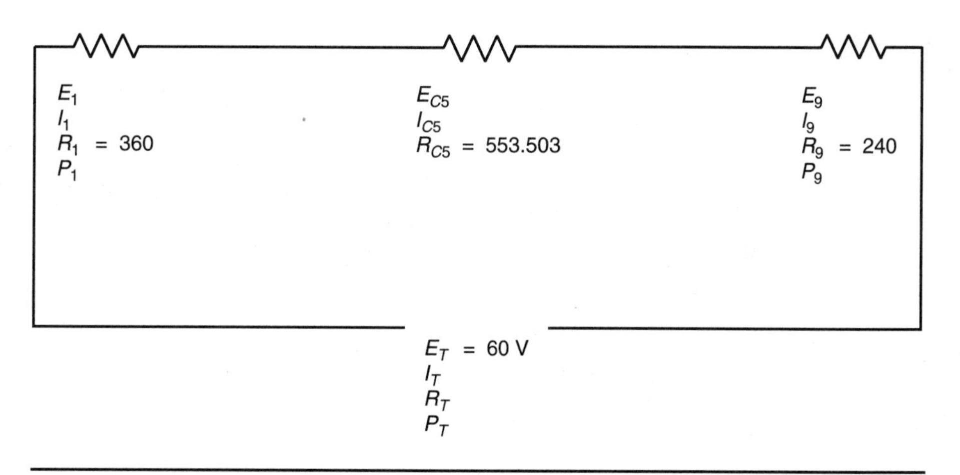

Figure 7–11 Resistors R_{C3} and R_{C4} are combined to form R_{C5}.

The circuit has now been reduced to a simple series circuit containing three resistors. The total resistance of the circuit can be computed by adding resistors R_1, R_{C5}, and R_9.

$$R_T = R_1 + R_{C5} + R_9$$

$$R_T = 360 + 553.503 + 240$$

$$R_T = 1153.503 \text{ ohms}$$

Now that the total resistance and total voltage are known, the total circuit current and power can be computed using Ohm's law.

$$I_T = \frac{E_T}{R_T}$$

$$I_T = \frac{60}{1153.503}$$

$$I_T = 0.052 \text{ amp}$$

$$P_T = E_T \times I_T$$

$$P_T = 60 \times 0.052$$

$$P_T = 3.12 \text{ watts}$$

Ohm's law can be used to find the missing values for resistors R_1, R_{C5}, and R_9.

$$E_1 = I_1 \times R_1$$

$$E_1 = 0.052 \times 360$$

$$E_1 = 18.72 \text{ volts}$$

$$P_1 = E_1 \times I_1$$

$$P_1 = 18.72 \times 0.052$$

$$P_1 = 0.973 \text{ watt}$$

$$E_{C5} = I_{C5} \times R_{C5}$$

$$E_{C5} = 0.052 \times 553.503$$

$$E_{C5} = 28.763 \text{ volts}$$

$$E_9 = I_9 \times R_9$$

$$E_9 = 0.052 \times 240$$

$$E_9 = 12.48 \text{ volts}$$

Resistor R_{C5} is actually the combination of resistors R_{C3} and R_{C4}. The values of R_{C5}, therefore, apply to resistors R_{C3} and R_{C4}. Since these two resistors are connected in parallel with each other, the voltage dropped across each will be the same. Each will have the same voltage drop as resistor R_{C5}, shown in Figure 7–12. Ohm's law can be used to find the remaining values of R_{C3} and R_{C4}.

$$I_{C4} = \frac{E_{C4}}{R_{C4}}$$

$$I_{C4} = \frac{28.763}{1260}$$

$$I_{C4} = 0.0228 \text{ amp}$$

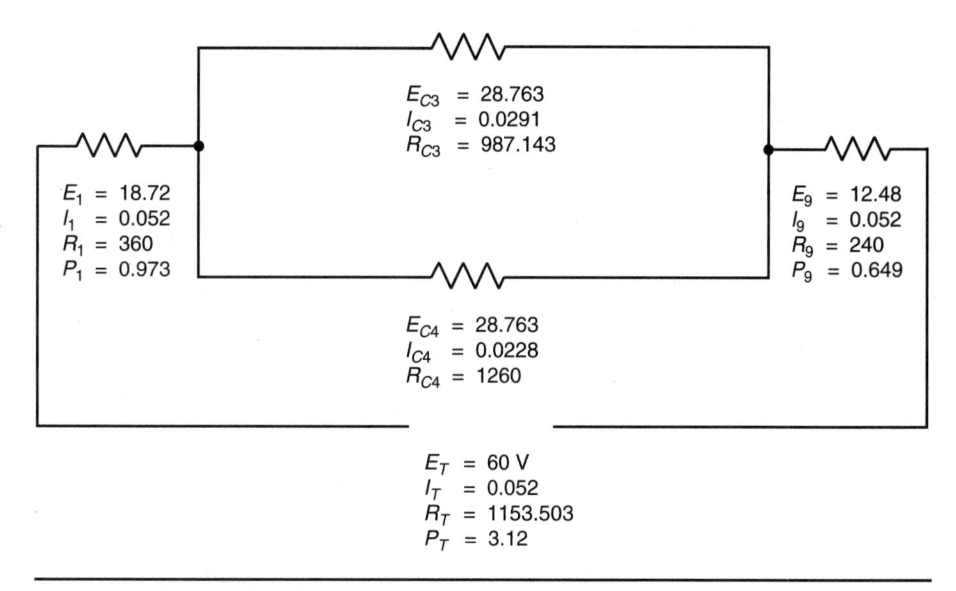

Figure 7–12 Solving the values for resistors R_{C3} and R_{C4}.

$$I_{C3} = \frac{E_{C3}}{R_{C3}}$$

$$I_{C3} = \frac{28.763}{987.143}$$

$$I_{C3} = 0.0291 \text{ amp}$$

Resistor R_{C4} is the combination of resistors R_7 and R_8. The values of resistor R_{C4} apply to resistors R_7 and R_8. Since resistors R_7 and R_8 are connected in series with each other, the current flow will be the same through both, shown in Figure 7–13. Ohm's law can be used to compute the remaining values for these two resistors.

$$E_7 = I_7 \times R_7$$

$$E_7 = 0.0228 \times 510$$

$$E_7 = 11.268 \text{ volts}$$

$$P_7 = E_7 \times I_7$$

$$P_7 = 11.268 \times 0.0228$$

$$P_7 = 0.265 \text{ watt}$$

$$E_8 = I_8 \times R_8$$

$$E_8 = 0.0228 \times 750$$

$$E_8 = 17.1 \text{ volts}$$

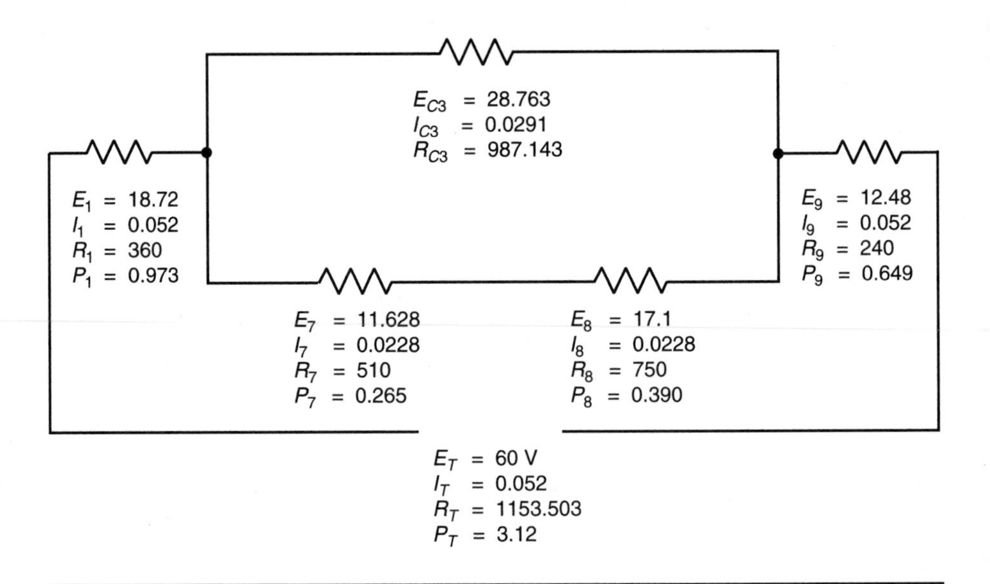

Figure 7–13 Solving values for resistors R_7 and R_8.

$$P_8 = E_8 \times I_8$$

$$P_8 = 17.1 \times 0.0228$$

$$P_8 = 0.390 \text{ watt}$$

Resistor R_{C3} is the combination of resistors R_2, R_{C2}, and R_5. Since these resistors are connected in series with each other, the current flow through each will be the same as the current flow through R_{C3}. The remaining values can be computed using Ohm's law (see Figure 7–14).

$$E_{C2} = I_{C2} \times R_{C2}$$

$$E_{C2} = 0.0291 \times 257.143$$

$$E_{C2} = 7.483 \text{ volts}$$

$$E_2 = I_2 \times R_2$$

$$E_2 = 0.0291 \times 300$$

$$E_2 = 8.73 \text{ volts}$$

$$P_2 = E_2 \times I_2$$

$$P_2 = 8.73 \times 0.291$$

$$P_2 = 0.254 \text{ watt}$$

$$E_5 = I_5 \times R_5$$

$$E_5 = 0.0291 \times 430$$

$$E_5 = 12.513 \text{ volts}$$

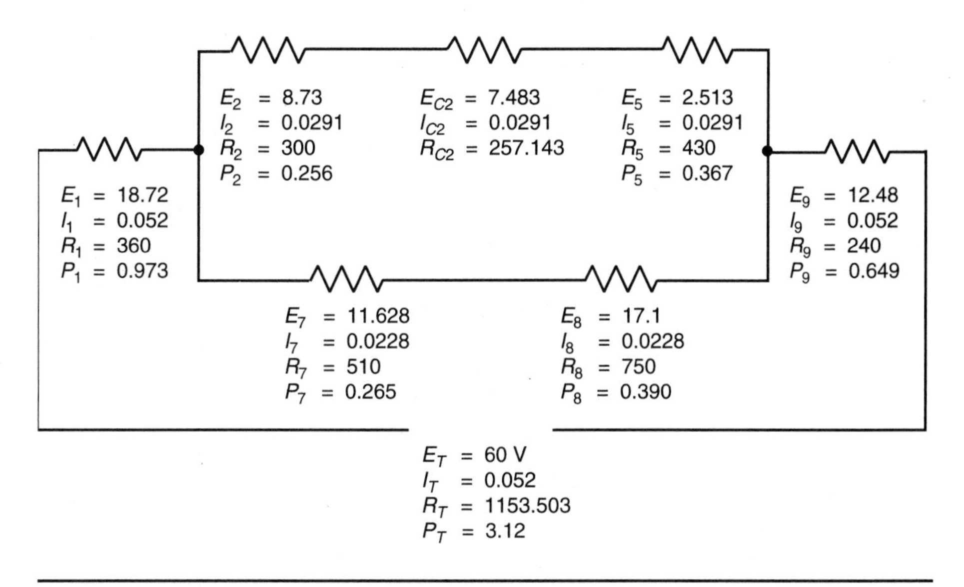

Figure 7–14 Determining values for R_2, R_{C2}, and R_5.

$$P_5 = E_5 \times I_5$$

$$P_5 = 12.513 \times 0.0291$$

$$P_5 = 0.364 \text{ watt}$$

Resistor R_{C2} is the combination of resistors R_{C1} and R_6. Resistors R_{C1} and R_6 are connected in parallel and will, therefore, have the same voltage drop as resistor R_{C2}. Ohm's law can be used to compute the remaining values for R_{C2} and R_6, shown in Figure 7–15.

$$I_{C1} = \frac{E_{C1}}{R_{C1}}$$

$$I_{C1} = \frac{7.483}{600}$$

$$I_{C1} = 0.0125 \text{ amp}$$

$$I_6 = \frac{E_6}{R_6}$$

$$I_6 = \frac{7.483}{450}$$

$$I_6 = 0.0166 \text{ amp}$$

$$P_6 = E_6 \times I_6$$

$$P_6 = 7.483 \times 0.0167$$

$$P_6 = 0.124 \text{ watt}$$

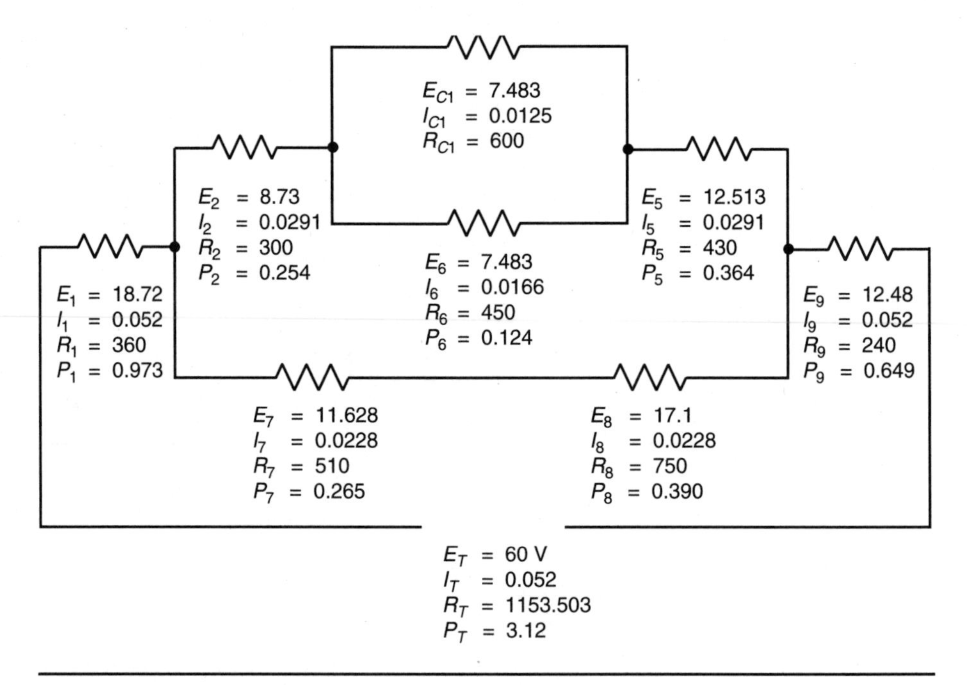

Figure 7–15 Computing the missing values for R_{C1} and R_6.

Resistor R_{C1} is the combination of resistors R_3 and R_4. Since these two resistors are connected in series, the amount of current flow through resistor R_{C1} will be the same as the flow through R_3 and R_4. The remaining values of the circuit can be found using Ohm's law (see Figure 7–16).

$$E_3 = I_3 \times R_3$$
$$E_3 = 0.0125 \times 270$$
$$E_3 = 3.375 \text{ volts}$$

$$P_3 = E_3 \times I_3$$
$$P_3 = 3.375 \times 0.0125$$
$$P_3 = 0.0423 \text{ watt}$$

$$E_4 = I_4 \times R_4$$
$$E_4 = 0.0125 \times 330$$
$$E_4 = 4.125 \text{ volts}$$

$$P_4 = E_4 \times I_4$$
$$P_4 = 4.125 \times 0.0125$$
$$P_4 = 0.0516 \text{ watt}$$

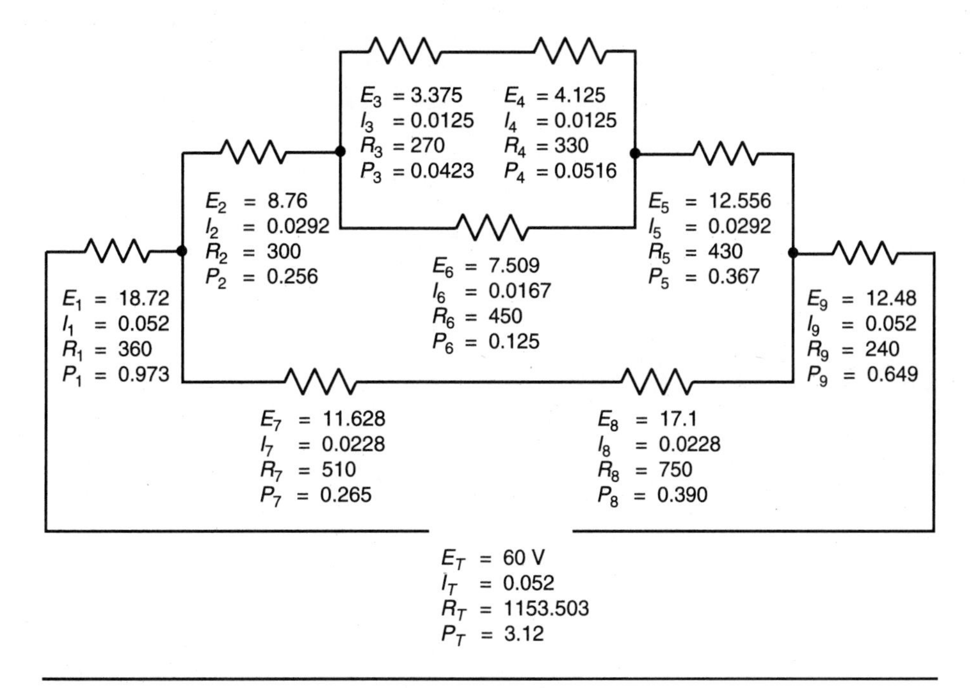

Figure 7–16 Computing the values for R_3 and R_4.

CHAPTER 8

Kirchoff's Law

Kirchoff's law is often employed to solve values of electric circuits that contain more than one power source. Kirchoff's law is based mainly on two rules for dealing with voltage and current in an electric circuit.

1. The algebraic sum of the voltage sources and voltage drops in a closed circuit must equal zero.
2. The algebraic sum of the currents entering and leaving a point must equal zero.

The first rule is the same as the series circuit rule that states that the sum of the voltage drops in a series circuit must equal the applied or source voltage. The second rule is the same as the parallel circuit rule that states that the total current will be the sum of the currents through all the circuit branches.

Kirchoff's Current Law

Kirchoff's current law states that the algebraic sum of the currents entering and leaving any particular point must equal zero. The proof of this law lies in the fact that if more current entered a point than left, some type of charge would be developed at that point. Consider the circuit shown in Figure 8–1. Four amperes of current flows through resistor R_1 to point *P*, and 6 amperes of current flows through resistor R_2 to point P. The current leaving point *P* is the sum of the two currents of 10 amperes (A). Kirchoff's law, however, states that the algebraic sum of the current must equal zero. When using Kirchoff's law, current entering a point is considered to be positive and current leaving point is considered to be negative.

$$+4\text{ A} +6\text{ A} -10\text{ A} = 0$$

Currents I_1 and I_2 are considered to be positive because they enter point P. Current I_3 is considered to be negative because it leaves point P.

A second circuit that illustrates Kirchoff's current law can be seen in Figure 8–2. Consider what happens to the current at point B. Two amperes of current flow into point B from resistor R_1. The current splits at point B, part flowing to resistor R_2 and part to resis-

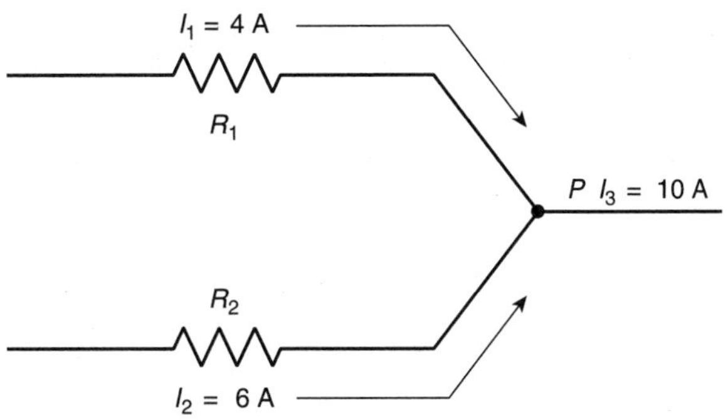

Figure 8–1 The algebraic sum of the currents entering and leaving a point must equal zero.

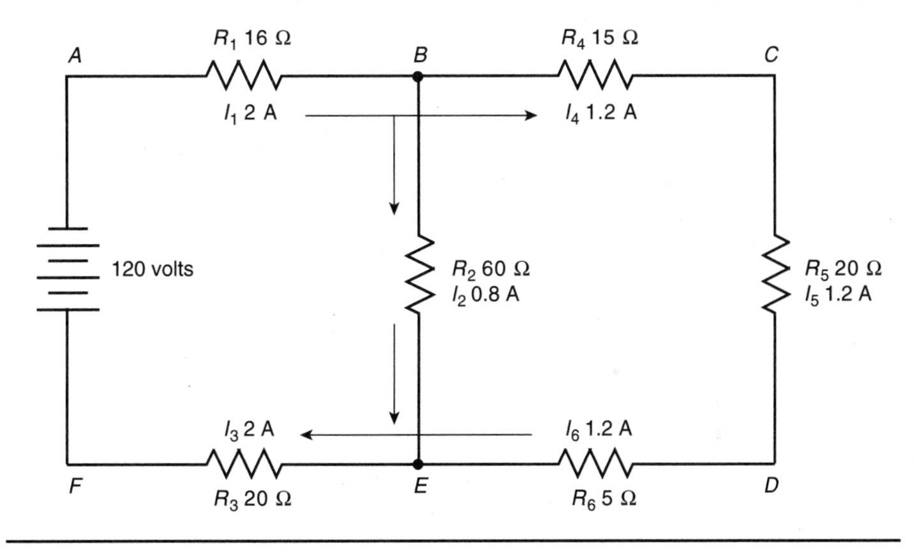

Figure 8–2 The current splits to separate branches.

tors R_4, R_5, and R_6. The current entering point B is positive and the two currents leaving point *B* are negative.

$$+I_1 - I_2 - I_4 - I_5 - I_6 = 0$$

$$+2\text{ A} - 0.8\text{ A} - 1.2\text{ A} = 0$$

(*Note:* I_4, I_5, and I_6 are in series and the current is the same for all.)

Now consider the currents at point E. There is 0.8 ampere of current entering point E from resistor R_2, and 1.2 amperes of current entering point *E* from resistors R_4, R_5, and R_6. Two amperes of current leave point E and flow through resistor R_3.

$$+0.8\text{ A} + 1.2\text{ A} - 2\text{ A} = 0$$

Kirchoff's Voltage Law

Kirchoff's voltage law is similar to the current law in that the algebraic sum of the voltages around any closed loop must equal zero. Before determining the algebraic sum of the voltages, it is necessary to first determine which end of the resistive element is positive and which is negative. To make this determination, assume a direction of current flow and mark the end of the resistive element where current enters and where current leaves. It is assumed that current flows from negative to positive. Therefore, the point at which current enters a resistor will be marked negative, and the point where current leaves the resistor will be marked positive. The voltage drops and the polarity marking have been added to each of the resistors in the example circuit shown in Figure 8–2. The amended circuit is shown in Figure 8–3.

To use Kirchoff's voltage law, start at some point and add the voltage drops around any closed loop. Be certain to return to the starting point. In the circuit shown in Figure 8–3, there are actually three separate closed loops. Loop ACDF contains voltage drops E_1, E_4, E_5, E_6, E_3, and E_T, (120 V source). Loop ABEF contains voltage drops E_1, E_2, E_3, and E_T. Loop BCDE contains voltage drops E_4, E_5, E_6, and E_2.

The voltage drops for the first loop are as follows:

$$-E_1 - E_4 - E_5 - E_6 - E_3 + E_T = 0$$

$$-32 - 18 - 24 - 6 - 40 + 120 = 0$$

The positive or negative sign for each number is determined by the assumed direction of current flow. In this example, it is assumed that current leaves point A and returns to point A. Current leaving point A enters resistor R_1 at the negative end. Therefore, the voltage is considered to be negative (–32 V). The same is true for resistors R_4, R_5, R_6, and R_3. The current enters the voltage source at the positive end, however. Therefore, E_T is assumed to be positive.

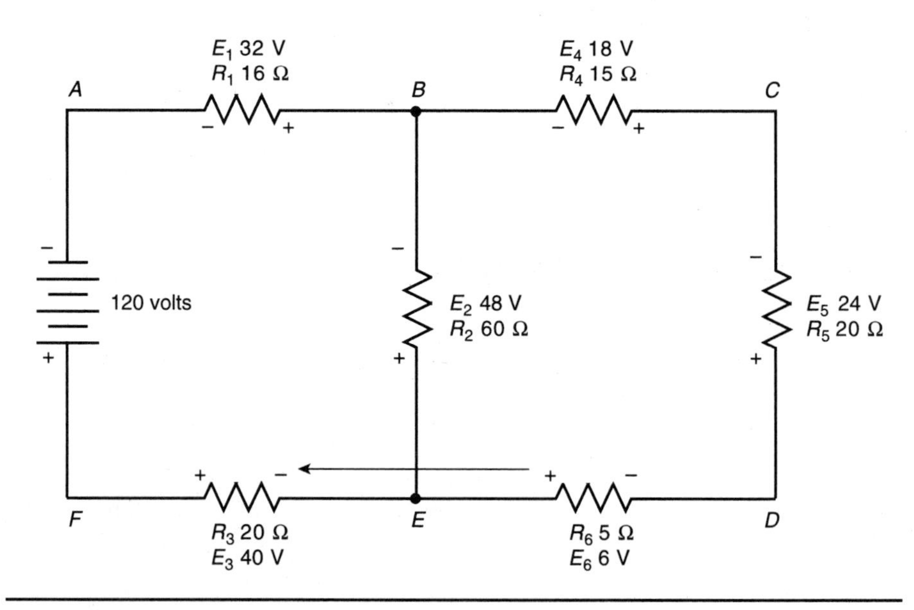

Figure 8–3 Marking resistive elements.

To determine the voltage drops around the second loop, it is assumed that the current leaves point A and returns to point A through resistors R_1, R_2, R_3, and the source voltage. The voltage drops are as follows.

$$-E_1 - E_2 - E_3 + E_T = 0$$

$$-32 - 48 - 40 + 120 = 0$$

The current path for the third loop assumes that current leaves point B and returns to point B. The current will flow through resistors R_4, R_5, R_6 and R_2.

$$-E_4 - E_5 - E_6 + E_2 = 0$$

$$-18 - 24 - 6 + 48 = 0$$

Using Kirchoff's Law with Circuits That Contain More Than One Power Source

The circuit shown in Figure 8–4 contains three resistors and two voltage sources. Although there are three separate loops in this circuit, only two are needed to find the missing values. The two loops used will be ABEF and CBED, as shown in Figure 8–4. The resistors have been marked positive and negative to correspond with the assumed direction of current flow. For the first loop it will be assumed that the current leaves point A and returns to point A. The equation for this loop is:

$$-E_1 - E_3 + E_{S1} = 0$$

$$-E_1 - E_3 + 60 = 0$$

For the second loop it is assumed that the current leaves point C and returns to point C. The equation for the second loop is:

$$-E_2 - E_3 + E_{S2} = 0$$

$$-E_2 - E_3 + 15 = 0$$

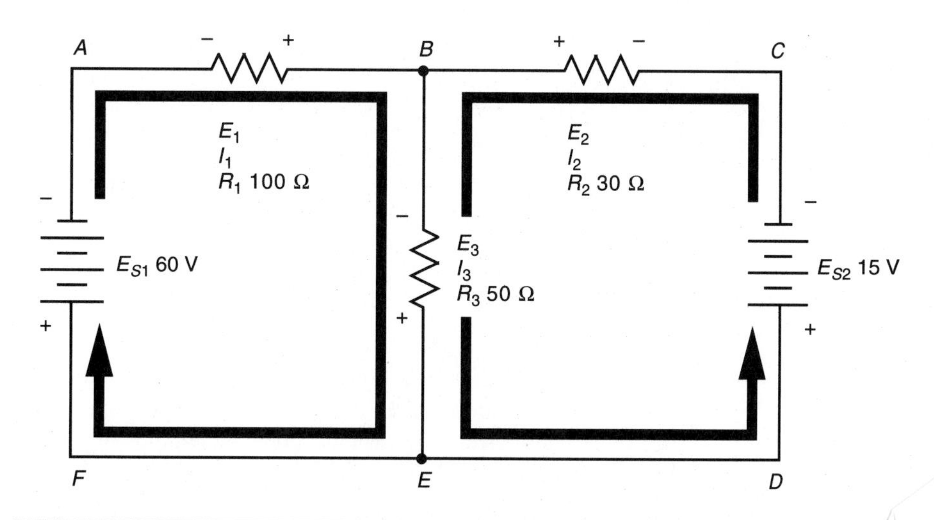

Figure 8–4 Finding circuit values with Kirchoff's law.

To simplify the two equations, the whole numbers must be moved to the other side of the equals sign. In the first equation, subtract 60 from both sides.

$$-E_1 - E_3 = -60$$

In the second equation, subtract 15 from both sides.

$$-E_2 - E_3 = -15$$

To further simplify the equations, multiply both by −1 to remove the negative signs. The new equations become:

$$E_1 + E_3 = 60$$

$$E_2 + E_3 = 15$$

According to Ohm's law, the voltage drop across any resistive element is equal to the amount of current flowing through the element times its resistance ($E = I \times R$). In order to solve these equations, it is necessary to change the values of E_1, E_2, and E_3 to their Ohm's law equivalents.

$$E_1 = I_1 \times R_1 = I_1 \times 100 = 100\ I_1$$

$$E_2 = I_2 \times R_2 = I_2 \times 30 = 30\ I_2$$

Although it is true that $E_3 = I_3 \times R_3$, it would produce three unknown currents in the equation. Since Kirchoff's current law states that the current entering a point must equal the current leaving a point, I_3 is actually the sum of currents I_1 and I_2. Therefore, the third voltage equation will be written:

$$E_3 = (I_1 + I_2) \times R_3 = (I_1 + I_2) \times 50 = 50(I_1 + I_2)$$

The two equations can now be written as:

$$100I_1 + 50(I_1 + I_2) = 60$$

$$30\ I_2 + 50(I_1 + I_2) = 15$$

The parentheses can be removed by multiplying I_1 and I_2 by 50. The equations now become:

$$100I_1 + 50I_1 + 50I_2 = 60$$

$$30I_2 + 50I_1 + 50I_2 = 15$$

The equations can be further simplified to:

$$150I_1 + 50I_2 = 60$$

$$50I_1 + 80I_2 = 15$$

In order to solve these equations, it is necessary to solve them as simultaneous equations. An equation cannot be solved if there is more than one unknown, so all but one unknown must be eliminated. This will be done by multiplying the bottom equation by −3. (Remember that if both sides of an equation are multiplied by the same number, the value of the unknown will not change.)

$$-3(50I_1 + 80I_2 = 15)$$

$$-150I_1 - 240I_2 = -45$$

The two equations can now be added.

$$150I_1 + 50I_2 = 60$$

$$-150I_1 - 240I_2 = -45$$

The positive $150I_1$ and the negative $150I_1$ will cancel each other, leaving $-190I_2$.

$$-190I_2 = 15$$

Dividing both sides of the equation by -190 will produce the answer for I_2.

$$I_2 = -0.0789 \text{ amp}$$

The negative answer for I_2 indicates that the assumed direction of current flow was incorrect. Current actually flows through the circuit as shown in Figure 8–5.

Now the value of I_2 is known, that answer can be substituted in either of the equations to find I_1.

$$150I_1 + 50(-0.0789) = 60$$

$$150I_1 - 3.945 = 60$$

Now add +3.945 to both sides of the equation.

$$150I_1 = 63.945$$

$$I_1 = 0.426 \text{ amp}$$

0.426 amp leaves point A, flows through resistor R_1, and enters point B. At point B, 0.0789 amp branches to point C through resistor R_2 and the remainder of the current branches to point E through resistor R_3. The value for I_3, therefore, can be found by subtracting 0.0789 amp from 0.426 amp.

$$I_3 = I_1 - I_2$$

$$I_3 = 0.426 - 0.0789$$

$$I_3 = 0.347 \text{ amp}$$

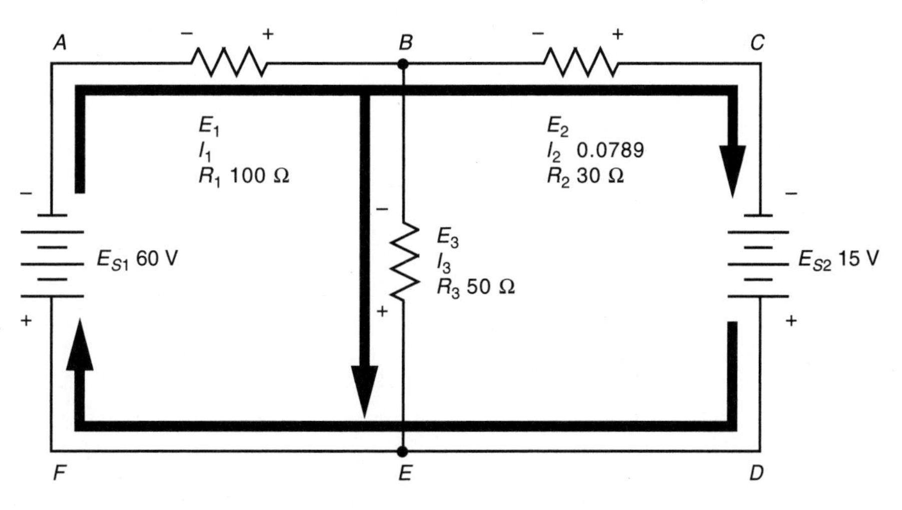

Figure 8–5 Actual direction of current flow.

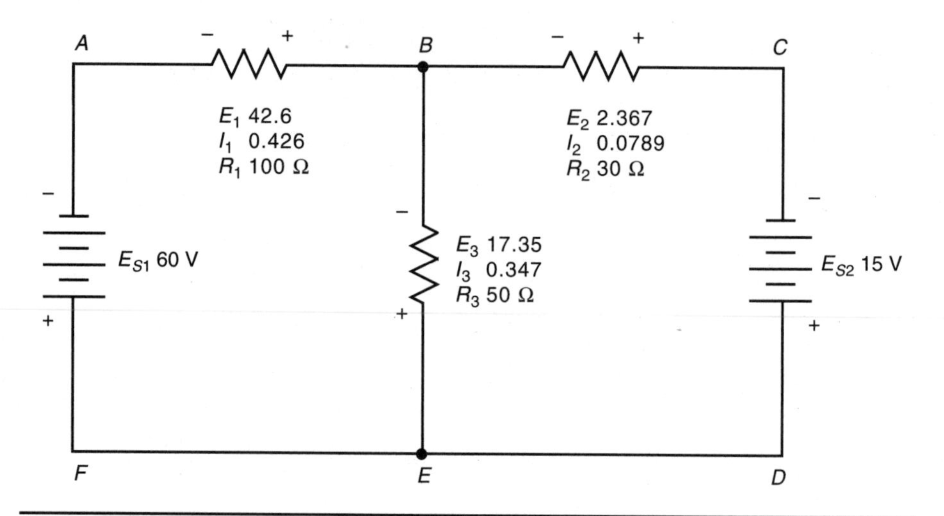

Figure 8–6 All circuit values have been computed.

The voltage drops across each resistor can now be determined using Ohm's law.

$$E_1 = 0.426 \times 100$$

$$E_1 = 42.6 \text{ volts}$$

$$E_2 = 0.0789 \times 30$$

$$E_2 = 2.367 \text{ volts}$$

$$E_3 = 0.347 \times 50$$

$$E_3 = 17.35 \text{ volts}$$

The values for the entire circuit are shown in Figure 8–6.

To check the answers, add the voltages around the loops. The answers should total zero. Loop BCDE will be added first.

$$-E_2 - E_{S2} + E_3 = 0$$

$$-2.367 - 15 + 17.35 = -0.017$$

(The slight negative voltage in the answer is caused by the rounding off of values.)

The second loop checked will be ABEF.

$$-E_1 - E_3 + E_{S1} = 0$$

$$-42.6 - 17.35 + 60 = 0.05$$

The third loop checked will be ACDF.

$$-E_1 - E_2 - E_{S2} + E_{S1} = 0$$

$$-42.6 - 2.367 - 15 + 60 = 0.033$$

CHAPTER 9

Thevenin's Theorem

Thevenin's theorem is used to simplify a circuit network into an equivalent circuit, which contains a single voltage source and series resistor (see Figure 9–1). Imagine a black box that contains an unknown circuit and two output terminals labeled A and B. The output terminals exhibit some amount of voltage and some amount of internal impedance.

Thevenin's theorem reduces the circuit inside the black box to a single source of power and a series resistor, which is equivalent to the internal impedance. The equivalent Thevenin circuit assumes the output voltage to be the open circuit voltage with no load connected. The equivalent Thevenin resistance is the open circuit resistance with no power source connected. Imagine the Thevenin circuit shown in Figure 9–1 with the power source removed and a single conductor between terminal B and the equivalent resistor (see Figure 9–2). If an ohmmeter were to be connected across terminals A and B, the equivalent Thevenin resistance would be measured.

Computing the Thevenin Values

The circuit shown in Figure 9–3 has a single power source of 24 volts and two resistors connected in series. Resistor R_1 has a value of 2 ohms and resistor R_2 has a value of 6 ohms. The Thevenin equivalent circuit will be computed across terminals A and B. To do this, determine

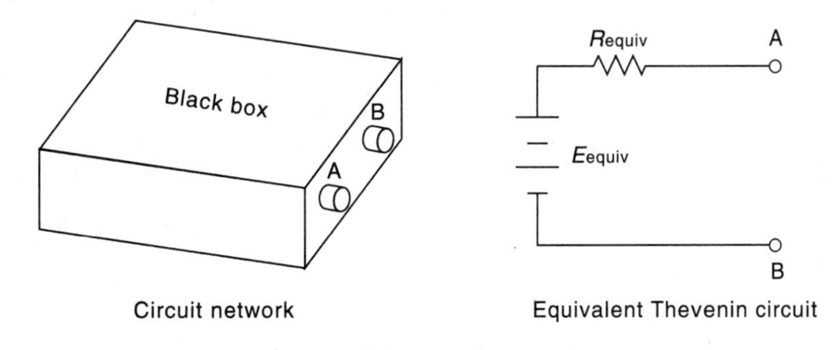

Figure 9–1 Thevenin's theorem reduces a circuit network to a single power source and a single series resistor.

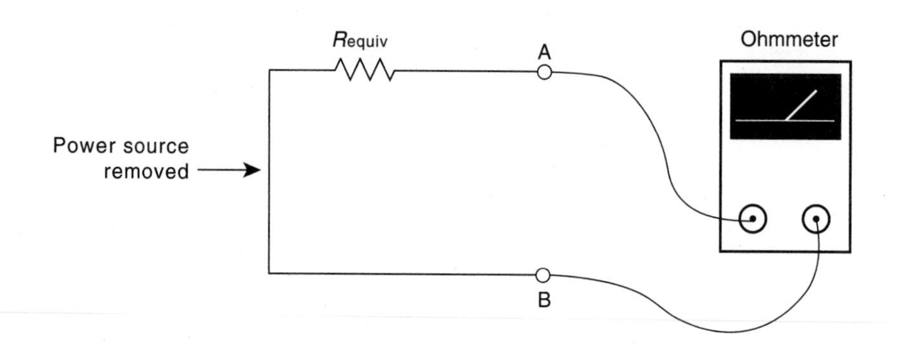

Figure 9–2 Thevenin equivalent resistance.

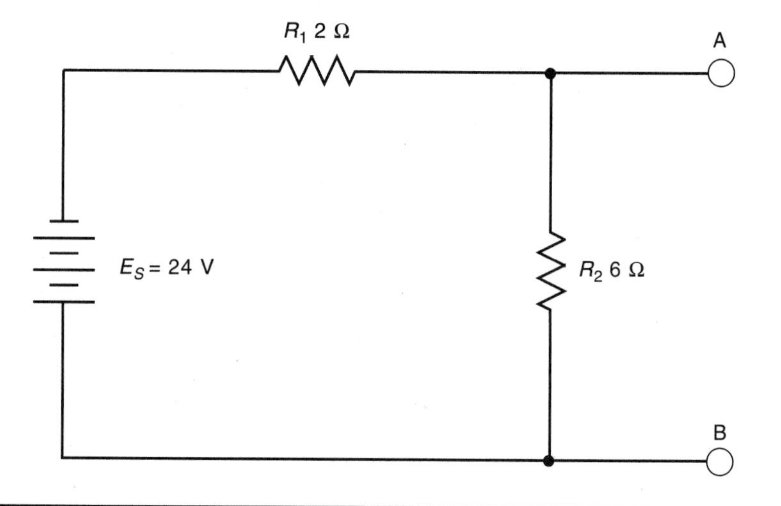

Figure 9–3 Determining the Thevenin equivalent circuit.

the voltage drop across resistor R_2 because it is connected directly across terminals A and B. The voltage drop across resistor R_2 will be the open circuit voltage of the equivalent Thevenin circuit when no load is connected across terminals A and B. Since resistors R_1 and R_2 form a series circuit, a total of 8 ohms is connected to the 24-volt power source. This will produce a current flow 3 amperes through resistors R_1 and R_2 (24 / 8 = 3). Since 3 amperes of current flows through resistor R_2, a voltage drop of 18 volts will appear across it (3 × 6 = 18); thus, 18 volts is the equivalent Thevenin voltage for this circuit.

To determine the equivalent Thevenin resistance, disconnect the power source and replace it with a conductor, as shown in Figure 9–4. In this circuit, resistors R_1 and R_2 are connected in parallel with each other. The total resistance can now be determined using one of the formulas for finding parallel resistance.

$$R_T = \frac{R_1 \times R_2}{R_1 + R_2}$$

$$R_T = \frac{2 \times 6}{2 + 6}$$

$$R_T = 1.5 \text{ ohms}$$

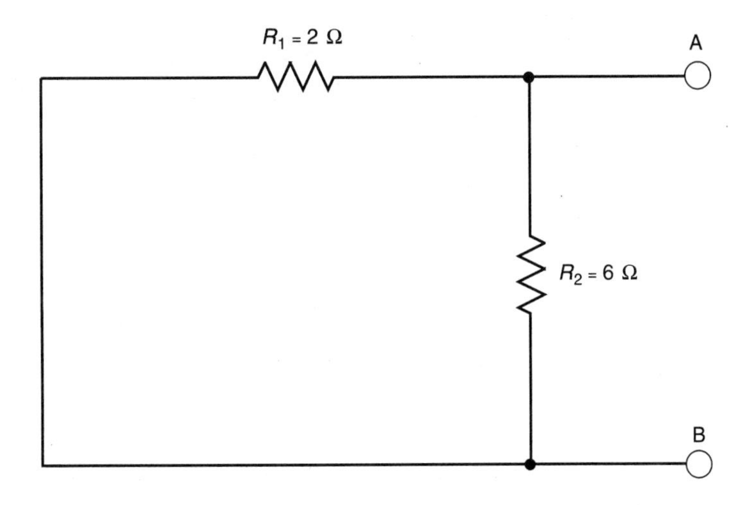

Figure 9–4 Determining the equivalent Thevenin resistance.

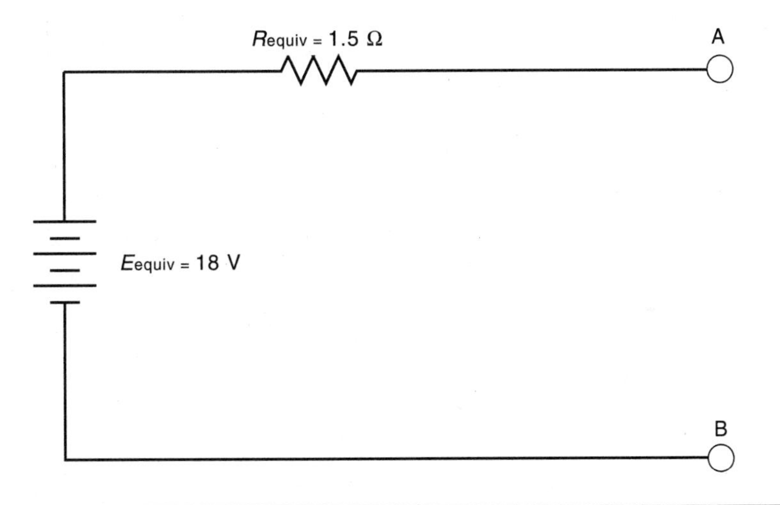

Figure 9–5 The Thevenin equivalent circuit.

The Thevenin equivalent circuit is shown in Figure 9–5.

Now that the Thevenin equivalent of the circuit is known, the voltage and current values for different load resistances can be quickly computed. Assume, for example, that a load resistance of 10 ohms is connected across terminals A and B. We can now compute the voltage and current value for the circuit. The total resistance of the circuit is 11.5 ohms (1.5 + 10). This produces a current flow of 1.565 amps (18 / 11.5) and a voltage drop of 15.65 volts (1.565 × 10) across the 10-ohm load resistor.

CHAPTER 10

Norton's Theorem

Norton's theorem is used to reduce a circuit network into a simple current source and a single parallel resistance. This is opposite to Thevenin's theorem, which reduces a circuit network into a simple voltage source and a single series resistor, as shown in Figure 10–1. Norton's theorem assumes a source of current that is divided among parallel branches. A source of current is often easier to work with—especially when calculating values for parallel circuits—than a voltage source, which drops voltages across series elements.

Current Sources

Power sources can be represented in one of two ways, as a voltage source or as a current source. Voltage sources are generally shown as a battery with a resistance connected in series with the circuit to represent the internal resistance of the source. This is the case when using Thevenin's theorem. Voltage sources are rated with some amount of voltage, such as 12 volts, 24 volts, and so on.

Power sources can also be represented by a current source connected to a parallel resistance that delivers a certain amount of current such as 1 amp, 2 amps, 3 amps, and so on. Assume that a current source is rated at 1.5 amps (Figure 10–2). This means that 1.5 amps

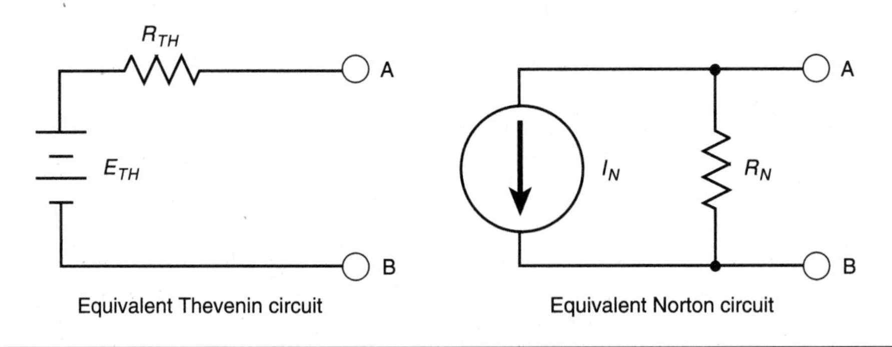

Figure 10–1 The Thevenin equivalent circuit contains a voltage source and series resistance. The Norton equivalent circuit contains a current source and parallel resistance.

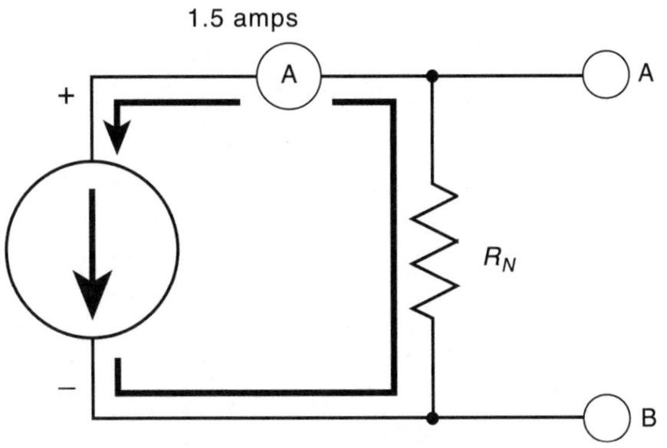

Figure 10–2 The current source supplies a continuous 1.5 amps.

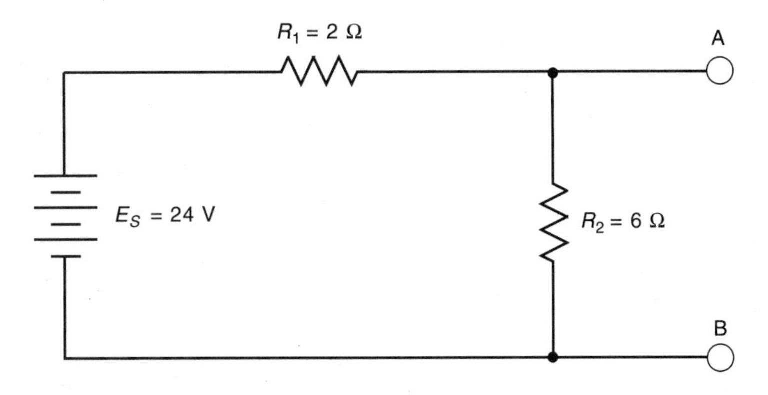

Figure 10–3 Determining the Norton equivalent circuit.

flow from the power source, regardless of the circuit connected. In the circuit shown in Figure 10–2, 1.5 amps flow through resistor R_N.

Determining the Norton Equivalent Circuit

The same circuit used previously to illustrate Thevenin's theorem will be used to illustrate Norton's theorem. Refer to the circuit shown in Figure 10–3. In this basic circuit, a 2-ohm and a 6-ohm resistor are connected in series with a 24-volt power source. To determine the Norton equivalent of this circuit, imagine a short circuit placed across terminals A and B (Figure 10–4). Since this places the short circuit directly across resistor R_2, that resistance is eliminated from the circuit and a resistance of 2 ohms is left connected in series with the voltage source. The next step is to determine the amount of current that can flow through this circuit. This current value will be known as I_N.

$$I_N = \frac{E}{R}$$

$$I_N = \frac{24}{2}$$

$$I_N = 12 \text{ amps}$$

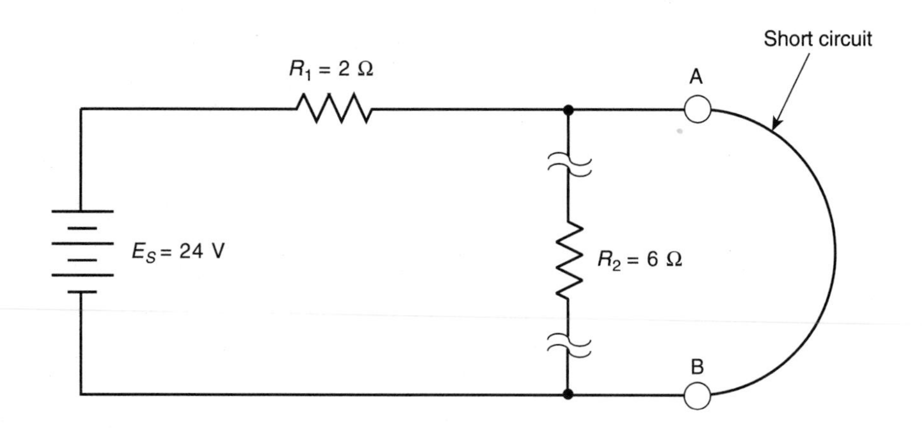

Figure 10–4 Shorting terminals A and B eliminates the 6-ohm resistor.

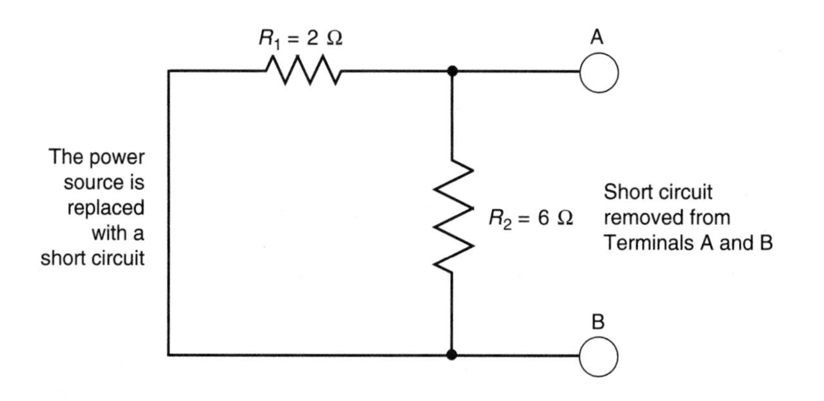

Figure 10–5 Determining the Norton equivalent resistance.

I_N, 12 amps, is the amount of current available in the Norton equivalent circuit.

The next step is to find the equivalent parallel resistance, R_N, connected across the current source. To do this, remove the short circuit across terminals A and B. Now replace the power source with a short circuit, just as was done in determining the Thevenin equivalent circuit (see Figure 10–5). The circuit now has a 2-ohm and a 6-ohm resistor connected in parallel. This produces an equivalent resistance of 1.5 ohms. The Norton equivalent circuit, shown in Figure 10–6, is a 1.5-ohm resistor connected in parallel with a 12-amp current source.

Now that the Norton equivalent for the circuit has been computed, any value of resistance can be connected across terminals A and B and the electrical values computed quickly. Assume that a 6-ohm load resistor, R_L, is connected across terminals A and B (see Figure 10–7). The 6-ohm load resistor is connected in parallel with the Norton equivalent resistance of 1.5 ohms. This produces a total resistance of 1.2 ohms for the circuit.

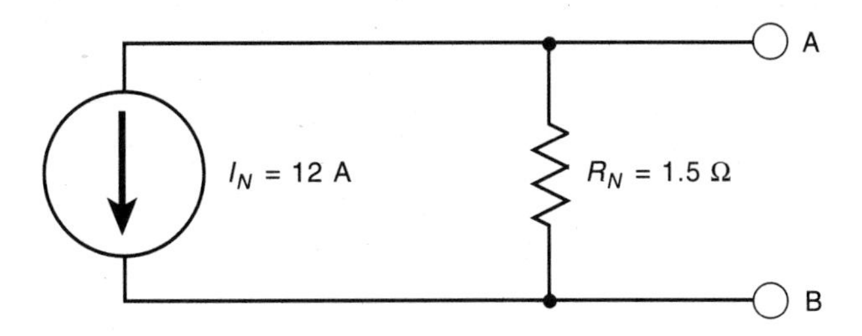

Figure 10–6 Equivalent Norton circuit.

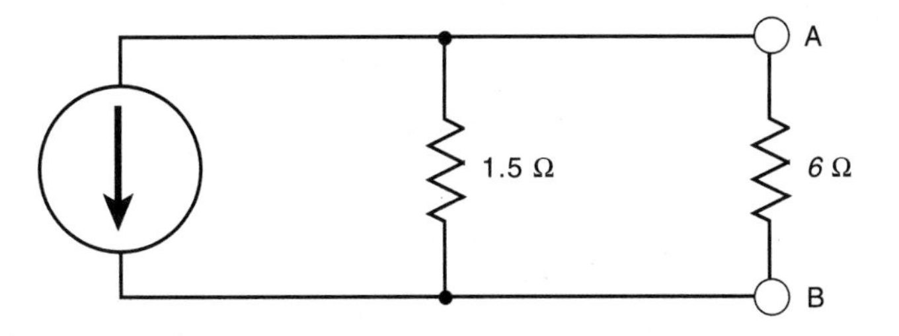

Figure 10–7 A 6 Ω load resistor is connected across the equivalent Norton circuit.

In a Norton equivalent circuit, assume that the Norton equivalent current, I_N, flows at all times. In this circuit the Norton equivalent current is 12 amps. Therefore, a current of 12 amps flows through the 1.2 ohms of resistance, producing a voltage drop of 14.4 volts across the resistance ($E = 12 \times 1.2$). Since the resistors shown in Figure 10–7 are connected in parallel, 14.4 volts is dropped across each. This produces a current flow of 9.6 amps through R_N (14.4 V / 1.5 Ω = 9.6 A) and a current flow of 2.4 amps through R_L (14.4 V / 6 Ω = 2.4 A).

CHAPTER 11

The Superposition Theorem

The superposition theorem is somewhat similar to combining Kirchoff's law with Thevenin and Norton's theorems. The superposition theorem can be used to find the current flow through any branch of a circuit containing more than one power source. The superposition theorem works on the principle that the current in any branch of a circuit supplied by a multipower source can be determined by finding the current produced in that particular branch by each of the individual power sources acting alone. All other power sources must be replaced by a resistance that is equivalent to their internal resistances. The total current flow through the branch will be the algebraic sum of the individual currents produced by each of the power sources.

Example Circuit #1

An example circuit is shown in Figure 11–1. In this example, the circuit contains two voltage sources, and the amount of current flowing through resistor R_2 will be determined using the superposition theorem. The circuit can be solved by following a procedure with several distinct steps.

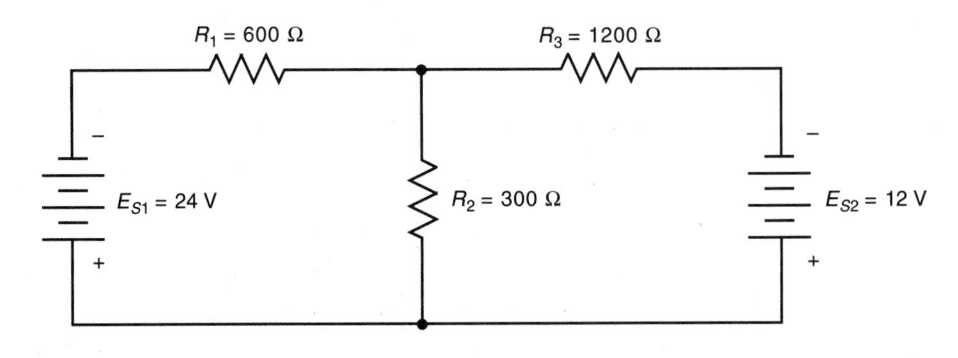

Figure 11–1 A circuit with two power sources.

Step #1

Reduce all but one of the voltage sources to zero by replacing it with a short circuit, leaving any internal series resistance. Reduce the current source to zero by replacing it with an open circuit, leaving any internal parallel resistance. Voltage source E_{S2} will be shorted, as shown in Figure 11–2. The circuit now exists as a simple combination circuit with resistor R_1 connected in series with resistors R_2 and R_3, which are in parallel with each other, shown in Figure 11–3. The total resistance of this circuit can now be found by finding the total resistance of the two resistors connected in parallel, R_2 and R_3, and adding them to R_1.

$$R_T = R_1 + \left(\frac{1}{\frac{1}{R_2} + \frac{1}{R_3}} \right)$$

$$R_T = 600 + \left(\frac{1}{\frac{1}{300} + \frac{1}{1200}} \right)$$

$$R_T = 840\ \Omega$$

Now that the total resistance is known, compute the total current flow.

$$I_T = \frac{E_T}{R_T}$$

$$I_T = \frac{24}{840}$$

$$I_T = 0.0286 \text{ amp}$$

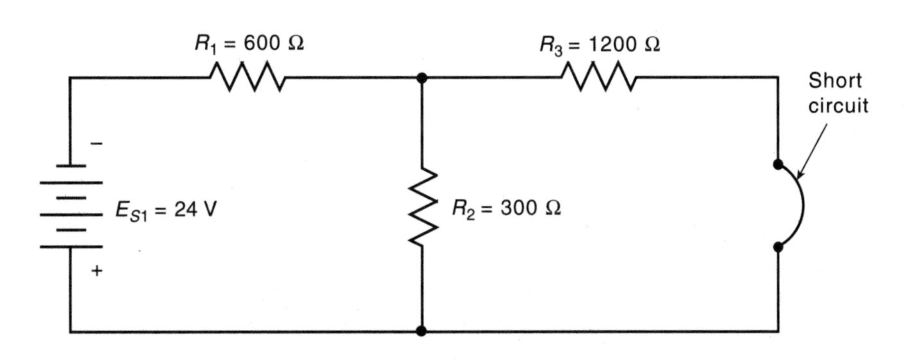

Figure 11–2 Voltage source E_{S2} is replaced with a short circuit.

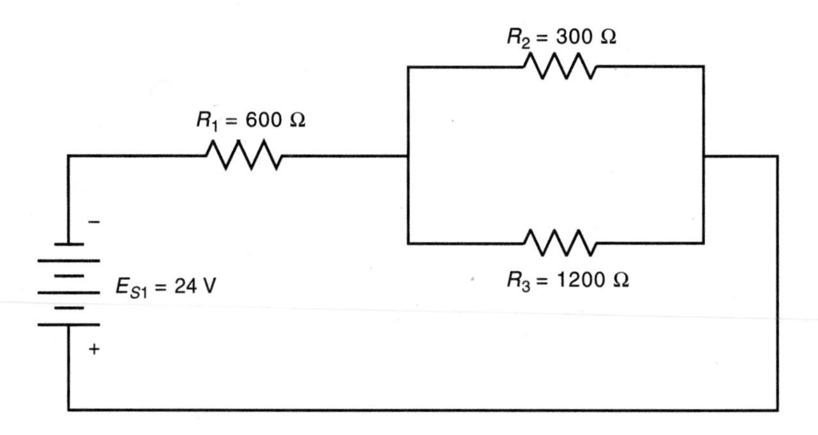

Figure 11–3 The circuit is reduced to a simple combination circuit.

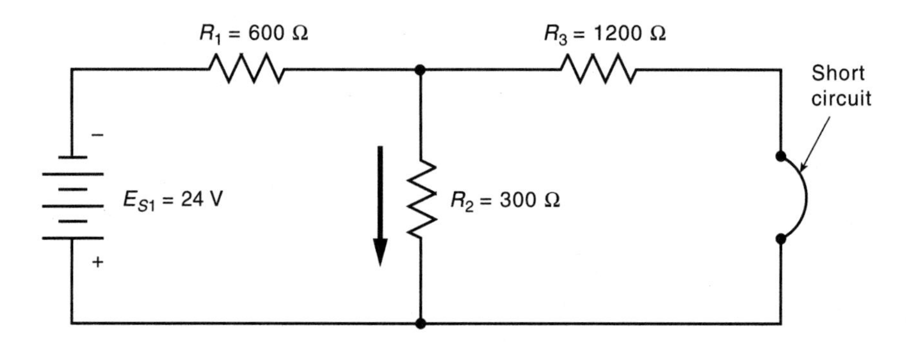

Figure 11–4 Current flows through the resistor in the direction shown.

The voltage drop across the parallel block can be computed using the total current and the combined resistance of resistors R_2 and R_3.

$$E_{\text{Combination}} = 0.0286 \times 240\ \Omega$$

$$E_{\text{Combination}} = 6.864 \text{ volts}$$

Compute the current flowing through resistor R_2.

$$I_2 = \frac{6.864}{300}$$

$$I_2 = 0.0229$$

Notice that the current is flowing through resistor R_2 in the direction of the arrow in Figure 11–4 .

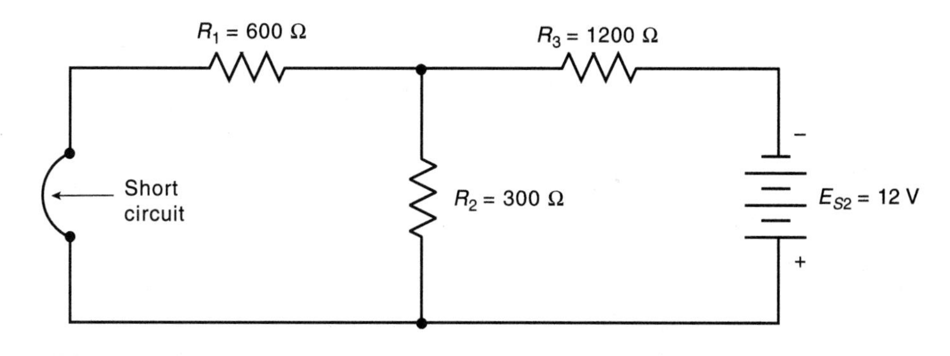

Figure 11–5 Voltage source E_{S1} is shorted.

Step #2

Find the current flow through resistor R_2 by shorting voltage source E_{S1}. Power is now supplied by voltage source E_{S2} (Figure 11–5). In this circuit, resistors R_1 and R_2 are in parallel with each other. Resistor R_3 is in series with R_1 and R_2.

$$R_T = R_3 + \left(\frac{1}{\frac{1}{R_1} + \frac{1}{R_2}} \right)$$

$$R_T = 1200 + \left(\frac{1}{\frac{1}{600} + \frac{1}{300}} \right)$$

$$R_T = 1200 + 200$$

$$R_T = 1400\ \Omega$$

The total current flow in the circuit can now be determined.

$$I_T = \frac{E_T}{R_T}$$

$$I_T = \frac{12}{1400}$$

$$I_T = 0.00857 \text{ amp}$$

Compute the amount of voltage drop across the parallel combination.

$$E_C = 0.00857 \times 200$$

$$E_C = 1.714$$

The amount of current flow through resistor R_2 can be computed using Ohm's law.

$$I_2 = \frac{1.714}{300}$$

$$I_2 = 0.00571 \text{ amp}$$

Notice that the current flowing through resistor R_2 is in the same direction as in the previous circuit, Figure 11–6.

Step #3

The next step is to find the algebraic sum of the two currents. Since both currents flow through resistor R_2 in the same direction, we can add the two currents.

$$I_{2(\text{Total})} = 0.0229 + 0.00571$$

$$I_{2(\text{Total})} = 0.0286 \text{ amp}$$

Example Circuit #2

Example circuit #2 is shown in Figure 11–7. This circuit contains a voltage source of 30 volts and a current source of 0.2 amp. We will compute the amount of current flowing through resistor R_2.

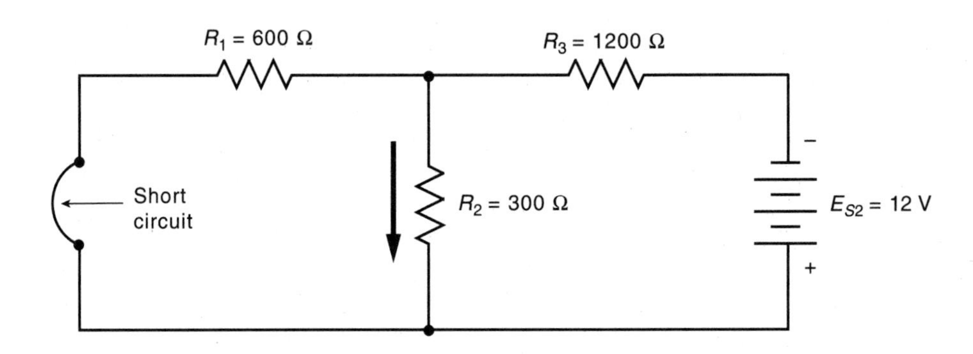

Figure 11–6 Current flows in the same direction.

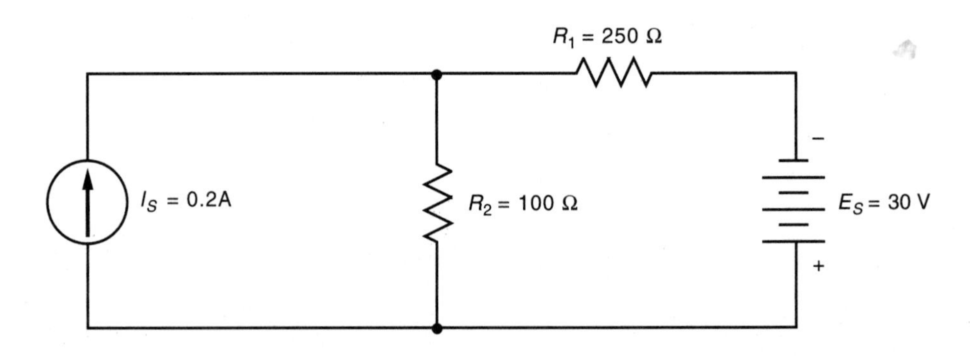

Figure 11–7 Example circuit #2 contains a current source and a voltage source.

Step #1

The first step will be to find the current flow through resistor R_2 using the voltage source only. This is done by replacing the current source with an open circuit, shown in Figure 11–8. Notice the direction of current flow through resistor R_2.

When the current source is replaced with an open circuit, resistors R_1 and R_2 become connected in series with each other. The total resistance is the sum of the two resistances.

$$R_T = R_1 + R_2$$

$$R_T = 250 + 100$$

$$R_T = 350\,\Omega$$

Now that the total resistance is known, we can find the total current flow in the circuit using Ohm's law.

$$I_T = \frac{30}{350}$$

$$I_T = 0.0857 \text{ amp}$$

Since the current flow must be the same in all points of a series circuit, the same amount of current flows through resistor R_2.

$$I_T = I_2$$

$$I_2 = 0.0857 \text{ amp}$$

Step #2

The next step will be to find the amount of current flow through resistor R_2 that would be supplied by the current source only. This can be done by replacing the voltage source with a short circuit, shown in Figure 11–9. Notice the direction of current flow through resistor R_2.

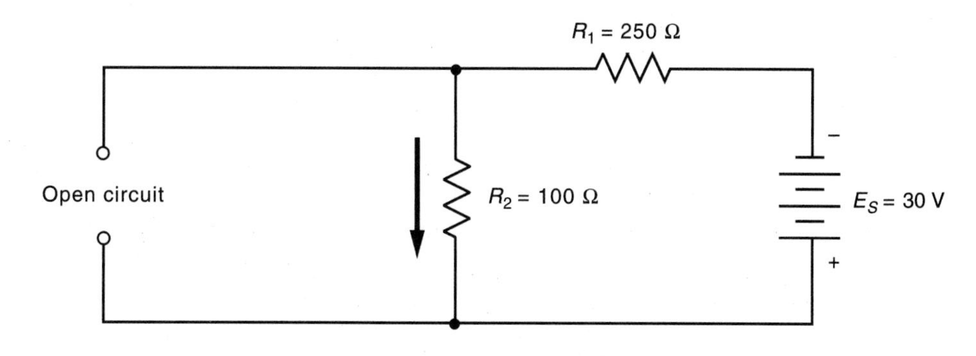

Figure 11–8 The current source is replaced with an open circuit.

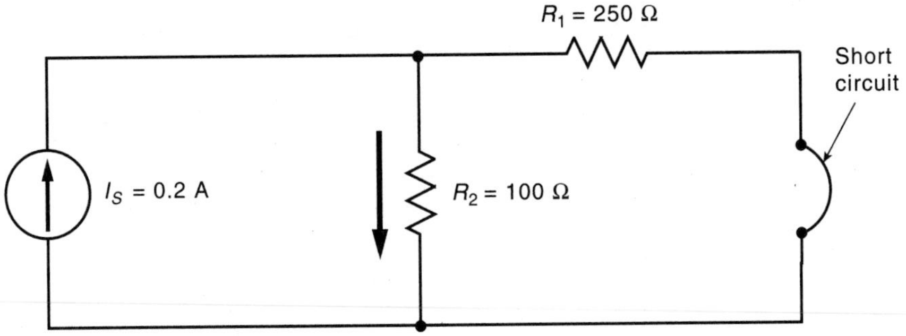

Figure 11–9 The voltage source is replaced with a short circuit.

When the voltage source is removed and replaced with a short circuit, resistors R_1 and R_2 become connected in parallel with each other. We can compute the current flow through resistor R_2 using the current divider formula.

$$I_2 = \left(\frac{R_1}{R_1 + R_2}\right) \times I_S$$

$$I_2 = \left(\frac{250}{350}\right) \times 0.2$$

$$I_2 = 0.143 \text{ amp}$$

Step #3

The total amount of current flow through resistor R_2 can now be determined by finding the algebraic sum of both currents.

$$I_{2\ (\text{Total})} = 0.0857 + 0.143$$

$$I_{2\ (\text{Total})} = 0.229 \text{ amp}$$

Example Circuit #3

In the third example, a circuit contains two resistors and two current sources (see Figure 11–10). The amount of current flowing through resistor R_2 will be determined.

Step #1

Remove one of the current sources from the circuit and replace it with an open circuit. In this example, current source I_{S2} will be removed first (see Figure 11–11). Resistor R_1 is now removed from the circuit, so the entire 0.25 amp of current flows through resistor R_2. Notice the direction of current flow through the resistor.

Step #2

The next step is to replace current source I_{S1} with an open circuit and determine the amount and direction of current flow through resistor R_2 produced by current source I_{S2}, shown in Figure 11–12.

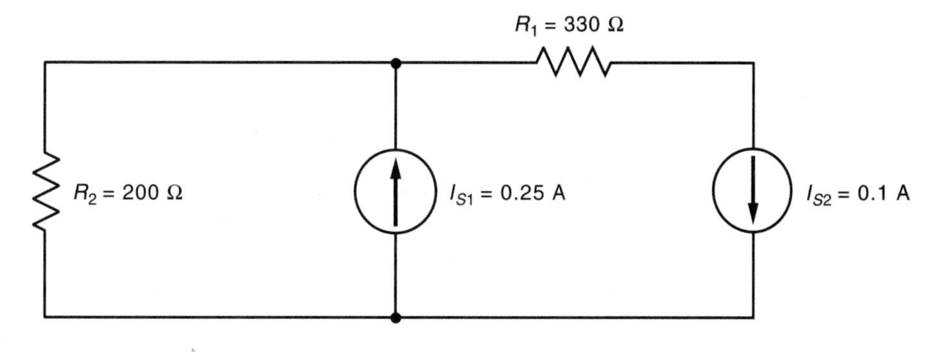

Figure 11–10 The circuit contains two current sources, I_{S1} and I_{S2}.

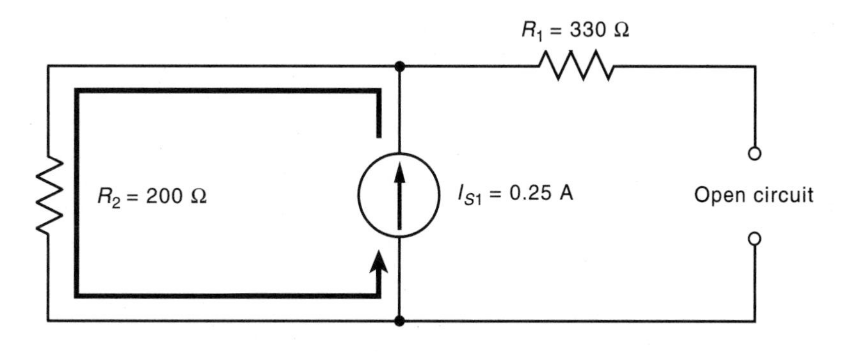

Figure 11–11 Current source I_{S2} is replaced with an open circuit.

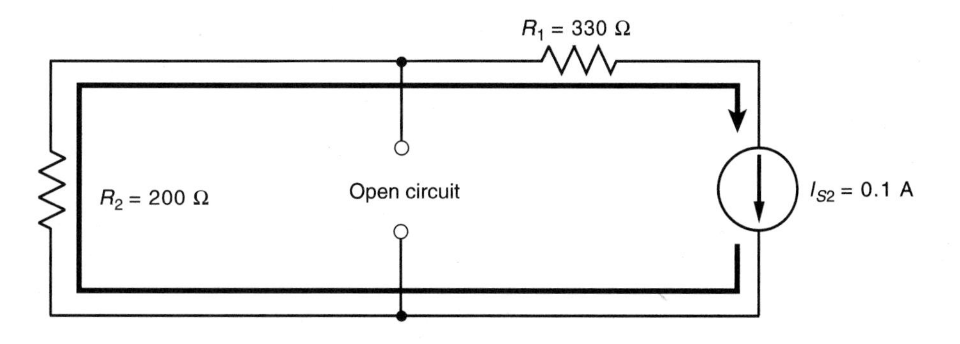

Figure 11–12 Current source I_{S1} is replaced with an open circuit.

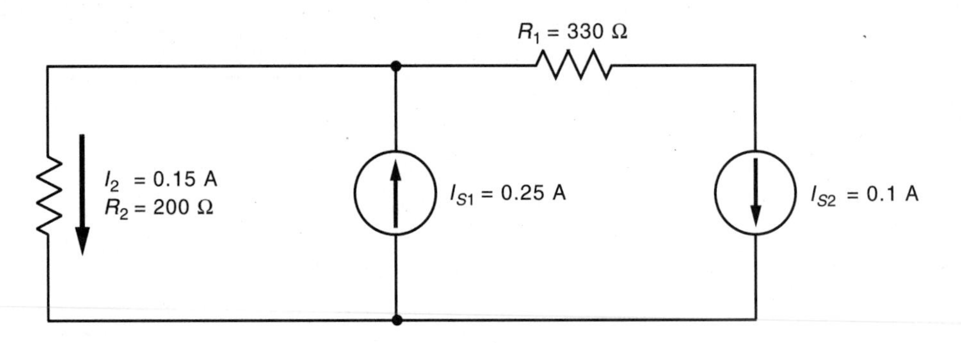

Figure 11–13 The amount and direction of current flow through resistor R_2 has been determined.

When I_{S1} is replaced with an open circuit, resistors R_1 and R_2 become connected in series with each other. Since the current is the same in a series circuit, both resistors have a current flow of 0.1 amp through them.

Step #3

Now that the amount and direction of current flow through resistor R_2 for both current sources is known, the total current flow will be the algebraic sum of the two currents. Since the currents flow in opposite directions, the algebraic sum will be the difference of the two currents, and the direction will be determined by the greater of the two currents, as shown in Figure 11–13.

$$I_{2(\text{Total})} = 0.25 - 0.1$$

$$I_{2(\text{Total})} = 0.15 \text{ amp}$$

CHAPTER 12

Voltage Dividers

Voltage dividers operate on the principle that the sum of the voltage drops in a series circuit must equal the applied or source voltage. Voltage dividers are used to provide different voltages between certain points (see Figure 12–1). If a voltmeter is connected between points A and B, a voltage of 20 volts will be seen. If the voltmeter is connected between points B and D, a voltage of 80 volts will be seen.

Voltage dividers can be constructed to provide any voltage desired. Assume that a voltage divider is connected to a source of 120 volts and is to provide voltage drops of 36 volts, 18 volts, and 66 volts. Notice that the sum of the voltage drops equals the applied voltage of 120 volts. The next step is to determine the amount of current that is to flow through the circuit. Since there is only one path for current flow, the current will be the same through each resistor. In this circuit, a current flow of 15 milliamperes (ma). (0.015 A) will be used. The resistance of each resistor can not be determined.

$$R_1 = \frac{36}{0.015}$$

$$R_1 = 2400\ \Omega$$

$$R_2 = \frac{18}{0.015}$$

$$R_2 = 1200\ \Omega$$

$$R_3 = \frac{66}{0.015}$$

$$R_3 = 4400\ \Omega$$

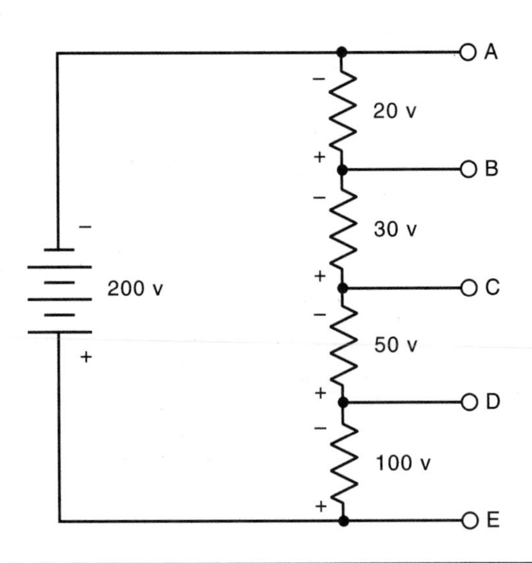

Figure 12–1 Series circuit used as a voltage divider.

The General Voltage Divider Formula

Another method of determining the voltage drop across series elements is to use the general voltage divider formula. The current flow through a series circuit is the same at all points in the circuit, so the voltage drop across any particular resistance is equal to the total circuit current (I_T) times the value of that resistor.

$$E_X = I_T \times R_X$$

The total circuit current is proportional to the source voltage (E_T) and the total resistance of the circuit.

$$I_T = \frac{E_T}{R_T}$$

If the value of I_T is a substitute for E_T/R_T in the previous formula, the expression now becomes

$$E_X = \left(\frac{E_T}{R_T}\right) R_X$$

If the formula is rearranged, it becomes what is known as the *general voltage divider formula.*

$$E_X = \left(\frac{R_X}{R_T}\right) E_T$$

We can compute the voltage drop across any series component (E_X) by substituting the value of R_X for the resistance value of that component when the source voltage and total resistance are known.

Example

Three resistors are connected in series to a 24-volt source. Resistor R_1 has a resistance of 200 ohms, resistor R_2 has a value of 300 ohms, and resistor R_3 has a value of 160 ohms. What is the voltage drop across each resistor?

Solution

Find the total resistance of the circuit.

$$R_T = R_1 + R_2 + R_3$$

$$R_T = 200 + 300 + 160$$

$$R_T = 660\,\Omega$$

Now use the voltage divider formula to compute the voltage drop across each resistor.

$$E_1 = \left(\frac{R_1}{R_T}\right)E_T$$

$$E_1 = \left(\frac{200}{660}\right)24$$

$$E_1 = 7.273 \text{ volts}$$

$$E_2 = \left(\frac{R_2}{R_T}\right)E_T$$

$$E_2 = \left(\frac{300}{660}\right)24$$

$$E_2 = 10.91 \text{ volts}$$

$$E_3 = \left(\frac{R_3}{R_T}\right)E_T$$

$$E_3 = \left(\frac{R_3}{R_T}\right)E_T$$

$$E_3 = 5.818 \text{ volts}$$

CHAPTER 13

Current Dividers

All parallel circuits are current dividers (see Figure 13–1). The sum of the currents in a parallel circuit must equal the total current. Assume that a current of 1 ampere enters the circuit at point A. This 1 ampere of current will divide between resistors R_1 and R_2, and will then recombine at point B. The amount of current that flows through each resistor is proportional to the resistance value. A greater amount of current will flow through a low-value resistor and less current will flow through a high-value resistor. In other words, the amount of current flowing through each resistor is inversely proportional to its resistance.

In a parallel circuit, the voltage drop across each branch must be equal. Therefore, we can compute the current flow through any branch by dividing the source voltage (E_T) by the resistance of that branch. The current flow through branch 1 can be computed using the formula:

$$I_1 = \frac{E_T}{R_1}$$

It is also true that the total circuit voltage is equal to the product of the total circuit current and the total circuit resistance.

$$E_T = I_T \times R_T$$

Substituting ($I_T \times R_T$) for E_T in the previous formula results in:

$$I_1 = \frac{I_T \times R_T}{R_1}$$

If the formula is rearranged, and the values of I_1 and R_1 are substituted for I_X and R_X, it becomes what is generally known as the *current divider formula.*

$$I_X = \left(\frac{R_T}{R_X}\right) I_T$$

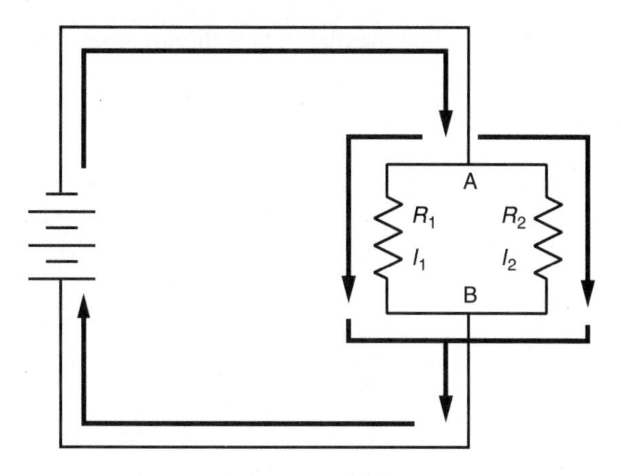

Figure 13–1 Parallel circuits are current dividers.

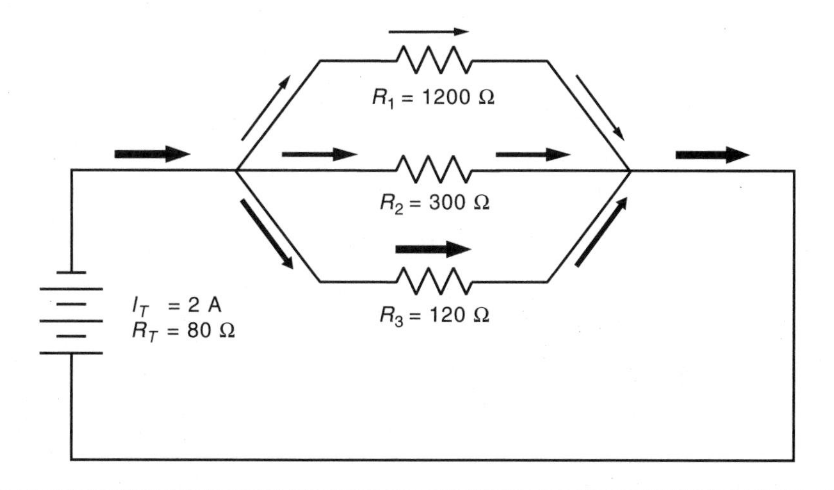

Figure 13–2 The current divides through each branch of a parallel circuit.

This formula can be used to compute the current flow through any branch by substituting the values of I_X and R_X for the branch values when the total circuit current and resistance are known. In the circuit shown in Figure 13–2, resistor R_1 has a value of 1200 ohms, resistor R_2 has a value of 300 ohms, and resistor R_3 has a value of 120 ohms, producing a total resistance of 80 ohms for the circuit. It is assumed that a total current of 2 amps flows in the circuit. We can find the amount of current flow through resistor R_1 using the formula:

$$I_1 = \left(\frac{R_T}{R_1}\right) I_T$$

$$I_1 = \left(\frac{80}{1200}\right) 2$$

$$I_1 = 0.133 \text{ amp}$$

We can find the current flow through each of the other resistors by substituting in the same formula.

$$I_2 = \left(\frac{R_T}{R_2}\right) I_T$$

$$I_2 = \left(\frac{80}{300}\right) 2$$

$$I_2 = 0.533 \text{ amp}$$

$$I_3 = \left(\frac{R_T}{R_3}\right) I_T$$

$$I_3 = \left(\frac{80}{120}\right) 2$$

$$I_3 = 1.333 \text{ amps}$$

CHAPTER 14

Alternating Current

Alternating current (AC) differs from direct current (DC) in that AC current reverses its direction of flow at periodic intervals. Alternating current waveforms can vary, depending on how the current is produced. One waveform frequently encountered is the square wave, shown in Figure 14–1. The oscilloscope in Figure 14-1 has been adjusted so that zero voltage is represented by the center horizontal line. The waveform shows that the voltage is in the positive direction for some length of time and then changes polarity. The voltage remains negative for some length of time and then changes back to positive again. Each time the voltage reverses polarity, the current flow through the circuit changes direction.

A square wave can be produced by a simple single-pole double throw switch connected to two batteries, as shown in Figure 14–2. Each time the switch position is changed, current flows through the resistor in a different direction. Although this circuit will produce a

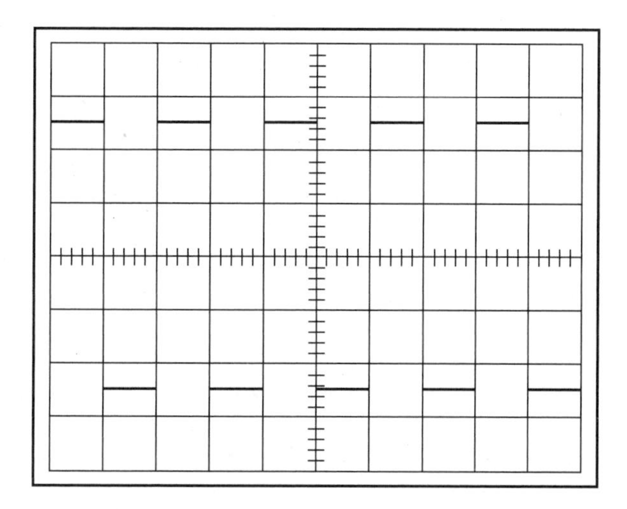

Figure 14–1 Oscilloscope shows square wave alternating current.

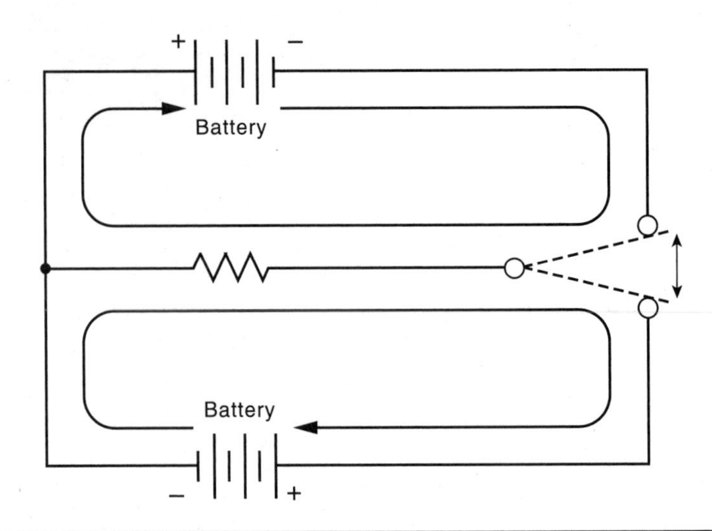

Figure 14–2 Square wave alternating current can be produced by a switch and two batteries.

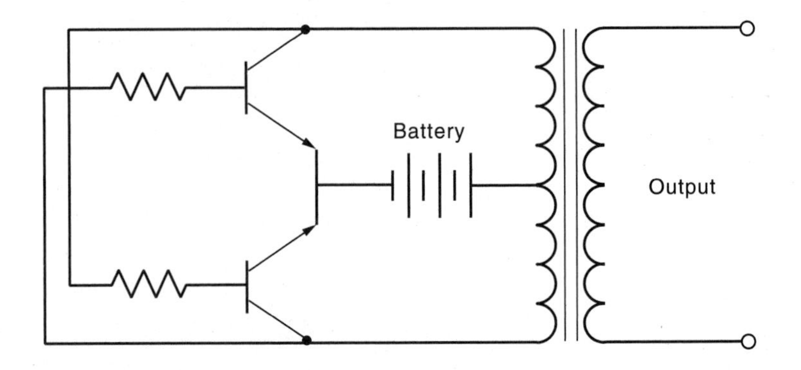

Figure 14–3 Schematic diagram of a square wave oscillator.

square wave alternating current, it is not practical. Square waves are generally produced by electronic devices called *oscillators.* The schematic diagram of a simple square-wave oscillator is shown in Figure 14–3. In this circuit, two bipolar transistors are used as switches to reverse the direction of current flow through the windings of the transformer. This type of oscillator is often used to change the 12-volt DC of an automobile battery into 120-volt AC to operate electric handtools such as drills and saws. This type of oscillator generally produces a large amount of electrical noise and voltage spikes. For this reason, electronic devices such as television sets should not be powered by it.

Another common AC waveform is the triangle wave, shown in Figure 14–4. The triangle wave is a linear wave. A linear wave is one in which the voltage rises at a constant rate with respect to time. Linear waves form straight lines when plotted on a graph. For example, assume the waveform shown in Figure 14–4 reaches a maximum positive value of 100 volts after 2 milliseconds. The voltage will be 25 volts after 0.5 millisecond, 50 volts after 1 millisecond, and 75 volts after 1.5 milliseconds.

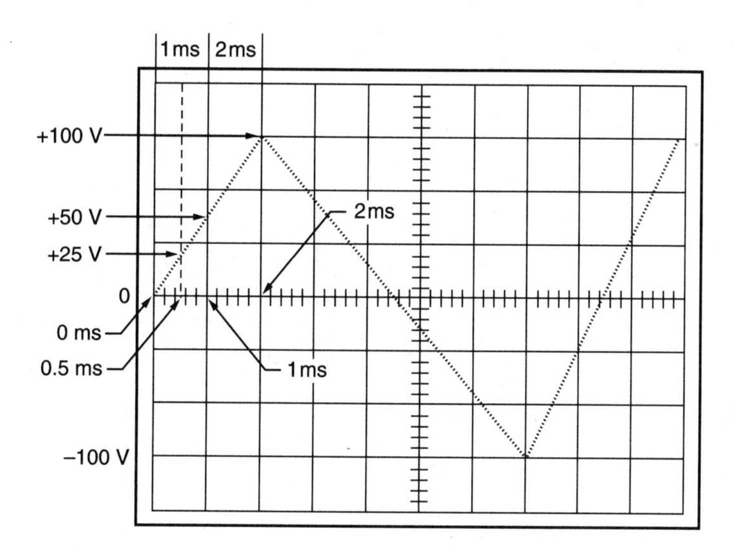

Figure 14–4 The triangle wave is a linear wave.

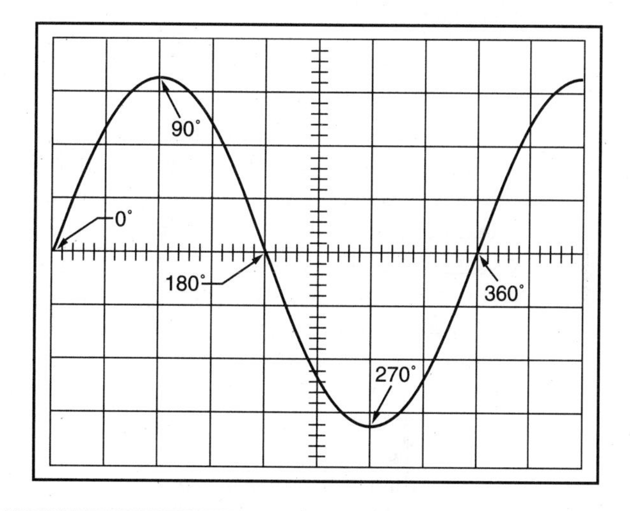

Figure 14–5 The sine wave is a cyclical wave.

Sine Waves

The most common of all AC waveforms is the sine wave, shown in Figure 14–5. Sine waves are produced by all rotating machines such as alternators, and therefore can be compared to circles. The sine wave contains a total of 360 electrical degrees (360°). It reaches its peak positive voltage at 90°, returns to a value of 0 volt at 180°, increases to its maximum negative voltage at 270°, and returns to 0 volt at 360°. Each complete waveform of 360° is called a cycle. The number of complete cycles that occur in one second is called the *frequency.* Frequency is measured in hertz (1 Hz = 1 cycle per second). The most common frequency in the United States and Canada is 60 hertz. This means that the voltage increases from 0 to its maximum value in the positive direction, returns to 0, increases to its maximum value in the negative direction, and returns to 0 sixty times each second.

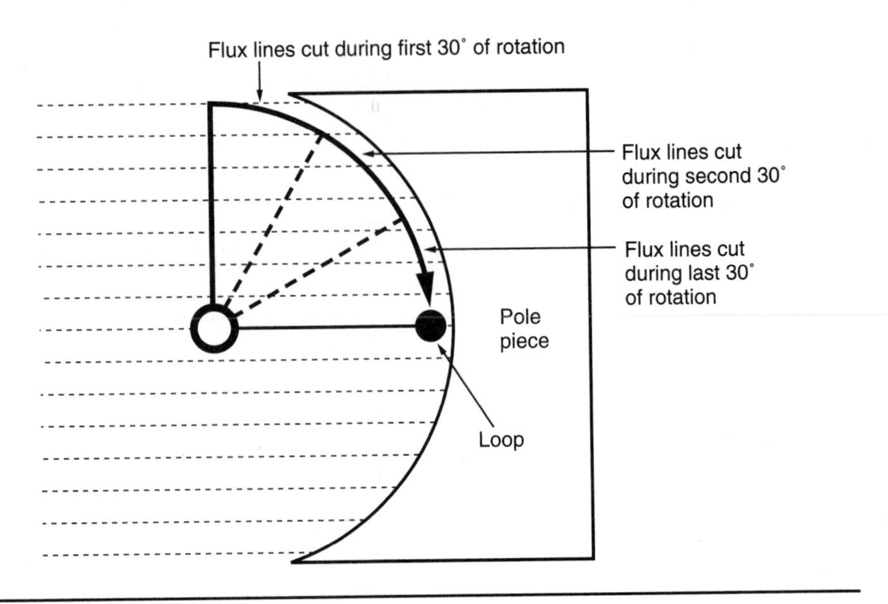

Figure 14–6 As the loop approaches 90 degrees of rotation the flux lines are cut at a faster rate.

Sine waves are so named because the voltage at any point along the waveform is equal to the maximum or peak value times the sine of the angle of rotation. Figure 14–6 illustrates one-half of a loop of wire cutting through lines of magnetic flux. The flux lines are shown with equal spacing between each line, and the arrow denotes the arc of the loop as it cuts through the lines of flux. Notice the number of flux lines that are cut by the loop during the first 30° of rotation. Now notice the number of flux lines cut during the second and third 30° of rotation. Because the loop is cutting the flux lines at an angle, it must travel a greater distance between flux lines during the first degrees of rotation. This means that less flux lines are cut per second, which results in a lower induced voltage. One volt is induced in a conductor when it cuts lines of magnetic flux at a rate of 1 *weber* per second. One weber is equal to 100 million lines of flux.

When the loop has rotated 90°, it is perpendicular to the flux lines and is cutting them at the maximum rate, which results in the highest or peak voltage being induced in the loop. The voltage at any point during the rotation is equal to the maximum induced voltage times the sine of the angle of rotation. For example, if the induced voltage after 90° of rotation is 100 volts, the voltage after 30° of rotation will be 50 volts because the sine of a 30° angle is 0.5, (100 × 0.5 = 50 volts). The induced voltage after 45° of rotation is 70.7 volts, because the sine of a 45° angle is 0.707 (100 × 0.707 = 70.7) volts. Figure 14–7 is a sine wave showing the instantaneous voltage values after different degrees of rotation. The instantaneous voltage value is the value of voltage at any instant on the waveform.

The following formula can be used to determine the instantaneous value at any point along the sine wave:

$$E_{(INST)} = E_{(MAX)} \times \sin \angle$$

where $E_{(INST)}$ = voltage at any point on the waveform
$E_{(MAX)}$ = maximum or peak voltage
$\sin \angle$ = sine of the angle of rotation

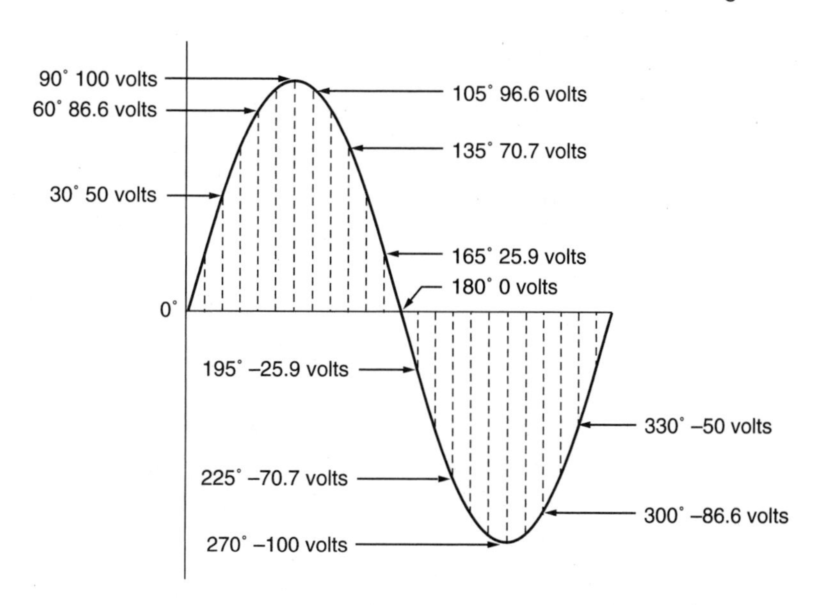

Figure 14–7 Instantaneous values of voltage and current.

Example #1

A sine wave has a maximum voltage of 138 volts. What is the voltage after 78° of rotation?

Solution

$$E_{(INST)} = E_{(MAX)} \times \sin \angle$$

$$E_{(INST)} = 138 \times 0.978 \text{ (sin of } 78°)$$

$$E_{(INST)} = 134.96 \text{ volts}$$

The formula can be changed to find the maximum value if the instantaneous value and the angle of rotation are known, or to find the angle if the maximum and instantaneous values are known.

$$E_{(MAX)} = \frac{E_{(INST)}}{\sin \angle}$$

Example #2

A sine wave has an instantaneous voltage of 246 volts after 53° of rotation. What is the maximum value the waveform will reach?

Solution

$$E_{(MAX)} = \frac{E_{(INST)}}{\sin \angle}$$

$$E_{(MAX)} = \frac{246}{0.799}$$

$$E_{(MAX)} = 307.88 \text{ volts}$$

Example #3

A sine wave has a maximum voltage of 350 volts. At what angle of rotation will the voltage reach 53 volts?

Solution

$$\sin \angle = \frac{E_{(INST)}}{E_{(MAX)}}$$

$$\sin \angle = \frac{53}{350}$$

$$\sin \angle = 0.151$$

Note: 0.151 is the sine of the angle, not the angle. To find what angle corresponds to a sine of 0.151 use a trigonometric table or a scientific calculator.

$$\angle = 8.71°$$

Sine Wave Values

There are several different measurements of voltage and current associated with sine waves: *peak-to-peak, peak, RMS,* and *average.* The peak-to-peak value is measured from the maximum value in the positive direction to the maximum value in the negative direction. The peak-to-peak value is often the simplest measurement to make when using an oscilloscope.

The peak value is measured from zero to the highest value obtained in either the positive or negative direction. The peak value is one-half of the peak-to-peak value.

RMS Values

In Figure 14–8, a 100-volt battery is connected to a 100-Ω resistor. This connection will produce 1 ampere of current flow and the resistor will dissipate 100 watts of power in the form of heat. An AC alternator that produces a peak voltage of 100 volts is also shown con-

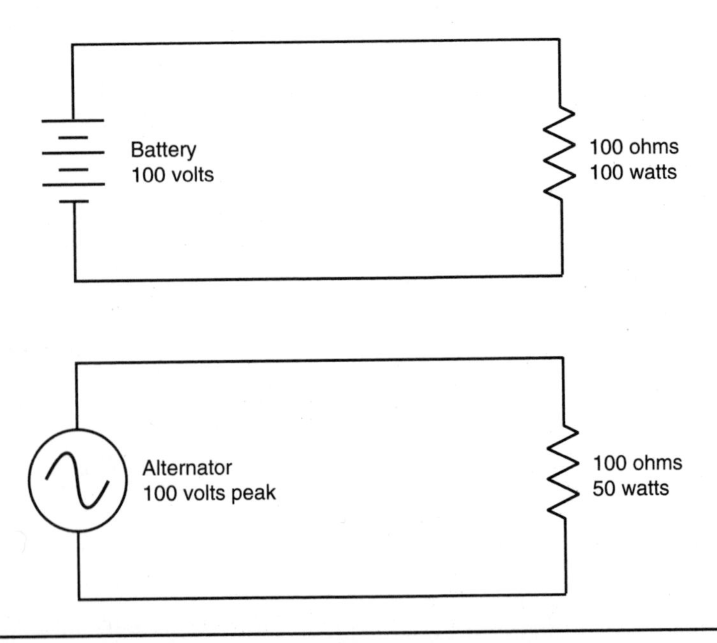

Figure 14–8 Direct current compared to a sine wave AC current.

nected to a 100-Ω resistor. A peak current of 1 ampere will flow in the circuit, but the resistor will dissipate only 50 watts in the form of heat. The reason for this is that the voltage produced by a pure source of direct current, such as a battery, is one continuous value. The sine wave, however, begins at zero, increases to the maximum value, and decreases back to zero during the same period of time. Since the sine wave has a value of 100 volts for only a short period of time, and is less than 100 volts during the rest of the half cycle, it cannot produce as much power as 100 volts of DC.

The solution to this problem is to use a value of AC voltage that will produce the same amount of power as a like value of DC voltage. This AC value is called the *RMS* or *effective* value, and is the value indicated by almost all AC voltmeters and ammeters. RMS stands for root-mean-square, which is an abbreviation for *"the square root of the mean of the square of the instantaneous currents."* The RMS value can be found by dividing the peak value by the $\sqrt{2}$ (1.414), or by multiplying the peak value by 0.707 (the reciprocal of 1.414). The formulas for determining the RMS or peak value are

$$\text{RMS} = \text{peak} \times 0.707 \qquad \text{peak} = \text{RMS} \times 1.414$$

Example #1

A sine wave has a peak value of 354 volts. What is the RMS value?

Solution

$$\text{RMS} = \text{peak} \times 0.707$$

$$\text{RMS} = 354 \times 0.707$$

$$\text{RMS} = 250.3 \text{ volts}$$

Example #2

An AC voltage has a value of 120 volts RMS. What is the peak value of voltage?

Solution

$$\text{peak} = \text{RMS} \times 1.414$$

$$\text{peak} = 120 \times 1.414$$

$$\text{peak} = 169.7 \text{ volts}$$

When the RMS value of voltage and current is used, it will produce the same amount of power as a like value of DC voltage or current. If 100 volts RMS is applied to a 100-Ω resistor, the resistor will produce 100 watts of heat. AC voltmeters and ammeters indicate the RMS value, not the peak value. All values of AC voltage and current used in this text will be RMS values unless otherwise stated.

Average Values

Average values of voltage and current are actually direct current values. The average value must be found when a sine wave AC voltage is changed into DC with a rectifier, shown in Figure 14–9. The rectifier shown is a bridge type, which produces full-wave rectification. This means that both the positive and negative half of the AC waveform is changed into DC. The average value is the amount of voltage that would be indicated by a DC voltmeter if it were connected across the load resistor. The average voltage is proportional to the peak

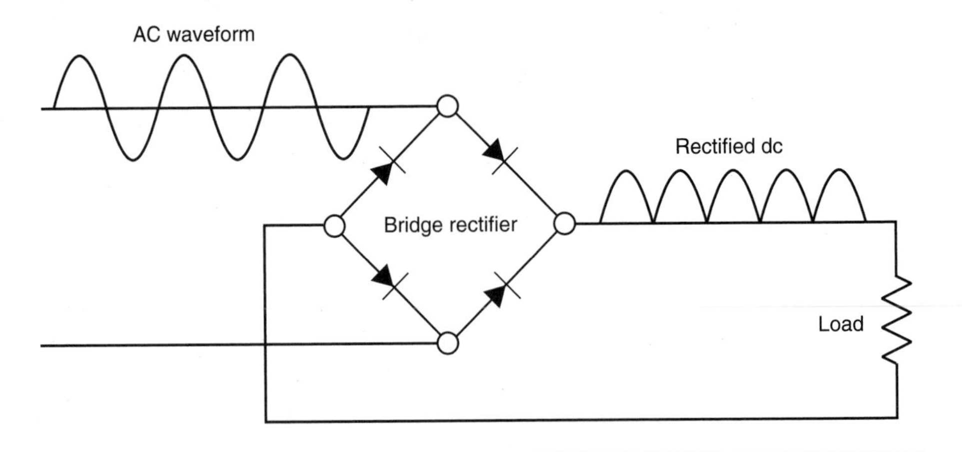

Figure 14–9 The bridge rectifier changes AC voltage into DC voltage.

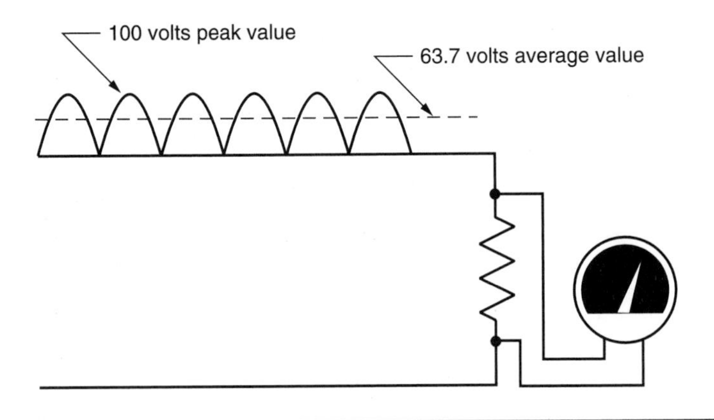

Figure 14–10 A DC voltmeter indicates the average value.

or maximum value of the waveform and the length of time it is on, as compared to the length of time it is off (see Figure 14–10). Notice in Figure 14–10 that the voltage waveform turns on and off, but it never changes polarity. The current, therefore, never reverses direction. This is called *pulsating direct current.* The pulses are often referred to as *ripple.* The average value of voltage will produce the same amount of power as a nonpulsating source of voltage such as a battery. For a sine wave, the average value of voltage is found by multiplying the peak value by 0.637, or by multiplying the RMS value by 0.9.

Example

An AC sine wave with an RMS value of 120 volts is connected to a full-wave rectifier. What is the average DC voltage?

Solution

The problem can actually be solved in one of two ways. The RMS value can be changed into peak, and then the peak value can be changed to the average value.

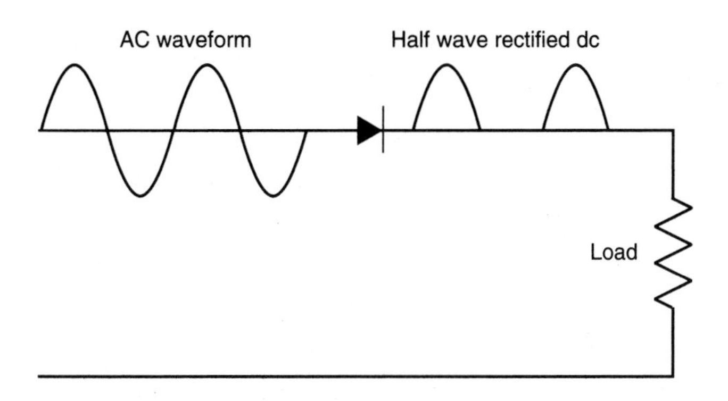

Figure 14–11 A half-wave rectifier converts only one-half of the AC waveform into DC.

$$\text{peak} = \text{RMS} \times 1.414$$

$$\text{peak} = 120 \times 1.414$$

$$\text{peak} = 169.7 \text{ volts}$$

$$\text{average} = \text{peak} \times 0.637$$

$$\text{average} = 169.7 \times 0.637$$

$$\text{average} = 108 \text{ volts}$$

The second method of determining the average value is to multiply the RMS value by 0.9.

$$\text{average} = \text{RMS} \times 0.9$$

$$\text{average} = 120 \times 0.9$$

$$\text{average} = 108 \text{ volts}$$

The conversion factors apply to full-wave rectification. If a half-wave rectifier is used, as in Figure 14–11, only one-half of the AC waveform is converted into DC. To determine the average voltage for a half-wave rectifier, multiply the peak value by 0.637 or the RMS value by 0.9 and then divide the product by two. Since only half of the AC waveform has been converted into direct current, the average voltage will be only half that of a full-wave rectifier.

Example

A half-wave rectifier is connected to 277 volts AC. What is the average DC voltage?

Solution

$$\text{average} = \text{RMS} \times 0.9/2$$

$$\text{average} = 277 \times 0.9/2$$

$$\text{average} = 124.6 \text{ volts}$$

Average Values for Three-Phase Rectifiers

There are two common types of rectifiers for three-phase systems, the half-wave and the full-wave bridge. The half-wave is constructed using a three-phase, four-wire wye or star connection, as shown in Figure 14–12. A single diode is connected into each of the three line leads and the center tap provides the return path for each phase. When this type of connection is used, the average voltage will be 0.827 of the peak value.

Example

A three-phase system has a peak value of 169.68 volts on each phase.

Solution

The average voltage value for a half-wave three-phase rectifier is:

$$169.68 \times 0.827 = 140.32 \text{ volts DC}$$

A three-phase bridge rectifier contains six diodes and can be connected to a delta or wye system, as in Figure 14–13. The average voltage value for the three-phase full-wave bridge rectifier is 0.955 of peak.

Example

A three-phase system has a peak value of 169.68 volts on each phase.

Solution

The average voltage value for a three-phase full-wave bridge rectifier is:

$$169.68 \times 0.955 = 162 \text{ volts DC}$$

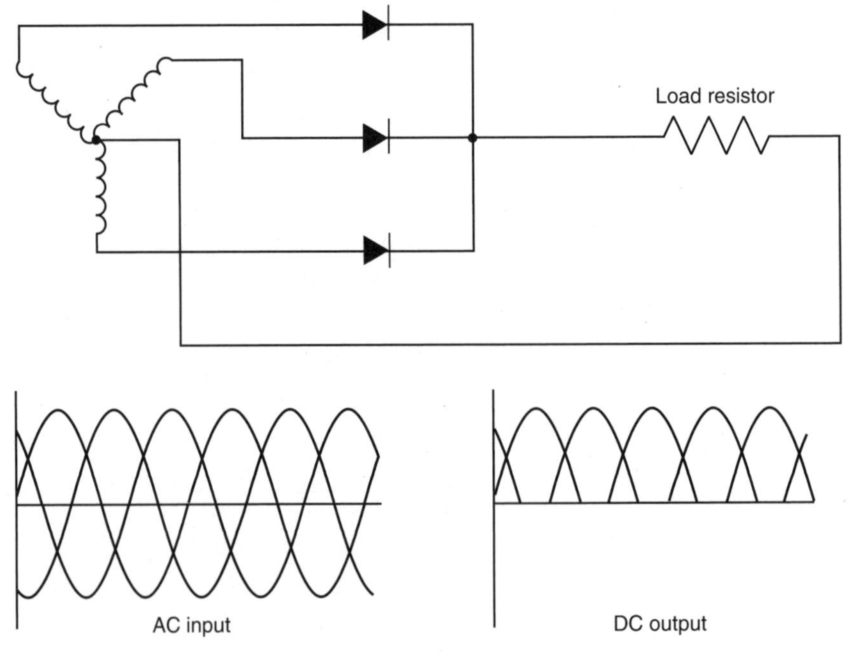

Figure 14–12 Three-phase half-wave rectifier.

Figure 14–14 lists conversion factors for peak-to-peak, peak RMS, and average values.

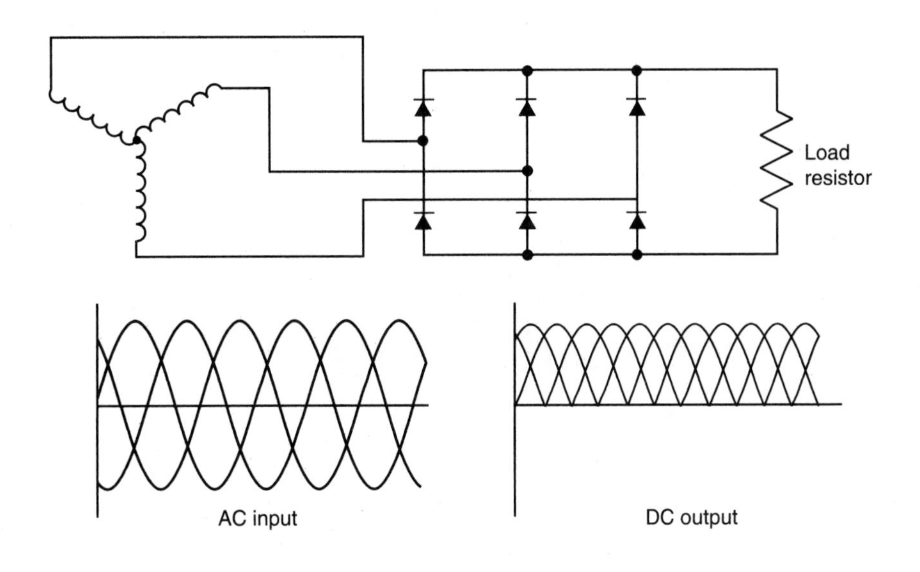

Figure 14–13 Three-phase bridge rectifier.

TO CHANGE	TO CHANGE	MULTIPLY BY
Peak-to-Peak	Peak	0.5
Peak	Peak-to-Peak	2
Peak	RMS	0.707
RMS	Peak	1.414
Peak	Average (Single-phase, full-wave unfiltered)	0.637
RMS	Average (Single-phase, full-wave unfiltered)	0.9
Peak	Average (Single-phase, half-wave unfiltered)	0.3185
Average (Single-phase, full-wave unfiltered)	Peak	1.567
Average (Single-phase, full-wave unfiltered)	RMS	1.111
Peak	Average (Three phase, half-wave unfiltered)	0.827
Peak	Average (Three phase, full-wave unfiltered)	0.955

Figure 14–14 Multiplication factors for peak-to-peak, peak RMS, and average voltage values.

CHAPTER 15

Alternating Current Circuits Containing Inductance

Inductance is one of the primary types of loads in alternating current circuits. Some amount of inductance is present in all alternating current circuits because of the continually changing magnetic field. The amount of inductance of a single conductor is extremely small, and in most instances is not considered in circuit calculations. Circuits are generally considered to contain inductance when any type of load that contains a coil is used. Loads such as motors, transformers, lighting ballast, and chokes all contain coils of wire.

Any time current flows through a coil of wire, a magnetic field is created around the wire. If the amount of current should decrease, the magnetic field will collapse. Concerning inductance:

1. When magnetic lines of flux cut through a coil, a voltage is induced in the coil.
2. An induced voltage is always opposite in polarity to the applied voltage.
3. The amount of induced voltage is proportional to the rate of change of current.
4. An inductor opposes a change of current.

Assume an inductor is connected to an alternating voltage. This causes the magnetic field to continually increase, decrease, and reverse polarity. Since the magnetic field continually changes magnitude and direction, a voltage is continually being induced in the coil. This induced voltage is 180 degrees out of phase with the applied voltage and is always in opposition to the applied voltage (see Figure 15–1). Since the induced voltage is always in opposition to the applied voltage, the applied voltage must overcome the induced voltage before current can flow through the circuit. For example, assume an inductor is connected to a 120-volt AC line. Now assume that the inductor has an induced voltage of 116 volts. Since an equal amount of applied voltage must be used to overcome the induced voltage, there will be only 4 volts to push current through the wire resistance of the coil ($120 - 116 = 4$).

Inductive Reactance

Notice that the induced voltage has the ability to limit the flow of current through the circuit in a manner similar to resistance. This induced voltage is **not** resistance, but it can limit the flow of current the way resistance does. This current-limiting property of the inductor is called *reactance* and is symbolized by the letter *X*. Since this reactance is caused by induc-

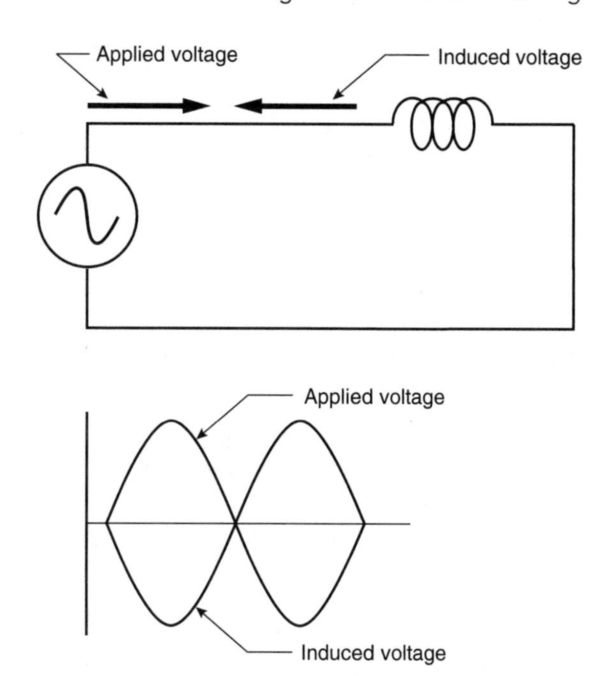

Figure 15–1 The applied voltage and induced voltage are 180 degrees out of phase with each other.

tance, it is called *inductive reactance* and is symbolized by X_L, pronounced "*X* sub *L*." Inductive reactance is measured in ohms the way resistance is, and it can be computed when the value of inductance and frequency are known. The formula shown can be used to find inductive reactance:

$$X_L = 2\pi FL$$

where X_L = inductive reactance
2 = constant
π = 3.1416
F = frequency in hertz
L = inductance in henrys

Inductive reactance is an inducted voltage and is, therefore, proportional to the three factors that determine induced voltage:

1. The number of turns of wire
2. The strength of the magnetic field
3. The speed of the cutting action (relative motion between the inductor and the magnetic lines of flux)

The number of turns of wire and strength of the magnetic field are determined by the physical construction of the inductor. Factors such as the size of wire used, the number of turns, how close the turns are to each other, and the type core material determine the amount of inductance (henrys) of the coil. The speed of the cutting action is proportional to the frequency (hertz). An increase of frequency will cause the magnetic lines of flux to cut the conductors at a faster rate. This will produce a higher induced voltage or more inductive reactance.

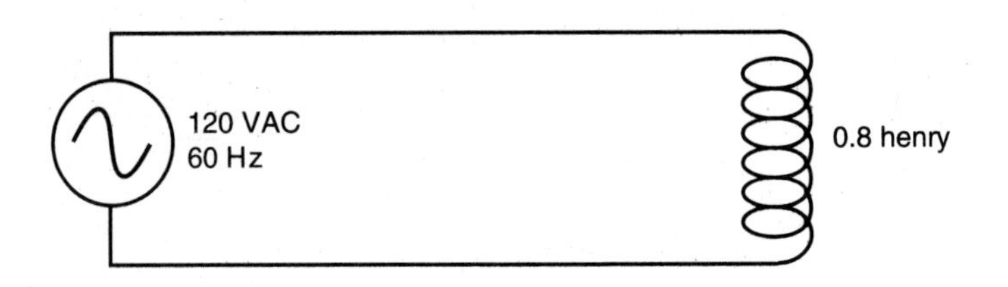

Figure 15–2 Circuit current is limited by inductive reactance.

Example #1

The inductor shown in Figure 15–2 has an inductance of 0.8 henry and is connected to a 120-volts, 60-Hz line. How much current will flow in this circuit if the wire resistance of the inductor is negligible?

Solution

The first step in solving this problem is to determine the amount of inductive reactance of the inductor.

$$X_L = 2\pi FL$$

$$X_L = 2 \times 3.1416 \times 60 \times 0.8$$

$$X_L = 301.6\Omega$$

Since inductive reactance is the current-limiting property of this circuit, it can be substituted for the value of R in an Ohm's law formula.

$$I = \frac{E}{X_L}$$

$$I = \frac{120}{301.6}$$

$$I = 0.398 \text{ amp}$$

If the amount of inductive reactance is known, the inductance of the coil can be determined using the formula:

$$L = \frac{X_L}{2\pi F}$$

Example #2

Assume an inductor with a negligible resistance is connected to a 36-volt, 400-Hz line. If the circuit has a current flow of 0.2 amp, what is the inductance of the inductor?

Solution

The first step is to determine the inductive reactance of the circuit.

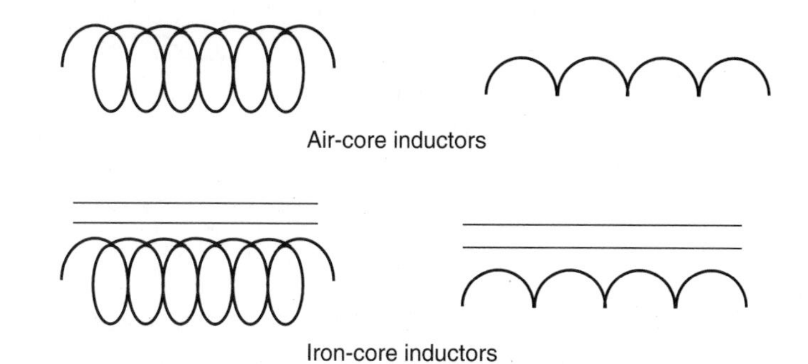

Figure 15–3 Schematic symbols for inductors.

$$X_L = \frac{E}{I}$$

$$X_L = \frac{36}{0.2}$$

$$X_L = 180\ \Omega$$

Now that the inductive reactance of the inductor is known, determine the inductance.

$$L = \frac{180}{2 \times 3.1416 \times 400}$$

$$L = 0.0716 \text{ henry}$$

The schematic symbol used to represent an inductor depicts a coil of wire. Several different symbols for inductors are shown in Figure 15–3. Although these symbols are different, they are similar. The symbols shown with the two parallel lines represent iron-core inductors, and the symbols without the parallel lines represent air-core inductors.

Inductors Connected in Series

When inductors are connected in series, as in Figure 15–4, the total inductance of the circuit will be equal to the sum of all the inductors.

$$L_T = L_1 + L_2 + L_3$$

When computing the inductive reactance of inductors connected in series, the total inductive reactance will equal the sum of the inductive reactance for each inductor.

$$X_{LT} = X_{L1} + X_{L2} + X_{L3}$$

Inductors Connected in Parallel

When inductors are connected in parallel, as in Figure 15–5, the total inductance can be found in a similar manner to finding the total resistance of a parallel circuit. The reciprocal of the total inductance is equal to the sum of the reciprocals of each inductor.

$$\frac{1}{L_T} = \frac{1}{L_1} + \frac{1}{L_2} + \frac{1}{L_3} + \frac{1}{L_N}$$

or

$$L_T = \frac{1}{\frac{1}{L_1} + \frac{1}{L_2} + \frac{1}{L_3} + \frac{1}{L_N}}$$

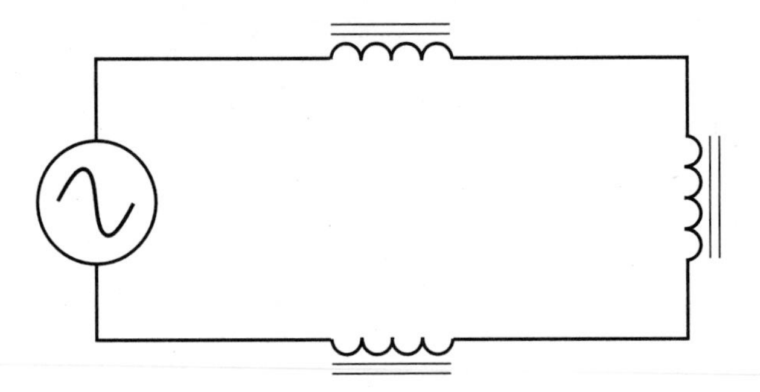

Figure 15–4 Inductors connected in series.

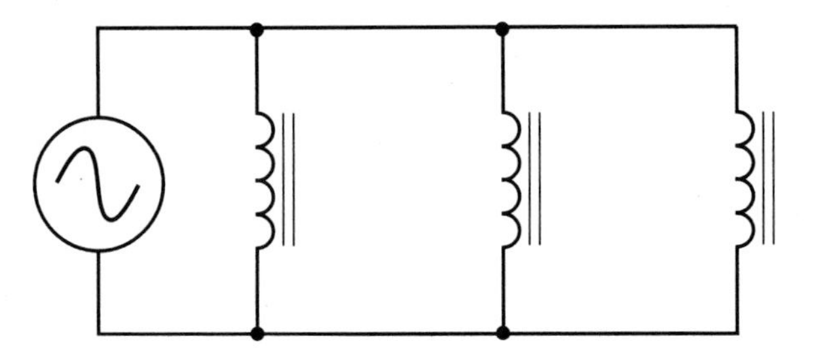

Figure 15–5 Inductors connected in parallel.

Another formula that can be used to find the total inductance of parallel inductors is the product-over-sum formula.

$$L_T = \frac{L_1 \times L_2}{L_1 + L_2}$$

If the values of all the inductors are the same, total inductance can be found by dividing the inductance of one inductor by the total number of inductors.

$$L_T = \frac{L}{N}$$

Similar formulas can be used to find the total inductive reactance of inductors connected in parallel.

$$\frac{1}{X_{LT}} = \frac{1}{X_{L1}} + \frac{1}{X_{L2}} + \frac{1}{X_{L3}} + \frac{1}{X_{LN}}$$

or

$$X_{LT} = \frac{1}{\frac{1}{X_{L1}} + \frac{1}{X_{L2}} + \frac{1}{X_{L3}} + \frac{1}{X_{LN}}}$$

or

$$X_{LT} = \frac{X_{L1} \times X_{L2}}{X_{L1} + X_{L2}}$$

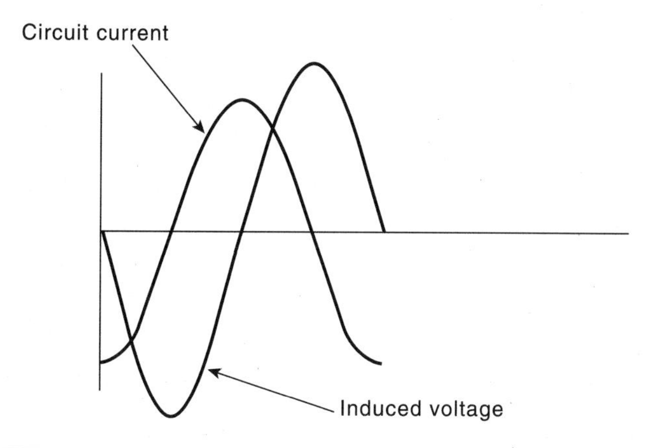

Figure 15–6 Induced voltage is proportional to the rate of change of current.

or

$$X_{LT} = \frac{X_L}{N}$$

Voltage and Current Relationships in an Inductive Circuit

When current flows through a pure resistive circuit, the current and voltage are in phase with each other. In a pure inductive circuit, the current lags the voltage by 90 degrees. At first this may seem to be an impossible condition, until the relationship of applied voltage and induced voltage is considered. To understand how the current and applied voltage can become 90 degrees out of phase with each other, compare the relationship of the current and induced voltage (see Figure 15–6). Recall that the induced voltage is proportional to the rate of change of the current (speed of cutting action). At the beginning of the waveform, the current is shown at its maximum value in the negative direction. At this point in time, the current is not changing, so induced voltage is zero. As the current begins to decrease in value, the magnetic field produced by the flow of current decreases or collapses and begins inducing a voltage into the coil as it cuts through the conductors.

The greatest rate of current change occurs when the current passes from negative, through zero, and begins to increase in the positive direction, shown in Figure 15–7. Because the current is changing at the greatest rate, the induced voltage is maximum. As current approaches its peak value in the positive direction, the rate of change decreases, causing a decrease in the induced voltage. The induced voltage will again be zero when the current reaches its peak value and the magnetic field stops expanding.

It can be seen that the current flowing through the inductor is leading the induced voltage by 90 degrees. Since the induced voltage is 180 degrees out of phase with the applied voltage, the current will lag the applied voltage by 90 degrees (see Figure 15–8).

Power in an Inductive Circuit

In a pure resistive circuit, the true power or watts is equal to the product of the voltage and current. In a pure inductive circuit, however, no true power or watts is produced. Voltage and current must both be either positive or negative before true power can be produced. Since the voltage and current are 90 degrees out of phase with each other in a pure inductive circuit, the current and voltage will be at different polarities 50 percent of the time and at the same polarity 50 percent of the time. During the period of time that the current and voltage have the same polarity, power is being given to the circuit in the form of creating a

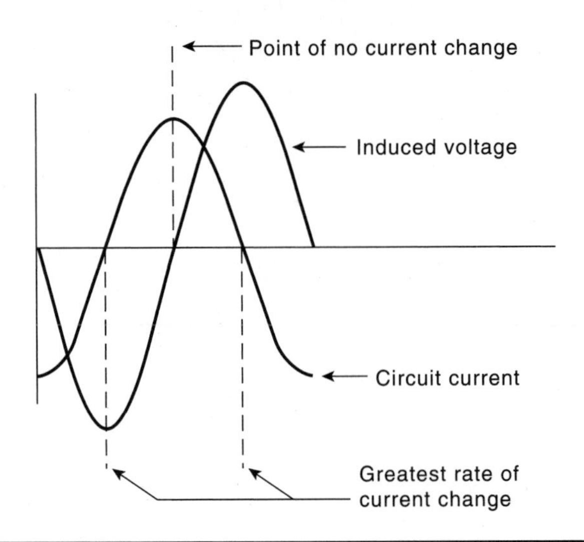

Figure 15–7 No voltage is induced when the current does not change.

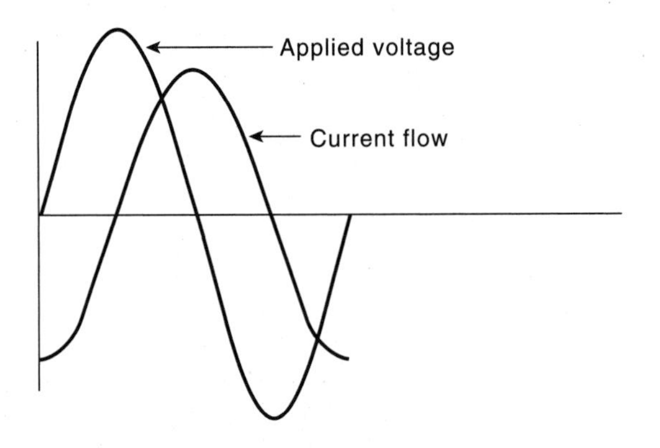

Figure 15–8 The current lags the applied voltage by 90 degrees.

magnetic field. When the current and voltage are opposite in polarity, power is being given back to the circuit as the magnetic field collapses and induces a voltage back into the circuit. Since power is stored in the form of a magnetic field and then given back, the inductor uses no power. Any power used in an inductor is caused by losses such as the resistance of the wire used to construct the coil, generally referred to as I^2R losses, eddy current losses, and hysteresis losses.

The current and voltage waveform in Figure 15–9 has been divided into four quadrants; A, B, C, and D.

A — During the first time period, indicated by A, the current is negative and the voltage is positive. During this period of time, energy is being given to the circuit as the magnetic field collapses.

B — During the second time period, section B, both the voltage and current are positive. Power is being used to produce the magnetic field.

C — In the third time period, indicated by C, the current is positive and the voltage is negative. Power is again being given back to the circuit as the field collapses.

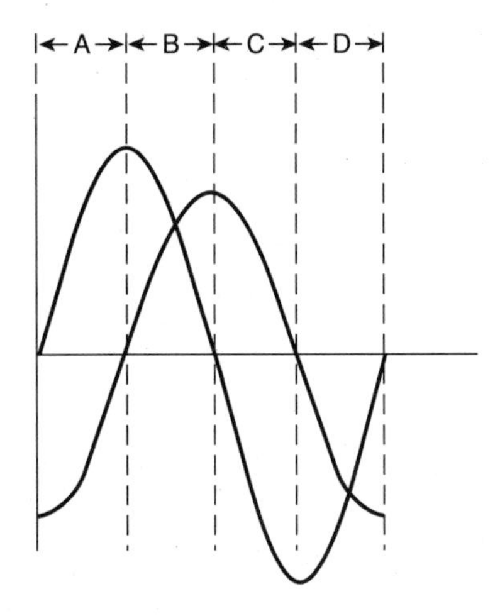

Figure 15–9 Voltage and current relationships during different parts of a cycle.

D During the fourth time period, indicated by D, both the voltage and current are negative. Power is again being used to produce the magnetic field.

If the amount of power used to produce the magnetic field is subtracted from the power given back, it will be seen that the result is zero.

Reactive Power (VARs)

Although there is essentially no true power being used, except by previously mentioned losses, an electrical measurement called **VARs** is used to measure the reactive power in a pure inductive circuit. VARs is an abbreviation for volt-amps-reactive. VARs can be computed in the same way as watts, except that inductive values are substituted for resistive values in the formulas. VARs is equal to the amount of current flowing through an inductive circuit times the voltage applied to the inductive part of the circuit. Several formulas for computing VARs are shown here:

$$VARs = E_L \times I_L$$

$$VARs = \frac{E_L^{\,2}}{X_L}$$

$$VARs = I_L^{\,2} \times X_L$$

where E_L = voltage applied to an inductor
I_L = current flow through an inductor
X_L = inductive reactance

Q of an Inductor

So far in this unit, it has been generally assumed that an inductor has no resistance and inductive reactance is the only current-limiting factor. In reality, this is not the case. Since inductors are actually coils of wire, they all contain some amount of internal resistance. Inductors actually appear to be a coil connected in series with some amount of resistance, as shown in Figure 15–10. The amount of resistance as compared to the inductive reactance determines the "Q" of the coil. The letter *Q* stands for quality. Inductors that have a higher ratio of inductive reactance as compared to resistance are considered to be inductors of higher quality. An inductor constructed with large wire will have a low wire resistance and, therefore, a higher *Q*. Inductors constructed with many turns of small wire have a

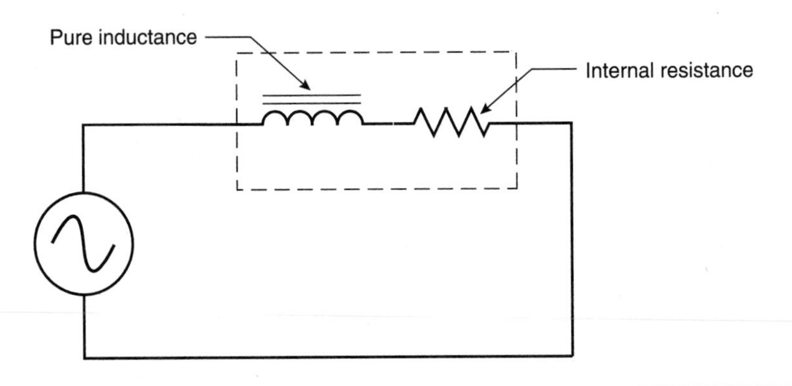

Figure 15–10 Inductors contain internal resistance.

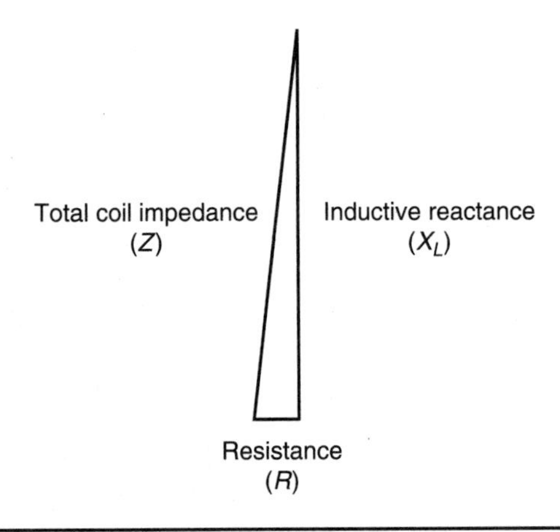

Figure 15–11 Coil impedance.

much higher resistance, and therefore, a lower *Q*. To determine the *Q* of an inductor, divide the inductive reactance by the resistance.

$$Q = \frac{X_L}{R}$$

Inductor Impedance

Although inductors have some amount of resistance, inductors that have a *Q* of 10 or greater are generally considered to be pure inductors. Once the ratio of inductive reactance becomes ten times greater than resistance, the amount of resistance is considered negligible. For example, assume that an inductor has an inductive reactance of 100 ohms and a wire resistance of 10 ohms. The inductive reactive component in the circuit is 90 degrees out phase with the resistive component. This relationship produces a right triangle, shown in Figure 15–11. The total current-limiting effect of the inductor is a combination of the inductive reactance and resistance. This total current-limiting effect is called *impedance* and is symbolized by the letter *Z*. The impedance of the circuit is represented by the hypotenuse of the right triangle formed by the inductive reactance, represented by the leg of the triangle, and the resistance, represented by the base of the triangle. To compute the value of impedance for the coil, it is necessary to add the inductive reactance and resistance together. Since these two components form the leg and base of a right triangle and the impedance forms the hypotenuse, the Pythagorean theorem is employed.

$$L = \frac{X_L}{2 \times \pi \times F} \qquad X_L = 2\pi FL$$

$$E_L = I_L \times X_L \qquad I_L = \frac{E_L}{X_L} \qquad X_L = \frac{E_L}{I_L} \qquad VARs_L = E_L \times I_L$$

$$E_L = \sqrt{VARs_L \times X_L} \qquad I_L = \frac{VARs_L}{E_L} \qquad X_L = \frac{{E_L}^2}{VARs_L} \qquad VARs_L = {I_L}^2 \times X_L$$

$$E_L = \frac{VARs_L}{I_L} \qquad I_L = \sqrt{\frac{VARs_L}{X_L}} \qquad X_L = \frac{VARs_L}{{I_L}^2} \qquad VARs_L = \frac{{E_L}^2}{X_L}$$

Series Inductive Circuits

$$E_{LT} = E_{L1} + E_{L2} + E_{L3}$$

$$X_{LT} = X_{L1} + X_{L2} + X_{L3}$$

$$I_{LT} = I_{L1} = I_{L2} = I_{L3}$$

$$VARs_{LT} = VARs_{L1} + VARs_{L2} + VARs_{L3}$$

$$L_T = L_1 + L_2 + L_3$$

Parallel Inductive Circuits

$$E_{LT} = E_{L1} = E_{L2} = E_{L3}$$

$$I_{LT} = I_{L1} + I_{L2} + I_{L3}$$

$$X_{LT} = \frac{X_{L1} \times X_{L2}}{X_{L1} + X_{L2}}$$

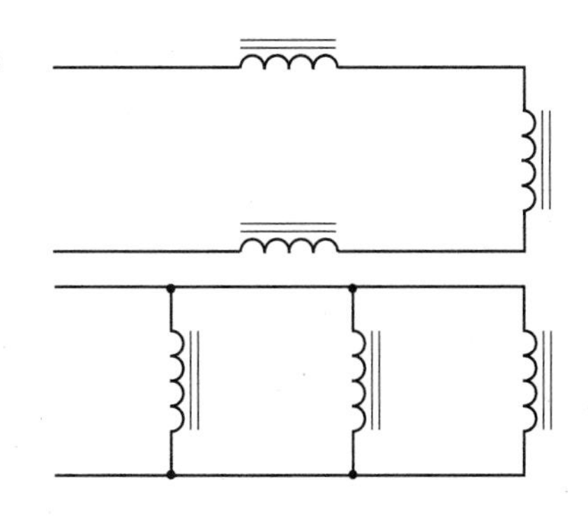

$$X_{LT} = \frac{1}{\frac{1}{X_{L1}} + \frac{1}{X_{L2}} + \frac{1}{X_{L3}}} \qquad L_T = \frac{1}{\frac{1}{L_1} + \frac{1}{L_2} + \frac{1}{L_3}} \qquad L_T = \frac{L_1 \times L_2}{L_1 + L_2}$$

$$VARsL_T = VARsL_1 + VARsL_2 + VARsL_3$$

Figure 15–12 Formulas for pure inductive circuits.

$$Z = \sqrt{R^2 + {X_L}^2}$$

$$Z = \sqrt{10^2 + 100^2}$$

$$Z = \sqrt{10100}$$

$$Z = 100.5\ \Omega$$

Notice that the value of total impedance for the inductor is only 0.5 ohms greater than the value of inductive reactance.

Pure Inductive Circuits

Formulas concerning alternating current circuits containing pure inductance are shown in Figure 15–12.

Note: In a pure inductive circuit, the current lags the voltage by 90 degrees. Therefore, there is no true power or watts and the power factor is 0. VARs is the inductive equivalent of watts.

CHAPTER 16

Resistive–Inductive Series Circuits

When a pure resistive load is connected to an alternating current circuit, the voltage and current are in phase with each other. When a pure inductive load is connected to an alternating current circuit, the voltage and current are 90 degrees out of phase with each other, as shown in Figure 16–1. When a circuit containing both resistance and inductance is connected to an alternating current circuit, the voltage and current will be out of phase with each other by some amount between 0 degrees and 90 degrees. The exact amount of phase angle difference is determined by the ratio of resistance as compared to inductance. In the following example, a series circuit containing 30 Ω of resistance and 40 Ω of inductive reactance is connected to a 240-volt, 60-Hz line, as shown in Figure 16–2. It is assumed the inductor has negligible resistance. The following unknown values will be computed:

Z = total circuit impedance
I_T = current flow
E_R = voltage drop across the resistor
P = watts (true power)
L = inductance of the inductor
E_L = voltage drop across the inductor
E_T = total voltage
$VARs$ = volt-amperes-reactive (reactive power)
VA = volt-amperes (apparent power)
PF = power factor
$\angle\varnothing$ = angle theta (indicates the angle the voltage and current are out of phase with each other)

Impedance (Z)

Impedance is a measure of the part of the circuit that impedes or hinders the flow of current. It is measured in ohms and is symbolized by the letter Z. In this circuit, impedance will be a combination of resistance and inductive reactance.

In a series circuit, the total resistance is equal to the sum of the individual resistors. In this instance, however, the impedance will be the sum of the resistance and the inductive

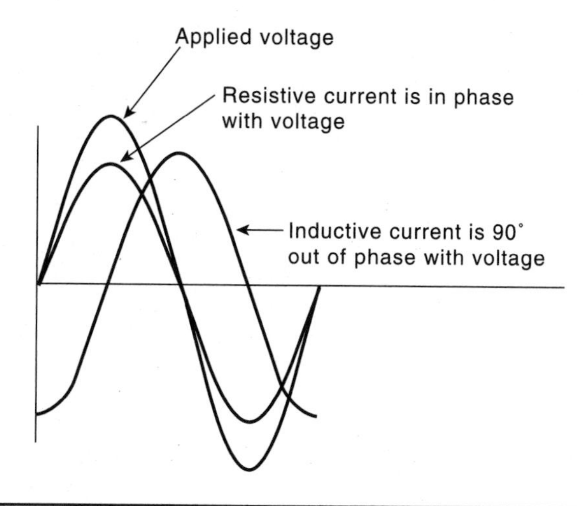

Figure 16–1 Relationship of resistive and inductive current with voltage.

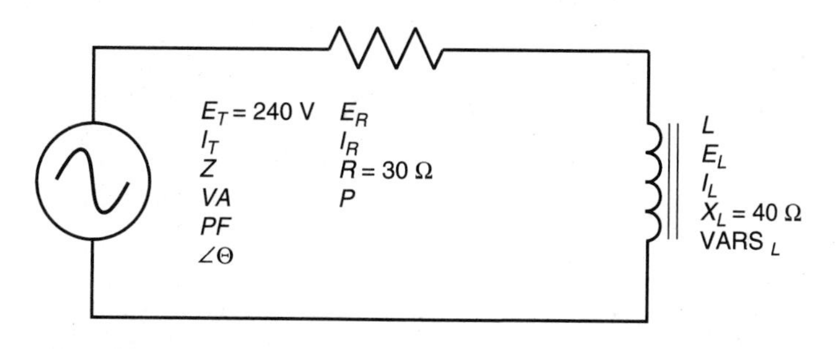

Figure 16–2 RL series circuit.

reactance. It would first appear that the sum of these two quantities should be 70 ohms (30 ohms + 40 ohms = 70 ohms). In practice, however, the resistive part of the circuit and the reactive part of the circuit are out of phase with each other by 90 degrees. To find the sum of these two quantities, vector addition must be used. Since these two quantities are 90 degrees out of phase with each other, the resistive and inductive reactance form the two legs of a right triangle and the impedance is the hypotenuse, as in Figure 16–3.

The impedance can be computed using the Pythagorean formula:

$$Z = \sqrt{R^2 + X_L{}^2}$$

$$Z = \sqrt{30^2 + 40^2}$$

$$Z = \sqrt{900 + 1600}$$

$$Z = \sqrt{2500}$$

$$Z = 50\ \Omega$$

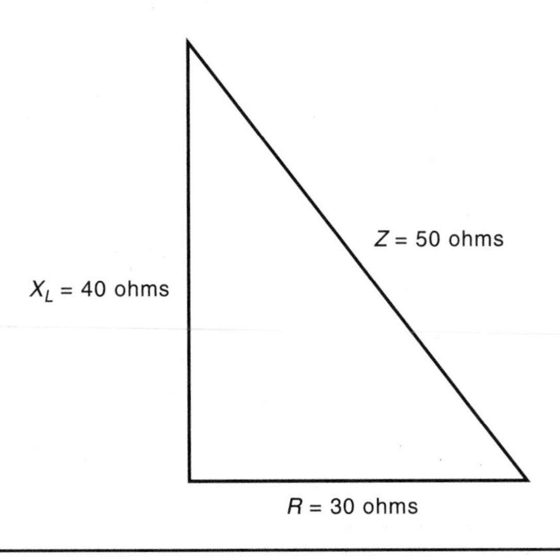

Figure 16–3 Impedance is the combination of resistance and inductive reactance.

Total Circuit Current (I_T)

One of the primary laws for series circuits is that the current must be the same in any part of the circuit. This law holds true of RL series circuits, also. Since the impedance is the total current-limiting component of the circuit, it can be used to replace R in an Ohm's law formula. The total current flow through the circuit can be computed by dividing the total applied voltage by the total current-limiting factor or impedance. Total current can be found by using the formula:

$$I_T = \frac{E}{Z}$$

$$I_T = \frac{240}{50}$$

$$I_T = 4.8 \text{ amps}$$

In a series circuit, the current is the same at any point in the circuit. Therefore, 4.8 amps of current flows through both the resistor and the inductor.

Voltage Drop Across the Resistor (E_R)

Now that the amount of current flow through the resistor is known, the voltage drop across the resistor can be computed using the following formula:

$$E_R = I_R \times R$$

$$E_R = 4.8 \times 30$$

$$E_R = 144 \text{ volts}$$

Notice that the amount of voltage dropped across the resistor was found using quantities that pertained only to the resistive part of the circuit.

Inductive reactance (X_L) is an inductive quantity, and impedance (Z) is a circuit total quantity. They can be used with vector addition to find like resistive quantities. For example, both inductive reactance and impedance are measured in ohms. The resistive quantity that is measured in ohms is resistance. If the impedance and inductive reactance of a circuit were known, they could be used with this formula to find the circuit resistance.

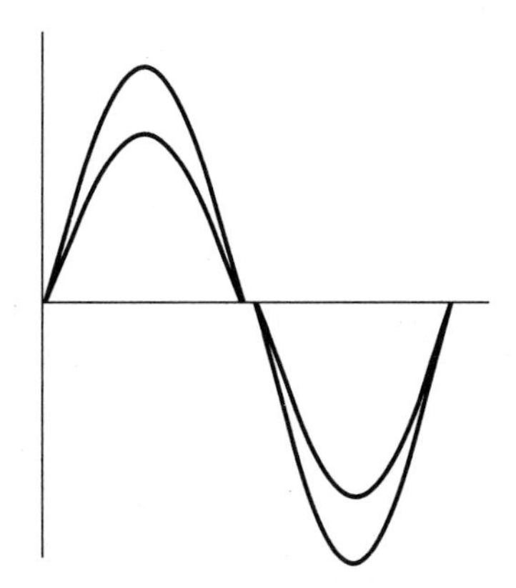

Voltage dropped across the resistor is in phase with the current.

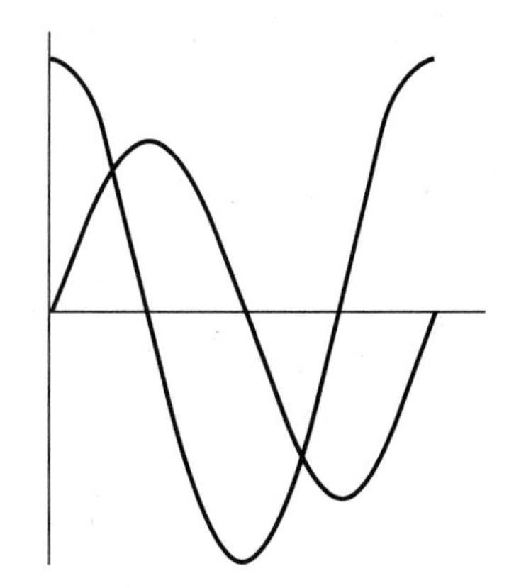

Voltage dropped across the inductor is 90° out of phase with the current.

Figure 16–4 Relationship of current and voltage in an RL series circuit.

$$R = \sqrt{Z^2 - X_L^{\ 2}}$$

Watts (*P*)

True power for the circuit can be computed by using any of the watts formulas with pure resistive parts of the circuit. Watts can be computed by multiplying the voltage dropped across the resistor, E_R, by the current flow through the resistor, I_R; or by squaring the voltage dropped across the resistor and dividing by the ohmic value of the resistor; or by squaring the current flow through the resistor and multiplying by the ohmic value of the resistor.

In an RL series circuit, the current is the same through both the resistor and the inductor. The voltage dropped across the resistor, however, is in phase with the current, and the voltage dropped across the inductor is 90 degrees out of phase with the current, as in Figure 16–4. True power or watts can be produced only when the current and voltage are both positive or both negative, so only resistive parts of the circuit can produce watts.

The formula used in this example will be:

$$P = E_R \times I_R$$

$$P = 144 \times 4.8$$

$$P = 691.2 \text{ watts}$$

Inductance (*L*) The amount of inductance can be computed using the formula:

$$L = \frac{X_L}{2\pi F}$$

$$L = \frac{40}{2 \times 3.1416 \times 60}$$

$$L = \frac{40}{377}$$

$$L = 0.106 \text{ henry}$$

Voltage Drop Across the Inductor (E_L)

The amount of voltage dropped across the inductor can be computed using the formula:

$$E_L = I_L \times X_L$$

$$E_L = 4.8 \times 40$$

$$E_L = 192 \text{ volts}$$

Notice that only inductive quantities were used to find the voltage drop across the inductor.

Total Voltage (E_T)

Although the total applied voltage was given at the beginning of this problem, it should be noted that the total voltage is also equal to the sum of the voltage drops, just as it is in any other series circuit. Since the voltage dropped across the resistor is in phase with the current and the voltage dropped across the inductor is 90 degrees out of phase with the current, vector addition must be used. The total voltage will be the hypotenuse of a right triangle and the resistive and inductive voltage drops will form the legs of the triangle. This relationship of voltage drops can also be represented using the parallelogram method of vector addition, as shown in Figure 16–5. The following formulas can be used to find total voltage or the voltage drops across the resistor or inductor if the other two voltage values are known.

$$E_T = \sqrt{E_R^2 + E_L^2}$$

$$E_R = \sqrt{E_T^2 + E_L^2}$$

$$E_L = \sqrt{E_T^2 + E_R^2}$$

Reactive Power (VARs)

VARs is an abbreviation for volt-amps-reactive, and is the amount of reactive power in the circuit. VARs should not be confused with watts, which is true power. VARs represents the product of the volts and amps that are 90 degrees out of phase with each other, such as the voltage dropped across the inductor and the current flowing through the inductor. Recall that true power can be produced only during periods of time that the voltage and current are both positive or both negative (see Figure 16–6). During these periods of time,

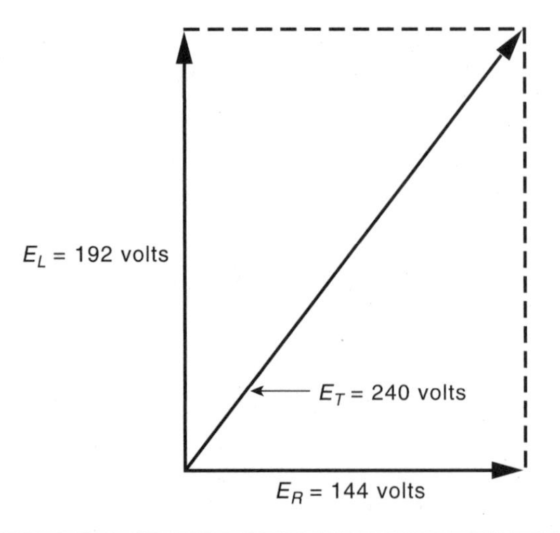

Figure 16–5 Graphic representation of voltage drops using the parallelogram method of vector addition.

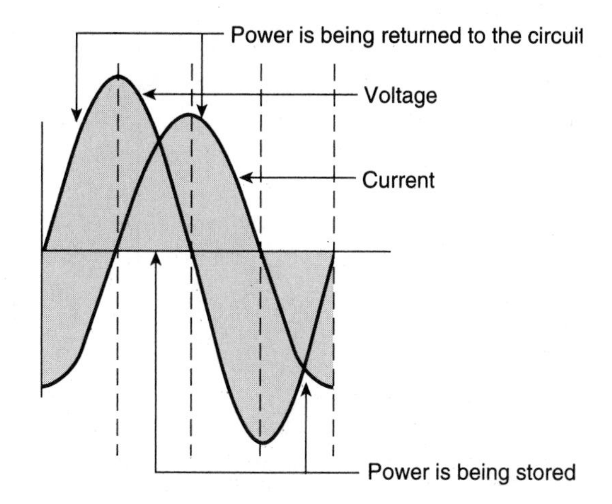

Figure 16–6 Power is stored and then returned to the circuit.

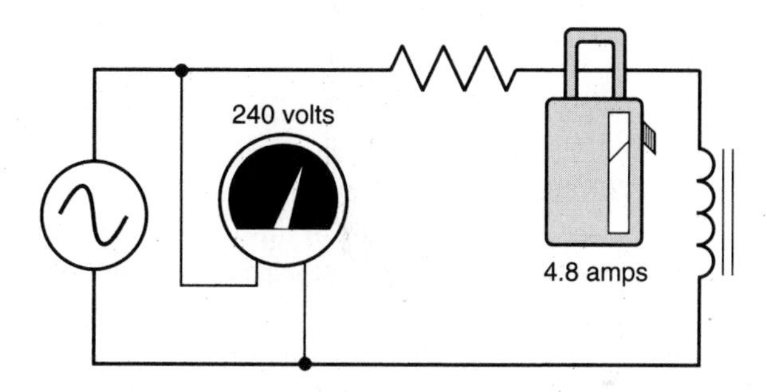

Figure 16–7 Apparent power is the product of measured values 240 × 4.8 = 1152 VA.

the power is being stored in the form of a magnetic field. During the periods of time that voltage and current have opposite signs, the power is returned to the circuit. For this reason, VARs is often referred to as quadrature power, or wattless power. It can be computed in a similar manner as watts, except that reactive values of voltage and current are used instead of resistive values. In this example the formula used will be:

$$VARs = I_L^2 \times X_L$$

$$VARs = 4.8^2 \times 40$$

$$VARs = 921.6$$

Apparent Power (*VA*)

Volt–amperes (VA) is the apparent power of the circuit. It can be computed in a similar manner as watts or VARs, except that total values of voltage and current are used. It is called apparent power because it is the value that would be found if a voltmeter and ammeter were used to measure the circuit voltage and current and then these measured values were multiplied together, as in Figure 16–7. In this example, the formula used will be:

$$VA = E_T \times I_T$$

$$VA = 240 \times 4.8$$

$$VA = 1152$$

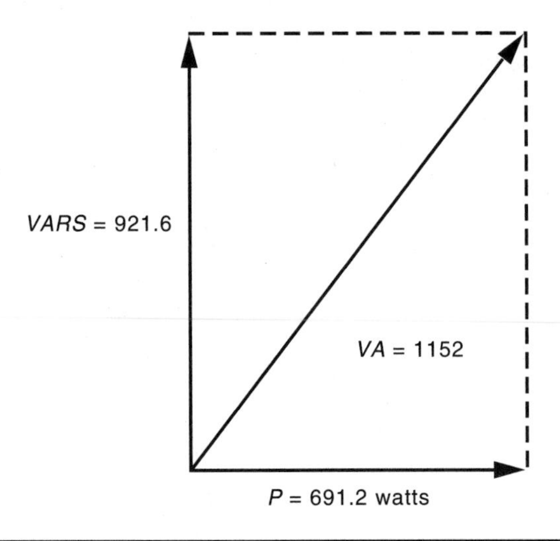

Figure 16–8 Using the parallelogram method to plot the relationship of volt amps, watts, and VARs.

The apparent power can also be found using vector addition in a similar manner as impedance or total voltage. Since true power or watts is a pure resistive component and VARs is a pure reactive component, they form the legs of a right triangle. The apparent power is the hypotenuse of this triangle. This relationship of the three power components can also be plotted using the parallelogram method, shown in Figure 16–8. These formulas can be used to compute the values of apparent power, true power, and reactive power when the other two values are known.

$$VA = \sqrt{P^2 + VARs_L{}^2}$$

$$P = \sqrt{VA^2 - VARs_L{}^2}$$

$$VARs_L = \sqrt{VA^2 - P^2}$$

Power Factor (*PF*)

Power factor is a ratio of the true power as compared to the apparent power. It can be computed by dividing any resistive value by its like total value. For example, power factor can be computed by dividing the voltage drop across the resistor by the total circuit voltage; or resistance divided by impedance; or watts divided by volt–amperes.

$$PF = \frac{E_R}{E_T} \qquad PF = \frac{R}{Z} \qquad PF = \frac{P}{VA} \qquad PF = \cos \angle Ø$$

Power factor is generally expressed as a percentage. The decimal fraction computed from the division will, therefore, be changed to a percent by multiplying it by 100. In this circuit the formula used will be:

$$PF = \frac{P}{VA}$$

$$PF = \frac{691.2}{1152}$$

$$PF = 0.6 \times 100, \text{ or } 60\%$$

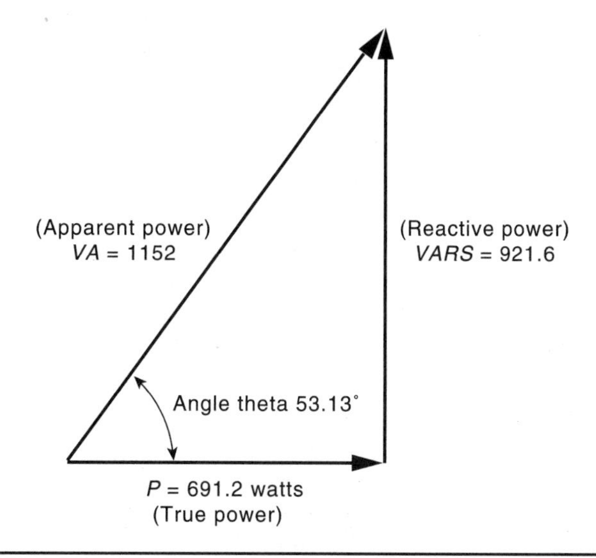

Figure 16–9 Angle theta is the relationship of true power to apparent power.

Note that in a series circuit, the power factor cannot be computed using current because current is the same in all parts of the circuit.

Power factor can become very important in an industrial application. Most power companies charge a substantial surcharge when the power factor drops below a certain percent. The reason for this is that electric power is sold on the basis of true power or watts consumed. The power company, however, must supply the apparent power. Assume that an industrial plant has a power factor of 60 percent and is consuming 5 megawatts of power per hour. At a power factor of 60 percent, the power company must actually supply 8.33 megavolt amps (5 megawatts / 0.6 = 8.33 megavolt amps) per hour. If the power factor were to be corrected to 95 percent, the power company would have to supply only 5.26 megavolt amps per hour to furnish the same amount of power to the plant.

Angle Theta

The angular displacement by which the voltage and current are out of phase with each other is called angle theta. Since the power factor is the ratio of true power as compared to apparent power, the phase angle of voltage and current is formed between the resistive leg of the right triangle and the hypotenuse, as shown in Figure 16–9. The resistive leg of the triangle is adjacent to the angle and the total or apparent power leg is the hypotenuse. The trigonometric function that corresponds to the adjacent side and the hypotenuse is the cosine. Angle theta will be the cosine of watts divided by volt amps. Watts divided by volt amps is also the power factor. Therefore, the cosine of angle theta is the power factor. You can get the value of the cosine from a trigonometric table or a calculator with trigonometric functions.

$$\cos \angle\varnothing = \text{PF}$$

$$\cos \angle\varnothing = 0.6$$

$$\angle\varnothing = 53.13°$$

The vectors formed using the parallelogram method of vector addition can also be used to find angle theta, as shown in Figure 16–10. Notice that the total quantity, volt amps, and the resistive quantity, watts, are again used to determine angle theta.

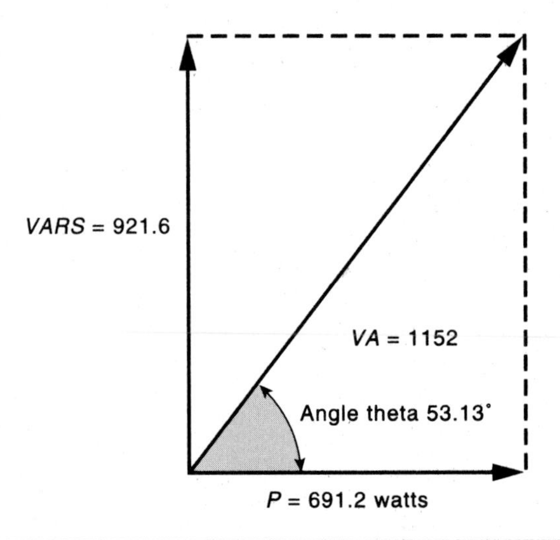

Figure 16–10 Angle theta can be found using vectors provided by the parallelogram method.

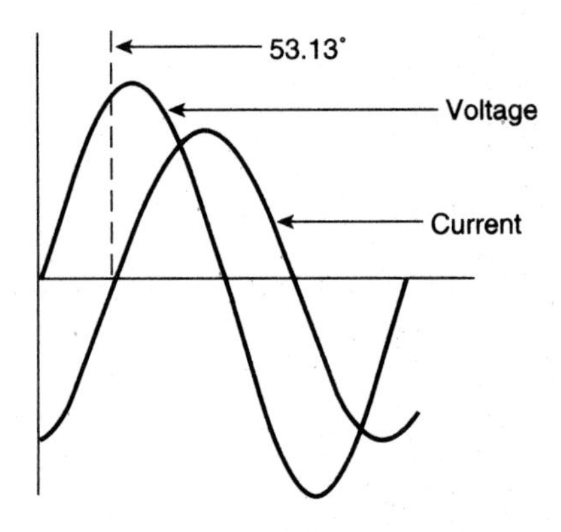

Figure 16–11 Current and voltage are 53.13 degrees out of phase with each other.

Since this circuit contains both resistance and inductance, the current is lagging the voltage by 53.13 degrees, shown in Figure 16–11. Angle theta can also be determined by using any of the other trigonometric functions.

$$\sin \angle\varnothing = \frac{VARs}{VA}$$

$$\tan \angle\varnothing = \frac{VARs}{P}$$

Example Circuit #1

In this example, two resistors and two inductors are connected in series, Figure 16–12. The circuit is connected to a 130-volt, 60-Hz line. The first resistor has a power dissipation of 56 watts, and the second resistor has a power dissipation of 44 watts. One inductor has a reactive power of 152 VARs and the second a reactive power of 88 VARs.

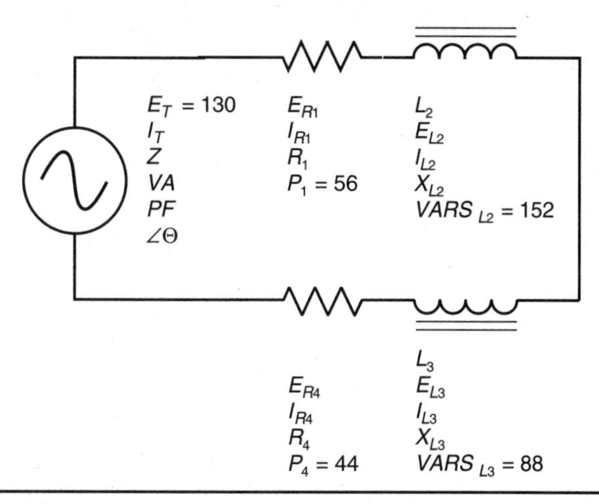

Figure 16–12 RL series circuit containing two resistors and two inductors.

The first step in finding the unknown values in this circuit is to find the total amount of true power and the total amount of reactive power. The total amount of true power can be computed by adding their values together.

$$P_T = P_1 + P_4$$

$$P_T = 56 + 44$$

$$P_T = 100 \text{ watts}$$

The total reactive power in the circuit can be found in the same manner.

$$VARs_T = VARs_{L2} + VARs_{L3}$$

$$VARs_T = 152 + 88$$

$$VARs_T = 240$$

Apparent Power

Now that the total amount of true power and the total amount of reactive power is known, the apparent power can be computed using vector addition.

$$VA = \sqrt{P^2 + VARs_L^{\ 2}}$$

$$VA = \sqrt{100^2 + 240^2}$$

$$VA = 260$$

The parallelogram method of vector addition is shown in Figure 16–13 for this calculation.

Total Circuit Current (I_T)

Now that the apparent power is known, the total circuit current can be found using the applied voltage and Ohm's law.

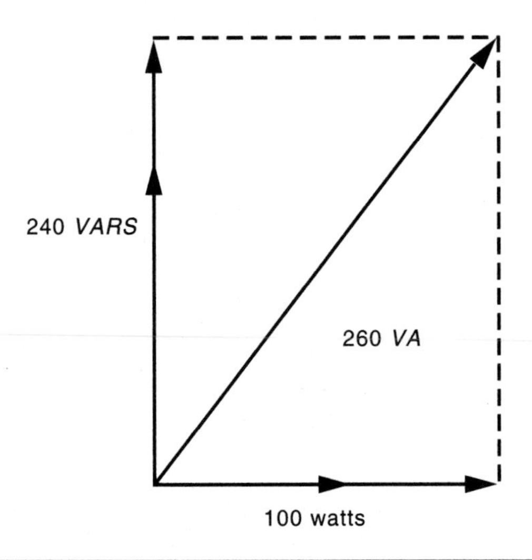

Figure 16–13 Power vector for circuit #2.

$$I_T = \frac{VA}{E_T}$$

$$I_T = \frac{260}{130}$$

$$I_T = 2 \text{ amps}$$

Since this is a series circuit, the current must be the same at all points in the circuit.

Now that the total circuit current has been found, other values can be computed using Ohm's law.

Impedance

$$Z = \frac{E_T}{I_T}$$

$$Z = \frac{130}{2}$$

$$Z = 65\,\Omega$$

Power Factor

$$PF = \frac{P}{VA}$$

$$PF = \frac{100}{260}$$

$$PF = 0.3846 \times 100, \text{ or } 38.46\%$$

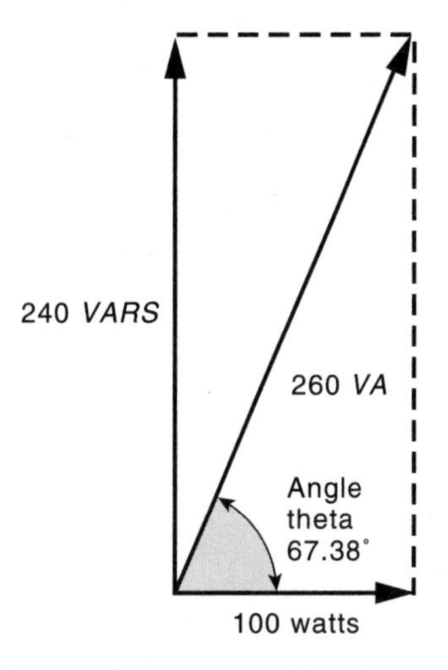

Figure 16–14 Angle theta for circuit #2.

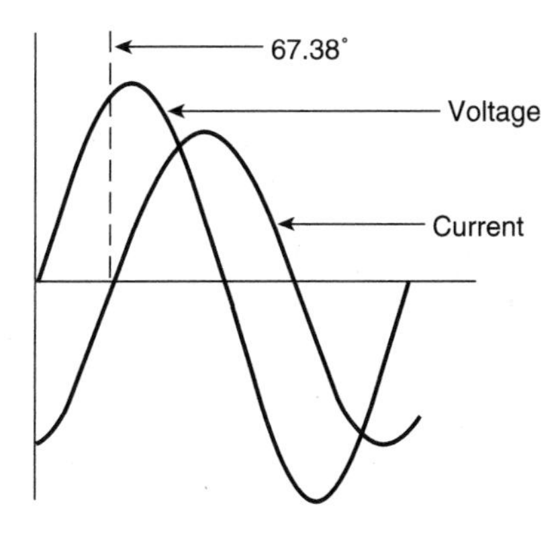

Figure 16–15 Voltage and current are 67.38 degrees out of phase with each other.

Angle Theta

Angle theta is the cosine of the power factor. Figure 16–14 is a vector diagram showing this relationship.

$$\cos \angle\varnothing = PF$$

$$\cos\angle\varnothing = 0.3846$$

$$\angle\varnothing = 67.38°$$

The relationship of voltage and current for this circuit is shown in Figure 16–15.

Voltage Across Resistor #1 (E_{R1})

$$E_{R1} = \frac{P_1}{I_{R1}}$$

$$E_{R1} = \frac{56}{2}$$

$$E_{R1} = 28 \text{ volts}$$

Resistance of Resistor #1 (R_1)

$$R_1 = \frac{E_{R1}}{I_{R1}}$$

$$R_1 = \frac{28}{2}$$

$$R_1 = 14\,\Omega$$

Voltage Across Inductor #2 (E_{L2})

$$E_{L2} = \frac{VARs_{L2}}{I_{L2}}$$

$$E_{L2} = \frac{152}{2}$$

$$E_{L2} = 76 \text{ volts}$$

Inductive Reactance of Inductor #2 (X_{L2})

$$X_{L2} = \frac{E_{L2}}{I_{L2}}$$

$$X_{L2} = \frac{76}{2}$$

$$X_{L2} = 38\,\Omega$$

Inductance of Inductor #2 (L_2)

$$L_2 = \frac{X_{L2}}{2\pi F}$$

$$L_2 = \frac{38}{377}$$

$$L_2 = 0.101 \text{ henry}$$

Voltage Across Inductor #3 (E_{L3})

$$E_{L3} = \frac{VARs_{L3}}{I_{L3}}$$

$$E_{L3} = \frac{88}{2}$$

$$E_{L3} = 44 \text{ volts}$$

Inductive Reactance of Inductor #3 (X_{L3})

$$X_{L3} = \frac{E_{L3}}{I_{L3}}$$

$$X_{L3} = \frac{44}{2}$$

$$X_{L3} = 22\ \Omega$$

Inductance of Inductor #3 (L_3)

$$L_3 = \frac{X_{L3}}{2\pi F}$$

$$L_3 = \frac{22}{377}$$

$$L_3 = 0.058 \text{ henry}$$

Voltage Across Resistor #4 (E_{R4})

$$E_{R4} = \frac{P_4}{I_{R4}}$$

$$E_{R4} = \frac{44}{2}$$

$$E_{R4} = 22 \text{ volts}$$

Resistance of Resistor #4 (R_4)

$$R_4 = \frac{E_{R4}}{I_{R4}}$$

$$R_4 = \frac{22}{2}$$

$$R_4 = 11\ \Omega$$

Figure 16–16 presents formulas for finding values in alternating current circuits containing resistance and inductance connected in series.

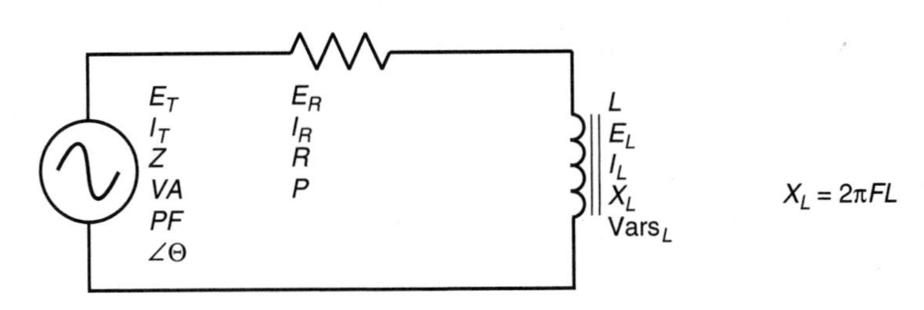

$$E_T = \sqrt{E_R^2 + E_L^2}$$
$$E_T = I_T \times Z$$
$$E_T = \frac{VA}{I_T}$$
$$E_T = \frac{E_R}{PF}$$

$$Z = \sqrt{R^2 + X_L^2}$$
$$Z = \frac{E_T}{I_T}$$
$$Z = \frac{VA}{I_T^2}$$
$$Z = \frac{R}{PF}$$

$$VA = E_T \times I_T$$
$$VA = I_T^2 \times Z$$
$$VA = \frac{E_T^2}{Z}$$
$$VA = \sqrt{P^2 + VARs_L^2}$$
$$Z = \frac{E_T^2}{VA}$$

$$L = \frac{X_L}{2 \times \pi \times F}$$
$$I_T = I_R = I_L$$
$$I_T = \frac{E_T}{Z}$$
$$I_T = \frac{VA}{E_T}$$
$$VA = \frac{P}{PF}$$

$$PF = \frac{R}{Z}$$
$$PF = \frac{P}{VA}$$
$$PF = \frac{E_R}{E_T}$$
$$PF = \cos\angle\theta$$

$$P = E_R \times I_R$$
$$P = \sqrt{VA^2 - VARs_L^2}$$
$$P = \frac{E_R^2}{R}$$
$$P = I_R^2 \times R$$
$$P = VA \times PF$$

$$E_R = I_R \times R$$
$$E_R = \sqrt{P \times R}$$
$$E_R = \frac{P}{I_R}$$
$$E_R = \sqrt{E_T^2 - E_L^2}$$
$$E_R = E_T \times PF$$

$$I_R = I_T = I_L$$
$$I_R = \frac{E_R}{R}$$
$$I_R = \frac{P}{E_R}$$
$$I_R = \sqrt{\frac{P}{R}}$$

$$R = \sqrt{Z^2 - X_L^2}$$
$$R = \frac{E_R}{I_R}$$
$$R = \frac{E_R^2}{P}$$
$$R = \frac{P}{I_R^2}$$
$$R = Z \times PF$$

$$E_L = I_L \times X_L$$
$$E_L = \sqrt{E_T^2 - E_R^2}$$
$$E_L = \sqrt{VARs_L \times X_L}$$
$$E_L = \frac{VARs_L}{I_L}$$

$$I_L = I_R = I_T$$
$$I_L = \frac{E_L}{X_L}$$
$$I_L = \frac{VARs_L}{E_L}$$
$$I_L = \sqrt{\frac{VARs_L}{X_L}}$$

$$X_L = \sqrt{Z^2 - R^2}$$
$$X_L = \frac{E_L}{I_L}$$
$$X_L = \frac{E_L^2}{VARs_L}$$
$$X_L = \frac{VARs_L}{I_L^2}$$

$$VARs_L = \sqrt{VA^2 - P^2} \qquad VARs_L = E_L \times I_L \qquad VARs_L = \frac{E_L^2}{X_L} \qquad VARs_L = I_L^2 \times X_L$$

Figure 16–16 Formulas for resistive–inductive series circuits.

CHAPTER 17

Resistive–Inductive Parallel Circuits

A circuit containing a resistor and inductor connected in parallel is shown in Figure 17–1. Since the voltage applied to any branch in parallel must be the same, the voltage applied to the resistor and inductor must be in phase and have the same value. The current flow through the inductor will be 90 degrees out of phase with the voltage and the current flow through the resistor will be in phase with the voltage, Figure 17–2. This produces a phase angle difference of 90 degrees between the current flow through a pure inductive load and a pure resistive load.

The amount of phase angle shift between the total circuit current and voltage is determined by the ratio of the amount of resistance as compared to the amount of inductance. The circuit power factor is still determined by the ratio of apparent power and true power.

Example Circuit #1

In the example circuit shown in Figure 17–3, a resistance of 15 Ω is connected in parallel with an inductive reactance of 20 Ω. The circuit is connected to a voltage of 240 volts AC and a frequency of 60 Hz. In this example problem, the following circuit values will be computed:

I_R = current flow through the resistor
P = watts (true power)
I_L = current flow through the inductor
$VARs$ = (reactive power)
L = inductance
I_T = total circuit current
Z = total impedance of the circuit
VA = volt-amperes (apparent power)
PF = power factor
$\angle\varnothing$ = angle theta

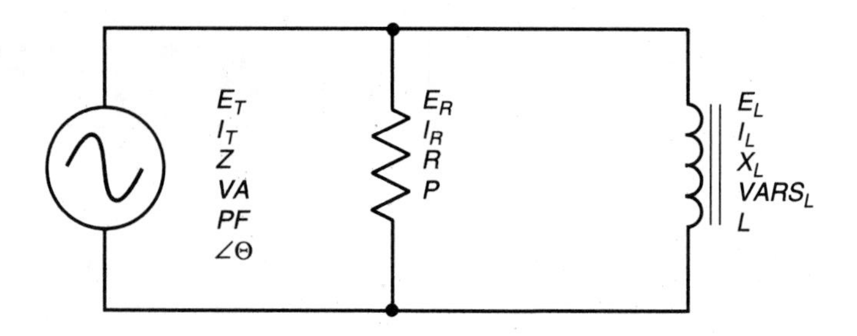

Figure 17–1 Resistive–inductive parallel circuit.

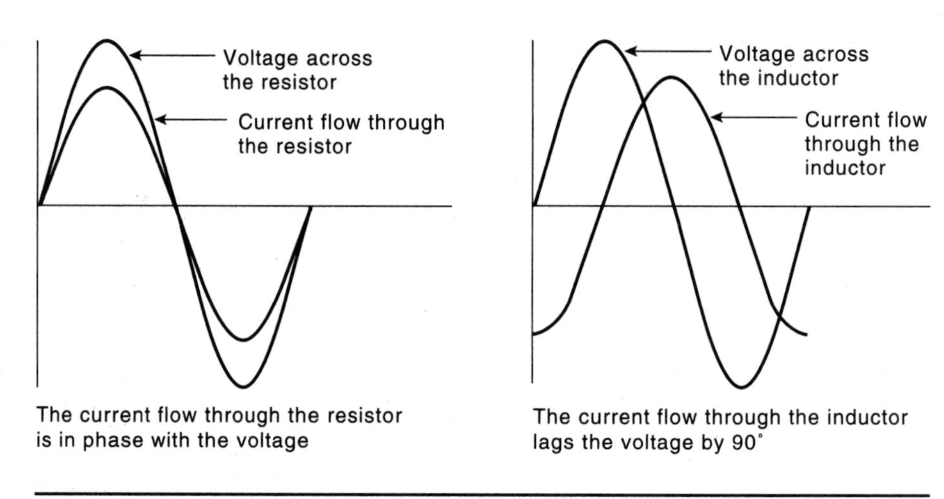

Figure 17–2 Relationship of voltage and current in an RL parallel circuit.

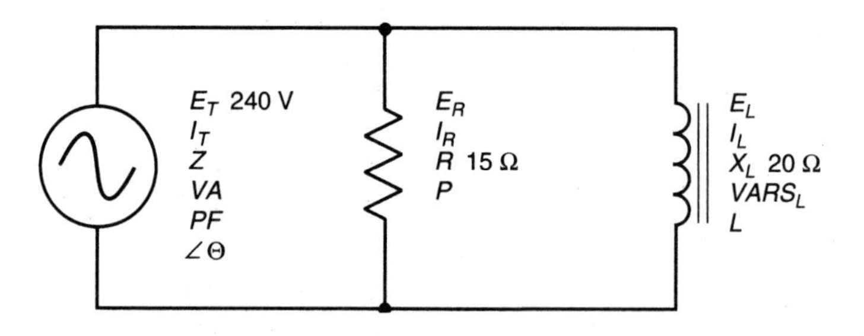

Figure 17–3 Typical RL parallel circuit.

Resistive Current (I_R)

In any parallel circuit, the voltage is the same across each branch in the circuit. Therefore, 240 volts is applied across both the resistor and the inductor. Since the amount of voltage applied to the resistor is known, the amount of current flow through the resistor can be computed by using the formula:

$$I_R = \frac{E}{R}$$

$$I_R = \frac{240}{15}$$

$$I_R = 16 \text{ amperes}$$

Watts (P)

True power or watts can be computed using any of the watts formulas and pure resistive values. The amount of true power in this circuit will be computed using the formula:

$$P = E_R \times I_R$$

$$P = 240 \times 16$$

$$P = 3840 \text{ watts}$$

Inductive Current (I_L)

Since the voltage applied to the inductor is known, we can find the current flow by dividing the voltage by the inductive reactance. The amount of current flow through the inductor will be computed using the formula:

$$I_L = \frac{E}{X_L}$$

$$I_L = \frac{240}{20}$$

$$I_L = 12 \text{ amps}$$

Reactive Power (VARs)

The amount of reactive power, VARs, will be computed using the formula:

$$VARs = E_L \times I_L$$

$$VARs = 240 \times 12$$

$$VARs = 2880$$

Inductance (L)

Since the frequency and the inductive reactance are known, the inductance of the coil can be found using the formula:

$$L = \frac{X_L}{2\pi F}$$

$$L = \frac{20}{2 \times 3.1416 \times 60}$$

$$L = 0.053 \text{ henry}$$

Total Current (I_T)

The total current flow through the circuit can be computed by adding the current flow through the resistor and the inductor. Since these two currents are 90 degrees out of phase with each other, vector addition will be used. Notice that the resistive and inductive currents form the legs of a right triangle and the total current forms the hypotenuse; therefore, the Pythagorean theorem can be used to add these currents together.

$$I_T = \sqrt{I_R^2 + I_L^2}$$

$$I_T = \sqrt{16^2 + 12^2}$$

$$I_T = \sqrt{256 + 144}$$

$$I_T = \sqrt{400}$$

$$I_T = 20 \text{ amps}$$

The parallelogram method for plotting the total current is shown in Figure 17–4.

Impedance (Z)

Now that the total current and total voltage are known, the impedance can be computed by substituting Z for R in an Ohm's law formula. The total impedance of the circuit can be computed using the formula:

$$Z = \frac{E_T}{I_T}$$

$$Z = \frac{240}{20}$$

$$Z = 12\ \Omega$$

The value of impedance can also be found if total current and voltage are not known. In a parallel circuit, the reciprocal of the total resistance is equal to the sum of the recipro-

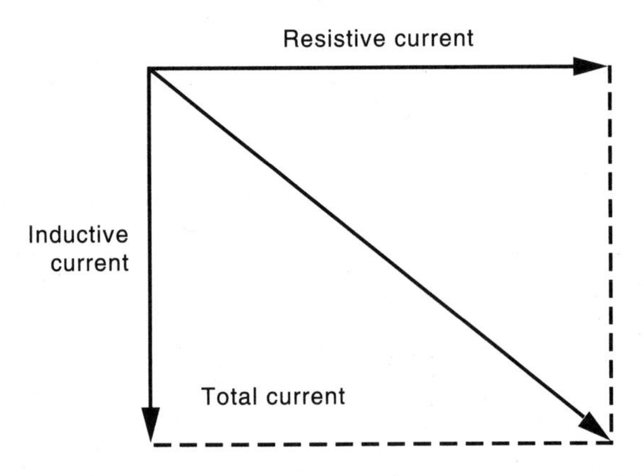

Figure 17–4 Plotting total current using the parallelogram method.

cals of each resistor. This same rule can be amended to permit a similar formula to be used in an RL parallel circuit. Since resistance and inductive reactance are 90 degrees out of phase with each other, vector addition must be used when the reciprocals are added. The initial formula is shown:

$$\left(\frac{1}{Z}\right)^2 = \left(\frac{1}{R}\right)^2 + \left(\frac{1}{X_L}\right)^2$$

This formula states that the square of the reciprocal of the impedance is equal to the sum of the squares of the reciprocals of resistance and inductive reactance. To remove the square from the reciprocal of the impedance, take the square root of both sides of the equation.

$$\frac{1}{Z} = \sqrt{\left(\frac{1}{R}\right)^2 + \left(\frac{1}{X_L}\right)^2}$$

The formula can now be used to find the reciprocal of the impedance, not the impedance. To change the formula so that it is equal to the impedance, take the reciprocal of both sides of the equation.

$$Z = \frac{1}{\sqrt{\left(\frac{1}{R}\right)^2 + \left(\frac{1}{X_L}\right)^2}}$$

Numeric values can now be substituted in the formula to find the impedance of the circuit.

$$Z = \frac{1}{\sqrt{\left(\frac{1}{15}\right)^2 + \left(\frac{1}{20}\right)^2}}$$

$$Z = \frac{1}{\sqrt{0.004444 + 0.0025}}$$

$$Z = \frac{1}{0.08333}$$

$$Z = 12\,\Omega$$

Apparent Power (*VA*)

The apparent power can be computed by multiplying the circuit voltage by the total current flow. The relationship of volt amps, watts, and VARs is the same for an RL parallel circuit as it is for an RL series circuit. The reason for this is that power adds in any type of circuit. Since the true power and reactive power are 90 degrees out of phase with each other, they form a right triangle with apparent power as the hypotenuse.

$$VA = E_T \times I_T$$

$$VA = 240 \times 20$$

$$VA = 4800$$

Power Factor (*PF*)

Power factor in an RL parallel circuit is the relationship of apparent power as compared to the true power, just as it was in the RL series circuit. There are some differences in the formulas used to compute power factor in a parallel circuit, however. In an RL series circuit, power factor could be computed by dividing the voltage dropped across the resistor by the total or applied voltage. In a parallel circuit, the voltage is the same, but the currents are different. Therefore, power factor can be computed by dividing the current flow through the resistive parts of the circuit by the total circuit current.

$$PF = \frac{I_R}{I_T}$$

$$PF = \frac{16\text{A}}{20\text{A}}$$

$$PF = 0.80, \text{ or } 80\%$$

Another formula that changes involves resistance and impedance. In a parallel circuit, the total circuit impedance will be less than the resistance. Therefore, if power factor is to be computed using impedance and resistance, the impedance must be divided by the resistance.

$$PF = \frac{Z}{R}$$

$$PF = \frac{12}{15}$$

$$PF = 0.80, \text{ or } 80\%$$

We can compute the circuit power factor in this example using the formula:

$$PF = \frac{P}{VA}$$

$$PF = \frac{3840}{4800} \times 100$$

$$PF = 0.80, \text{ or } 80\%$$

Angle Theta

Angle theta is the cosine of the power factor.

$$\cos \angle\varnothing = 0.80$$

$$\angle\varnothing = 36.87°$$

A vector diagram using apparent power, true power, and reactive power is shown in Figure 17–5. Notice that angle theta is the angle produced by the apparent power and the true power. The relationship of current and voltage for this circuit is shown in Figure 17–6.

Example Circuit #2

In this circuit, one resistor is connected in parallel with two inductors, Figure 17–7. The frequency is 60 hertz. The circuit has an apparent power of 6120 volt amps, the resistor has a resistance of 45 Ω, the first inductor has an inductive reactance of 40 Ω, and the second

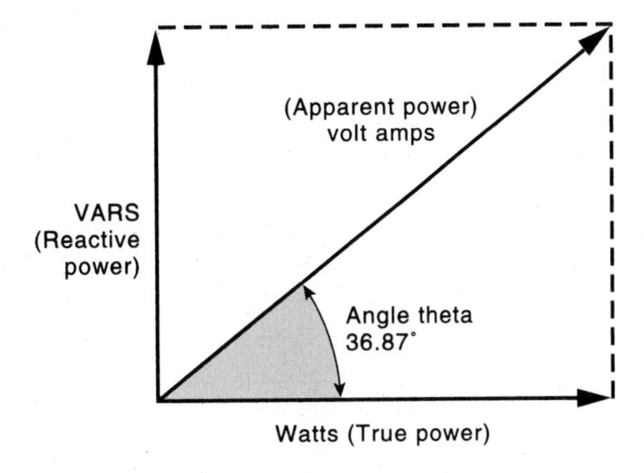

Figure 17–5 Angle theta.

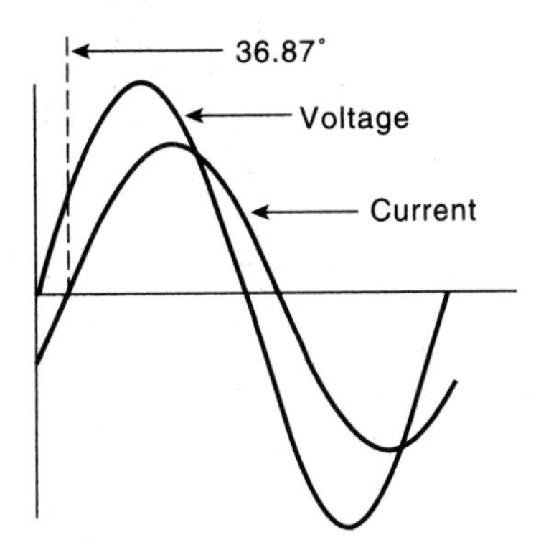

Figure 17–6 The current is 36.87 degrees out of phase with the voltage.

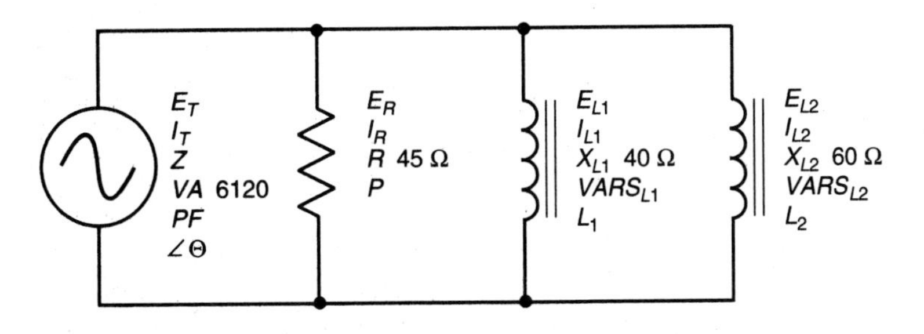

Figure 17–7 Example circuit #2.

inductor has an inductive reactance of 60 Ω. It is assumed that both inductors have a Q greater than 10 and their resistance is negligible. The following missing values will be found:

Z = circuit impedance
I_T = total current in the circuit
E_T = applied voltage
E_R = voltage drop across the resistor
I_R = current flow through the resistor
P = true power
E_{L1} = voltage drop across the first inductor
I_{L1} = current flow through the first inductor
$VARs_{L1}$ = reactive power of the first inductor
L_1 = inductance of the first inductor
E_{L2} = voltage drop across the second inductor
I_{L2} = current flow through the second inductor
$VARs_{L2}$ = reactive power of the second inductor
L_2 = inductance of the second inductor

Impedance (Z)

Before it is possible to compute the impedance of the circuit, the total amount of inductive reactance (X_{LT})for the circuit must be found. Since these two inductors are connected in parallel, the reciprocal of their inductive reactances must be added. This will give the reciprocal of the total inductive reactance:

$$\frac{1}{X_{LT}} = \frac{1}{X_{L1}} + \frac{1}{X_{L2}}$$

To find the total inductive reactance, take the reciprocal of both sides of the equation:

$$X_{LT} = \frac{1}{\frac{1}{X_{L1}} + \frac{1}{X_{L2}}}$$

Refer to the formulas for pure inductive circuits. Numeric values can now be substituted in the formula to find the total inductive reactance.

$$X_{LT} = \frac{1}{\frac{1}{40} + \frac{1}{60}}$$

$$X_{LT} = \frac{1}{\frac{1}{0.025} + \frac{1}{0.01667}}$$

$$X_{LT} = 24\ \Omega$$

Now that the total amount of inductive reactance for the circuit is known, we can compute the impedance using the formula:

$$Z = \frac{1}{\sqrt{\left(\frac{1}{R}\right)^2 + \left(\frac{1}{X_{LT}}\right)^2}}$$

$$Z = \frac{1}{\sqrt{\left(\frac{1}{45}\right)^2 + \left(\frac{1}{24}\right)^2}}$$

$$Z = 21.176\ \Omega$$

Applied Voltage (E_T)

Now that the circuit impedance and the apparent power are known, the applied voltage can be computed using the formula:

$$E_T = \sqrt{VA \times Z}$$

$$E_T = \sqrt{6220 \times 21.176}$$

$$E_T = 360 \text{ volts}$$

Voltage Drop (E_R, E_{L1}, and E_{L2})

In a parallel circuit, the voltage must be the same across any leg or branch. Therefore, 360 volts is dropped across the resistor, the first inductor, and the second inductor.

$$E_R = 360 \text{ volts} \qquad E_{L1} = 360 \text{ volts} \qquad E_{L2} = 360 \text{ volts}$$

Total Circuit Current (I_T)

The total current of the circuit can be computed using the formula:

$$I_T = \frac{E_T}{Z}$$

$$I_T = \frac{360}{21.176}$$

$$I_T = 17 \text{ amps}$$

The remaining values of the circuit can be found using Ohm's law. Refer to the formulas for circuits contains resistance and inductance connected in parallel as shown in Figure 17–8.

Resistant Current (I_R)

$$I_R = \frac{E_R}{R}$$

$$I_R = \frac{360}{45}$$

$$I_R = 8 \text{ amps}$$

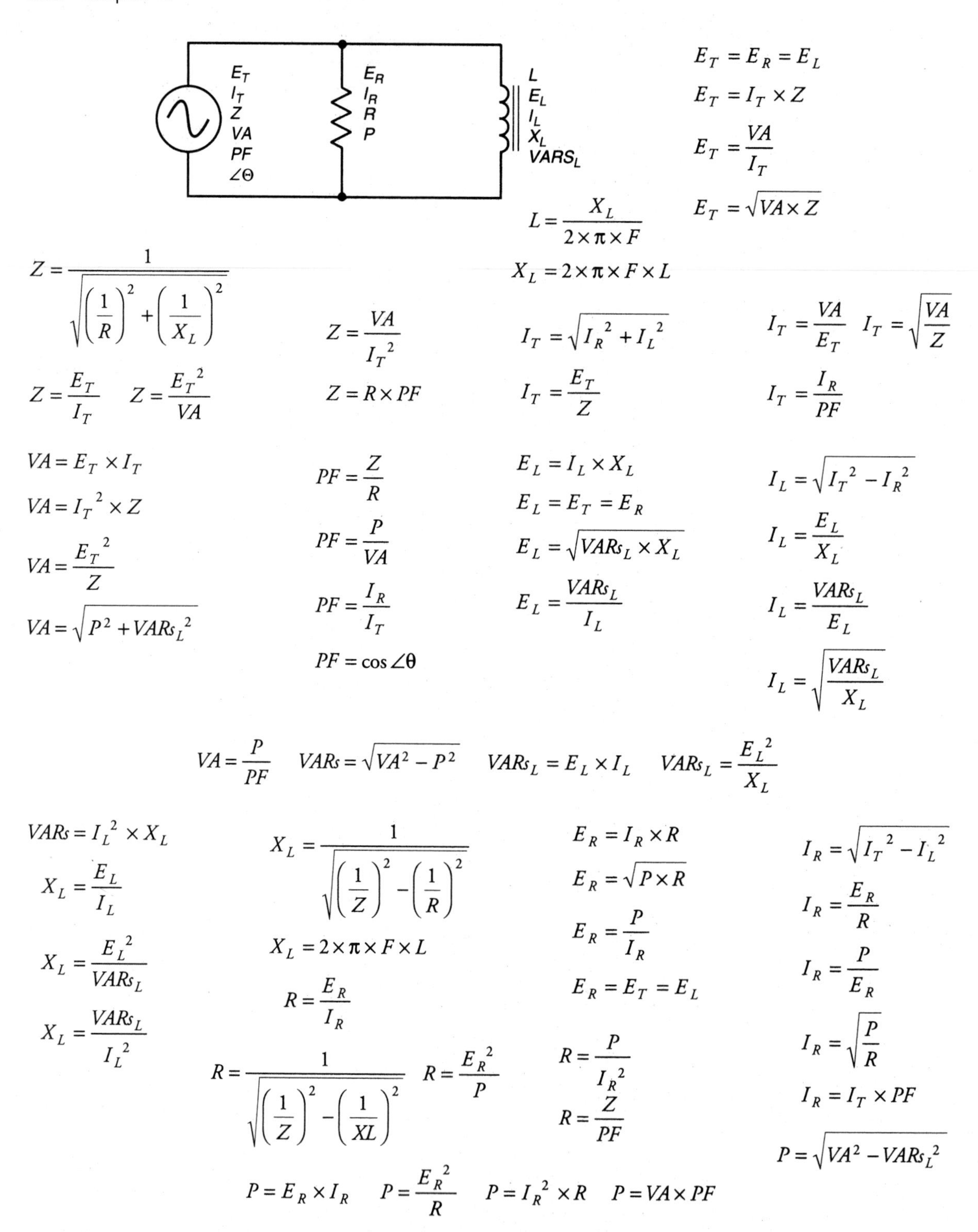

Figure 17–8 Formulas for resistive–inductive parallel circuits.

True Power (P)

$$P = E_R \times I_R$$

$$P = 360 \times 8$$

$$P = 2880 \text{ watts}$$

Current Flow through First Inductor (I_{L1})

$$I_{L1} = \frac{E_{L1}}{X_{L1}}$$

$$I_{L1} = \frac{360}{40}$$

$$I_{L1} = 9 \text{ amps}$$

Reactive Power of First Inductor ($VARs_{L1}$)

$$VARs_{L1} = E_{L1} \times I_{L1}$$

$$VARs_{L1} = 360 \times 9$$

$$VARs_{L1} = 3240$$

Inductance of First Inductor (L_1)

$$L_1 = \frac{X_{L1}}{2\pi F}$$

$$L_1 = \frac{40}{377}$$

$$L_1 = 0.106 \text{ henry}$$

Current Flow through Second Inductor (I_{L2})

$$I_{L2} = \frac{E_{L2}}{X_{L2}}$$

$$I_{L2} = \frac{360}{60}$$

$$I_{L2} = 6 \text{ amps}$$

Reactive Power of Second Inductor ($VARs_{L2}$)

$$VARs_{L2} = E_{L2} \times I_{L2}$$

$$VARs_{L2} = 360 \times 6$$

$$VARs_{L2} = 2160$$

Inductance of Second Inductor (L_2)

$$L_2 = \frac{X_{L2}}{2\pi F}$$

$$L_2 = \frac{60}{377}$$

$$L_2 = 0.159 \text{ henry}$$

Power Factor (PF)

$$PF = \frac{P}{VA}$$

$$PF = \frac{2880}{6120} \times 100$$

$$PF = 47.06\%$$

Angle Theta ($\angle\varnothing$)

$$\cos \angle\varnothing = PF$$

$$\cos \angle\varnothing = 0.4706$$

$$\angle\varnothing = 61.93°$$

Figure 17–9 is a vector diagram showing angle theta. The vectors used are those for apparent power, true power, and reactive power. The phase relationship of voltage and current for this circuit are shown in Figure 17–10. Formulas for circuits containing resistance and inductance connected in parallel are shown in Figure 17–8.

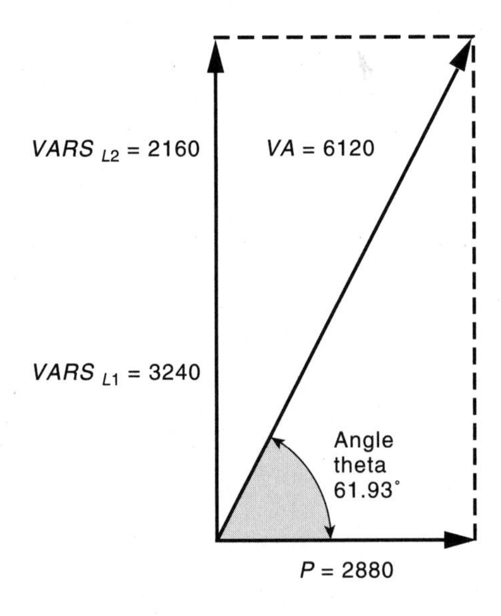

Figure 17–9 Angle theta determined by power vectors.

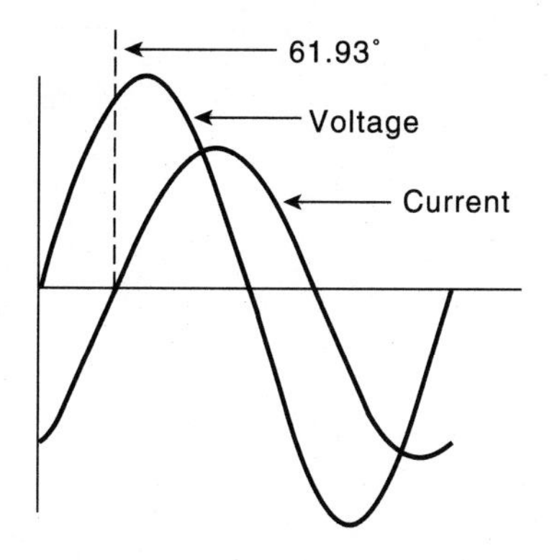

Figure 17–10 Voltage and current are 61.93 degrees out of phase with each other.

CHAPTER 18

Capacitors

Capacitors perform a variety of jobs. For example, they correct the power factor where large inductive loads are present and store an electrical charge to produce a large current pulse. They are used in timing circuits and in the start windings of some AC motors, and they serve as electronic filters. Capacitors can be nonpolarized or polarized, depending on the application. Nonpolarized capacitors can be used in both AC and DC circuits, while polarized capacitors can be used in DC circuits only. Both types will be discussed.

Capacitors are devices that oppose a change of voltage. The simplest type of capacitor is constructed by separating two metal plates by some type of insulating material called the *dielectric,* shown in Figure 18–1. Three factors determine the capacitance of a capacitor:

1. The area of the plates
2. The distance between the plates
3. The type of dielectric used

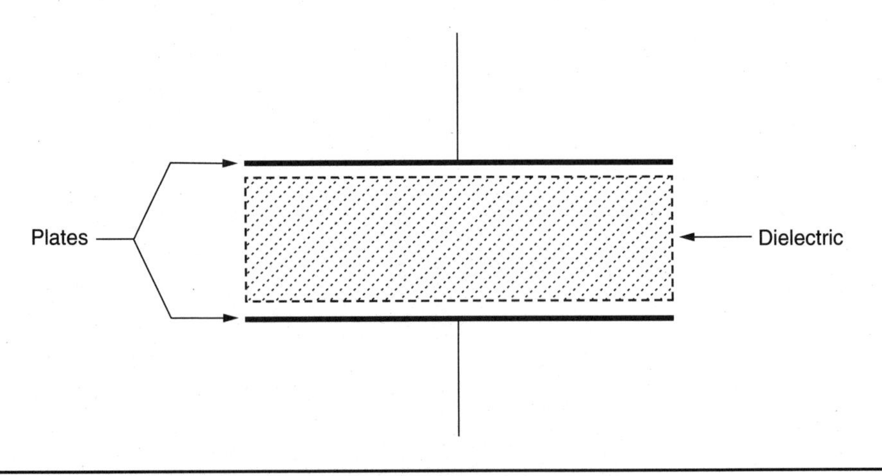

Figure 18–1 A capacitor is made by separating two metal plates by a dielectric.

The greater the surface area of the plates, the more capacitance a capacitor will have. If a capacitor is charged by connecting it to a source of direct current, as in Figure 18–2, electrons are removed from the plate connected to the positive battery terminal and deposited on the plate connected to the negative terminal. This flow of current will continue until a voltage is established across the plates of the capacitor that is equal to the battery voltage. When these two voltages become equal, the flow of electrons will stop. The capacitor is now charged. If the battery is disconnected from the capacitor, the capacitor will remain charged as long as there is no path by which the electrons can move from one plate to the other. A good rule to remember concerning a capacitor and current flow is that current can flow only during the period of time that a capacitor is either charging or discharging.

In theory, it should be possible for a capacitor to remain in a charged condition forever. In actual practice, however, this is not the case. No dielectric is a perfect insulator, and electrons eventually move through the dielectric from the negative plate to the positive, causing the capacitor to discharge, as shown in Figure 18–3. This current flow through the dielectric is called leakage current, and is proportional to the resistance of the dielectric and the charge across the plates. If the dielectric of a capacitor should become weak, it will permit an excessive amount of leakage current to flow. A capacitor in this condition is often called a *leaky capacitor.*

Electrostatic Charge

Two other factors that determine capacitance is the type of dielectric used and the distance between the plates. To understand these concepts, it is necessary to understand how a capacitor stores energy. An inductor stores energy in the form of an electromagnetic field. A capacitor stores energy in an electrostatic field.

When a capacitor is not charged, the atoms of the dielectric are uniform, as shown in Figure 18–4. The valence electrons orbit the nucleus in a circular pattern. When the capacitor becomes charged, however, a potential exists between the plates of the capacitor. The plate with the lack of electrons has a positive charge and the plate with the excess of electrons has a negative charge. Since electrons are negative particles, they are repelled from the negative plate and attracted to the positive plate. This causes the electron orbit to become

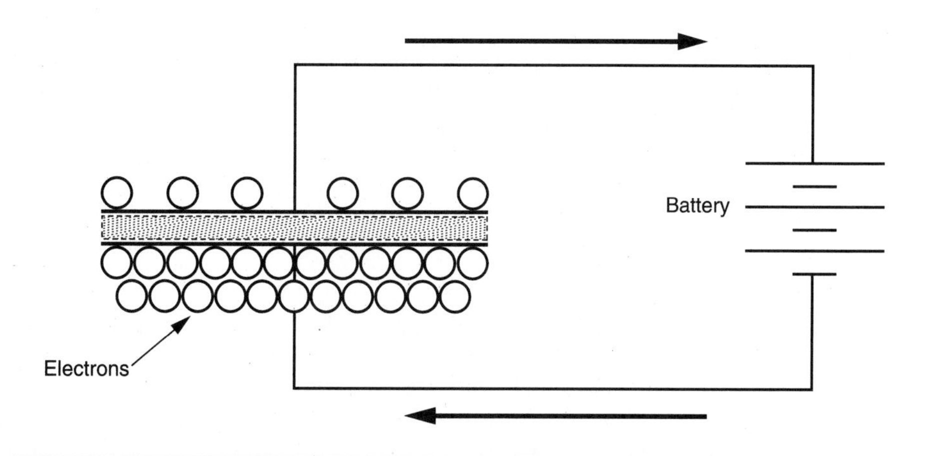

Figure 18–2 A capacitor can be charged by removing electrons from one plate and depositing electrons on the other plate.

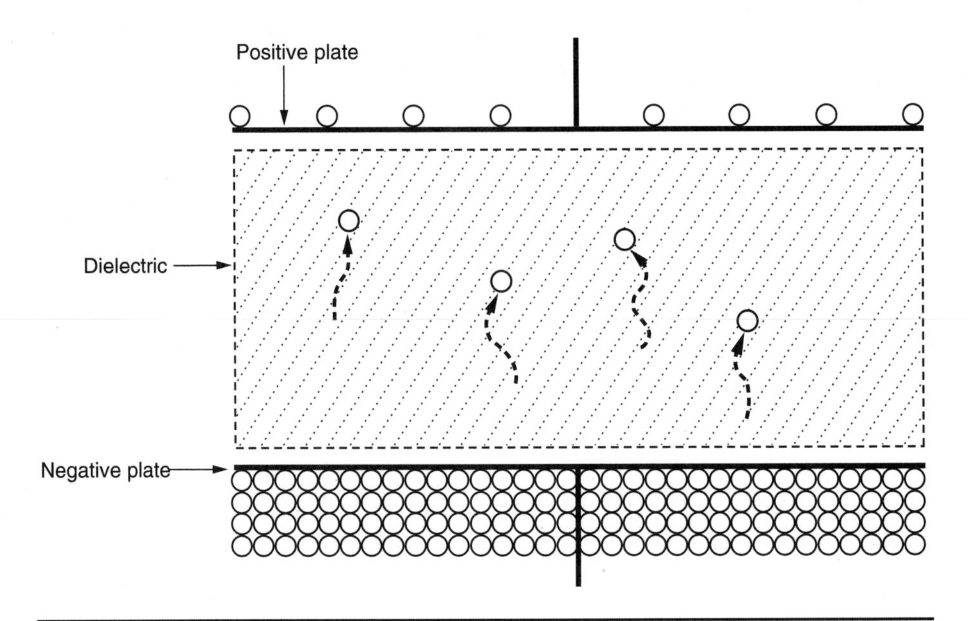

Figure 18–3 Electrons eventually leak through the dielectric. This flow of electrons is known as leakage current.

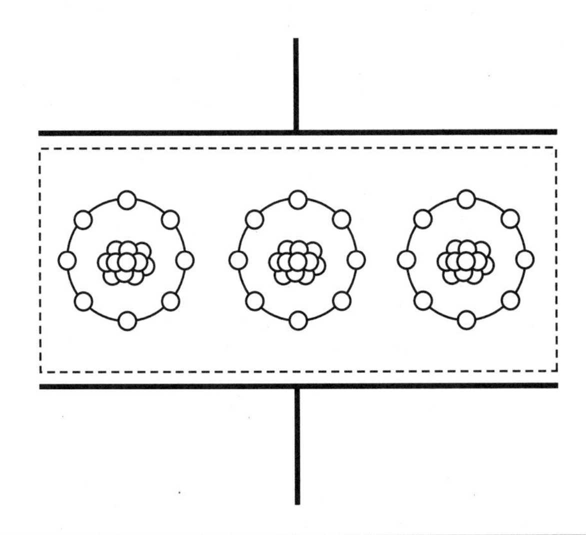

Figure 18–4 Atoms of the dielectric in an uncharged capacitor are uniform.

stretched as shown in Figure 18–5. This stretching of the dielectric atoms is called dielectric stress. Placing the atoms of the dielectric under stress has the same effect as drawing back a bow and arrow, and holding it.

The amount of dielectric stress is proportional to the voltage difference between the plates. The greater the voltage, the greater the dielectric stress. If the voltage becomes too great, the dielectric will break down and permit current to flow between the plates. At this point the capacitor becomes shorted. Capacitors have voltage rating that should not be exceeded. This voltage rating may be expressed as peak or RMS, depending on the capacitor.

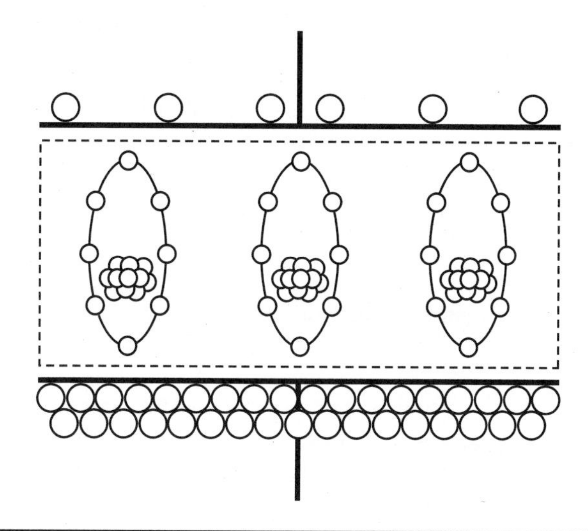

Figure 18–5 Atoms of the dielectric in a charged capacitor are stretched.

Capacitors that give the voltage rating in volts DC (VDC) or working volts DC (WVDC) indicate the peak value of voltage that can be applied. Capacitors that indicate the voltage value in volts AC indicate the RMS value. The voltage rating indicates the maximum amount of voltage the dielectric is intended to withstand without breaking down. The amount of voltage applied to a capacitor is critical to its life span. Capacitors operated above their voltage rating will fail relatively quickly. Many years ago, the United States military studied the voltage rating of a capacitor as compared to its life span. The results showed that a capacitor operated at one-half its rated voltage will have a life span approximately eight times longer than a capacitor operated at the rated voltage.

The energy of the capacitor is stored in the dielectric in the form of an electrostatic charge. It is this electrostatic charge that permits the capacitor to produce extremely high currents under certain conditions. If the leads of a capacitor are shorted together, it has the effect of releasing the drawn-back bow. When the bow string is released, the arrow will be propelled forward at a high rate of speed. The same is true for the capacitor. When the leads are shorted, the atoms of the dielectric snap back to their normal position. This causes the electrons on the negative plate to be literally blown off and attracted to the positive plate. Capacitors can produce currents of thousands of amperes for short periods of time.

This principle is used to operate the electronic flash of many cameras. Electronic flash attachments contain a small glass tube filled with a gas called Xenon. Xenon produces a very bright white light similar to sunlight when the gas is ionized. A large amount of current flow is required, however, to produce a bright flash. A battery capable of directly ionizing the Xenon would be very large, expensive, and have a potential of about 500 volts. The simple circuit shown in Figure 18–6 can be used to overcome the problem. In this circuit, two small 1.5-volt batteries are connected to an oscillator. The oscillator changes the direct current of the batteries into square wave alternating current. The alternating current is then connected to a transformer, and the voltage is increased to about 500 volts peak. A diode changes the AC voltage back into DC and charges a capacitor. The capacitor charges to the peak value of the voltage waveform. When the switch is closed, the capacitor suddenly discharges through the Xenon tube and supplies the power needed to ionize the gas. It may take several seconds to store enough energy in the capacitor to ionize the gas in the tube, but the capacitor can release the stored energy in a fraction of a second.

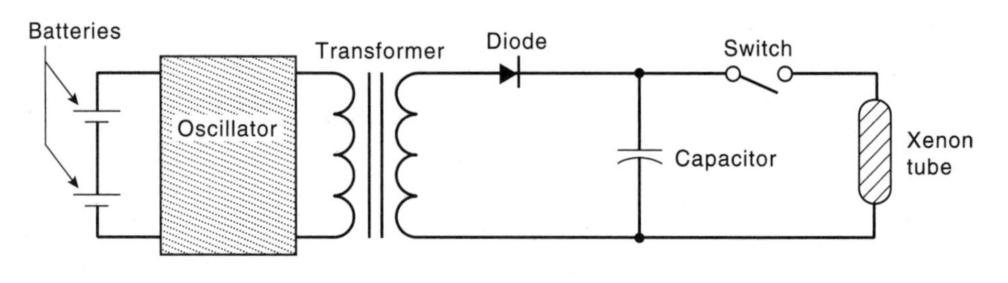

Figure 18–6 Energy is stored in a capacitor.

To understand how the capacitor can supply the energy needed, consider the amount of gun powder contained in a 0.357 cartridge. If the powder were to be removed from the cartridge and burned in the open air, the actual amount of energy contained in the powder would be very small. This amount of energy would not even be able to raise the temperature by a noticeable amount in a small enclosed room. If this same amount of energy is converted into heat in a fraction of a second, however, enough force is developed to propel a heavy projectile with great force. This same principle is at work when a capacitor is charged over some period of time and then discharged in a fraction of a second. Recall that 1 ampere is equal to 1 coulomb per second. If 1coulomb of electrons can be made to flow in 1/1000 of a second, there is a current flow of 1000 amperes.

Dielectric Constants

Since the energy of the capacitor is stored in the dielectric, the type of dielectric used is extremely important in determining the amount of capacitance a capacitor will have. Different materials are assigned a number called the dielectric constant. Air is assigned the number 1 and is used as a reference for comparison. For example, assume a capacitor uses air as the dielectric and its capacitance value is found to be 1 microfarad. Now assume that some dielectric material is placed between the plates without changing the spacing and the capacitance value becomes 5 microfarads. This material has a dielectric constant of 5. A chart showing the dielectric constant of different materials is shown in Figure 18–7.

Capacitor Ratings

The basic unit of capacitance is the farad and is symbolized by the letter F. It receives its name from a famous scientist named Michael Faraday.

A capacitor has a capacitance of one farad when a change of one volt across its plates results in a movement of one coulomb.

$$Q = C \times V$$

where Q = charge in coulombs
C = capacitance in farads
V = charging voltage

Although the farad is the basic unit of capacitance, it is seldom used because it is an extremely large amount of capacitance. The following formula can be used to determine the capacitance of a capacitor when the area of the plates, the dielectric constant, and the distance between the plates is known.

$$C = \frac{K \times A}{4.45\ D}$$

Material	Dielectric Constant
Air	1
Bakelite	4.0 to 10.0
Castor oil	4.3 to 4.7
Cellulose acetate	7.0
Ceramic	1200
Dry paper	3.5
Hard rubber	2.8
Insulating oils	2.2 to 4.6
Lucite	2.4 to 3.0
Mica	6.4 to 7.0
Mycalex	8.0
Paraffin	1.9 to 2.2
Porcelain	5.5
Pure water	81
Pyrex glass	4.1 to 4.9
Rubber compounds	3.0 to 7.0
Teflon	2
Titanium dioxide compounds	90 to 170

Figure 18–7 Dielectric constant of different materials.

where C = capacitance in pF (picofarads)
K = dielectric constant
A = area of one plate in square inches (both plates have equal area)
D = distance between the plates in inches

Example #1

What would be the plate area of a one-farad capacitor if air is used as the dielectric and the plates are separated by a distance of one inch?

Solution

The first step is to convert the above formula to solve for area.

$$A = \frac{C \times 4.45D}{K}$$

$$A = \frac{1{,}000{,}000{,}000{,}000 \times 4.45 \times 1}{1}$$

$$A = 4{,}450{,}000{,}000{,}000 \text{ square inches}$$

$$A = 1108.5 \text{ square miles}$$

Since the basic unit of capacitance is so large, other units such as the microfarad (μF), nanofarad (nF), and picofarad (pF) are generally used,

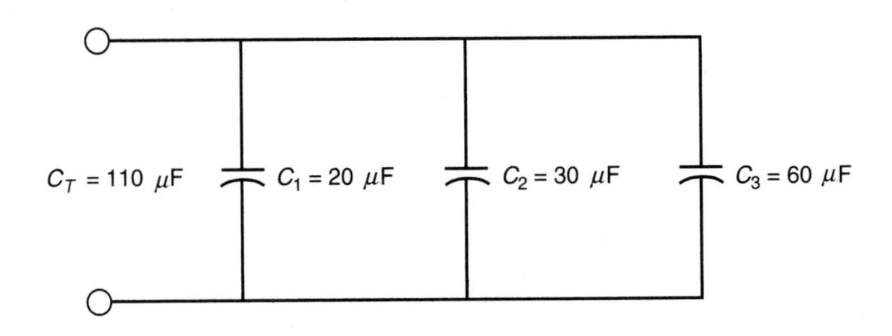

Figure 18–8 Capacitors connected in parallel.

$$\text{where } \mu = \frac{1}{1{,}000{,}000} \text{ of a farad } (1\times10^{-6})$$

$$\text{n} = \frac{1}{1{,}000{,}000} \text{ of a farad } (1\times10^{-9})$$

$$\text{p} = \frac{1}{1{,}000{,}000{,}000{,}000} \text{ of a farad } (1\times10^{-12})$$

The picofarad is sometimes referred to as a micro-microfarad and is symbolized by μμF.

Capacitors Connected in Parallel

When capacitors are connected in parallel, as in Figure 18–8, it has the same effect as increasing the plate area of one capacitor. In the example shown, three capacitors having a capacitance of 20 μF, 30 μF, and 60 μF are connected in parallel. The total capacitance of this connection is:

$$C_T = C_1 + C_2 + C_3$$

$$C_T = 20 + 30 + 60$$

$$C_T = 110\,\mu\text{F}$$

Capacitors Connected in Series

When capacitors are connected in series, as in Figure 18–9, it has the effect of increasing the distance between the plates. This reduces the total capacitance of the circuit. The total capacitance can be computed in a similar manner to computing parallel resistance. The following formulas can be used to find the total capacitance when capacitors are connected in series.

$$C_T = \frac{1}{\frac{1}{C_1} + \frac{1}{C_2} + \frac{1}{C_3} + \frac{1}{C_N}}$$

or

$$C_T = \frac{C_1 \times C_2}{C_1 + C_2}$$

or

$$C_T = \frac{C}{N}$$

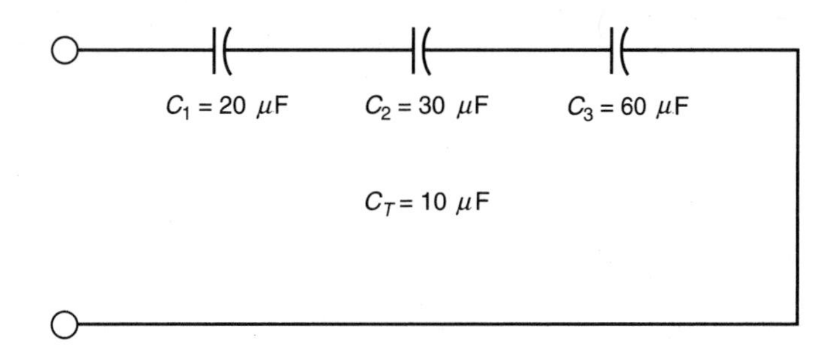

Figure 18–9 Capacitors connected in series.

Note The last formula can be used only when all the capacitors connected in series are the same value.

where C = capacitance of one capacitor
N = number of capacitors connected in series

Example #2

What is the total capacitance of three capacitors connected in series if C_1 has a capacitance of 20 μF, C_2 has a capacitance of 30 μF, and C_3 has a capacitance of 60 μF?

Solution

$$C_T = \frac{1}{\frac{1}{C_1} + \frac{1}{C_2} + \frac{1}{C_3} + \frac{1}{C_N}}$$

$$C_T = \frac{1}{\frac{1}{20} + \frac{1}{30} + \frac{1}{60}}$$

$$C_T = \frac{1}{0.05 + 0.0333 + 0.0166}$$

$$C_T = \frac{1}{0.1}$$

$$C_T = 10\ \mu\text{F}$$

Capacitive Charge and Discharge Rates

Capacitors charge and discharge at an exponential rate. A charge curve for a capacitor is shown in Figure 18–10. The curve is divided into five time constants, and each time constant is equal to 63.2 percent of the whole value. In Figure 18–10 it is assumed that a capacitor is to be charged to a total of 100 volts. At the end of the first time constant the voltage has reached 63.2 percent of 100, or 63.2 volts. At the end of the second time constant, the voltage reaches 63.2 percent of the remaining voltage, or 86.4 volts. This continues until the capacitor has been charged to 100 volts.

The capacitor discharges in the same manner (see Figure 18–11. At the end of the first time constant, the voltage will decrease to 63.2 percent of its charged value. In this example the voltage will decrease from 100 volts to 36.8 volts in the first time constant. At the

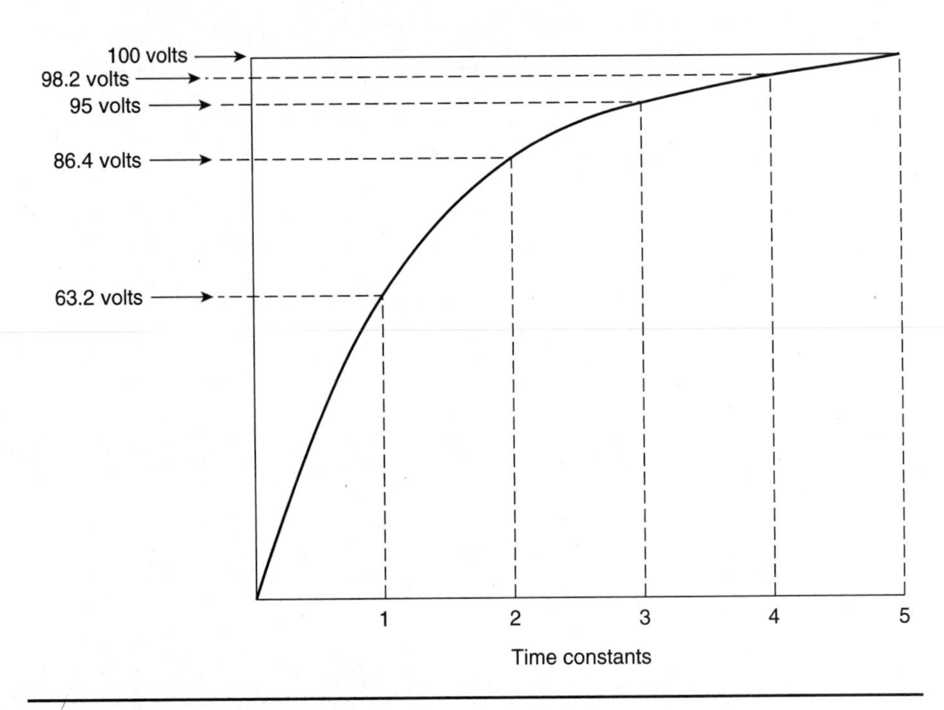

Figure 18–10 Capacitors charge at an exponential rate.

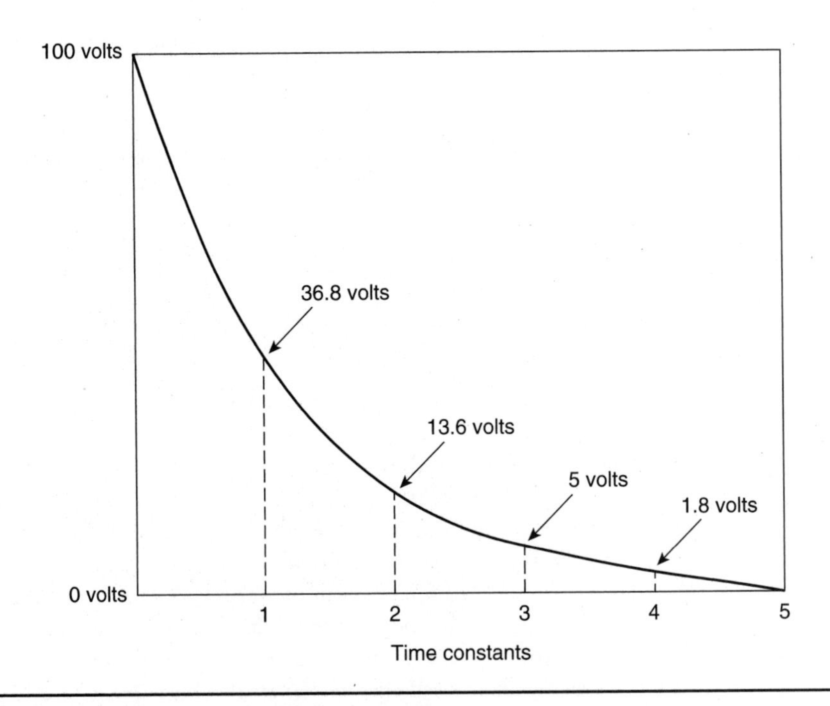

Figure 18–11 Capacitor discharge curve.

end of the second time constant, the voltage will drop to 13.6 volts, and by the end of the third time constant the voltage has dropped to 5 volts. The voltage will continue to drop at this rate until it reaches approximately 0 after five time constants.

RC Time Constants

When a capacitor is connected in a circuit with a resistor, the amount of time needed to charge the capacitor can be accurately determined (see Figure 18–12). The formula for determining charge time is:

$$\tau = R \times C$$

where τ = time for one time constant in seconds
R = resistance in ohms
C = capacitance in farads

Note: The Greek letter τ (tau) is used to represent the time for one time constant. It is not unusual, however, for the letter T to be used to represent time.

Example #3

How long will it take the capacitor shown in Figure 18–12 to charge if it has a value of 50 μF and the resistor has a value of 100,000 ohms (100K Ω)?

Solution

$$\tau = R \times C$$

$$\tau = .000{,}050 \text{ farads} \times 100{,}000 \text{ ohms}$$

$$\tau = 5 \text{ seconds}$$

The formula is used to find the time for one time constant. Five time constants are required to charge the capacitor.

$$\text{Total } \tau = 5 \text{ seconds} \times 5 \text{ time constants}$$

$$\text{Total } \tau = 25 \text{ seconds}$$

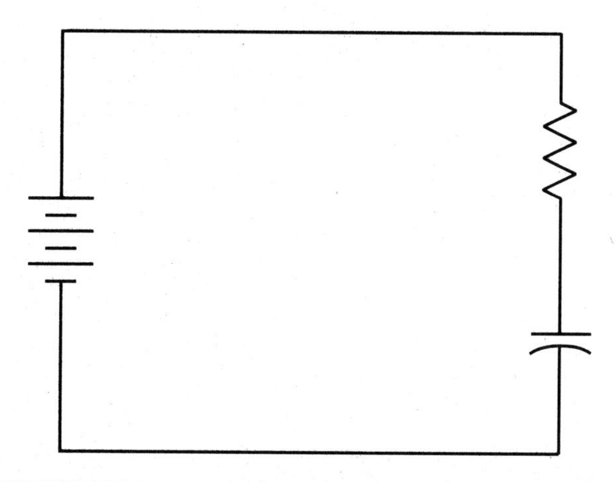

Figure 18–12 The charge time of the capacitor can be determined accurately.

Example #4

How much resistance should be connected in series with a 100-pF capacitor to give it a total charge time of 0.2 second?

Solution

Change the previous formula to solve for resistance.

$$R = \frac{\tau}{C}$$

The total charge time is 0.2 second. The value of τ is therefore 0.2/5 = 0.04 second. Substitute these values in the formula

$$R = \frac{0.04}{100^{\times 10^{-12}}}$$

$$R = 400 \text{ mega ohms}$$

Example #5

A 500K Ω resistor is connected in series with a capacitor. The total charge time of the capacitor is 15 seconds. What is the capacitance of the capacitor?

Solution

Change the base formula to solve for the value of capacitance.

$$C = \frac{\tau}{R}$$

Since the total charge time is 15 seconds, the time of one time constant will be 3 seconds (15/5 = 3).

$$C = \frac{3}{500{,}000}$$

$$C = 0.000006 \text{ farads}$$

or

$$C = 6\ \mu\text{F}$$

Applications for Capacitors

Capacitors are among the most used of electrical components. They are used for power factor correction in industrial applications, in the start windings of many single-phase AC motors, to produce phase shifts for SCR and Triac circuits (SCRs and Triacs are solid-state electronic devices used throughout industry to control high current circuits), to filter pulsating DC, and in RC timing circuits. Capacitors are used extensively in electronic circuits for control of frequency and pulse generation. The type of capacitor used is dictated by the circuit application.

Nonpolarized Capacitors

Capacitors can be divided into two basic groups, nonpolarized and polarized. Nonpolarized capacitors are often referred to as AC capacitors. The reason for this is that they are not sensitive to polarity connection. Nonpolarized capacitors can be connected to either

DC or AC circuits without harm to the capacitor. Nonpolarized capacitors are constructed by separating metal plates by some type of dielectric, as was shown earlier in Figure 18–1.

A common type of AC capacitor—called the paper capacitor or oil-filled capacitor—is often used in motor circuits and for power factor correction (see Figure 18–13). This capacitor derives its name for the type of dielectric used. It is constructed by separating metal-foil plates with thin sheets of paper soaked in a dielectric oil, as shown in Figure 18–14. These capacitors are often used as the run or starting capacitor for single-phase motors. Many manufacturers of oil-filled capacitors will identify one terminal with an arrow or a painted dot, or by stamping a dash in the capacitor can, as shown in Figure 18–15. This identified terminal marks the connection to the plate that is located nearer to the metal container or can. It has long been known that when a capacitor's dielectric breaks down and permits a short circuit to ground, it is most often the plate located nearer to the outside case that becomes grounded. For this reason, it is generally desirable to connect the identified capacitor terminal to the line side instead of to the motor start winding.

In Figure 18–16, the run capacitor has been connected in such a manner that the identified terminal is connected to the start winding of a single-phase motor. If the capacitor should become shorted to ground, a current path exists through the motor start winding. The start winding is an inductive type load, and inductive reactance will limit the value of current flow to ground. Since the flow of current is limited, it will take the circuit breaker or fuse time to open the circuit and disconnect the motor from the power line. This time delay can permit the start winding to overheat and become damaged.

Figure 18–13 An oil-filled paper capacitor is often used in motor circuits and for power factor correction.

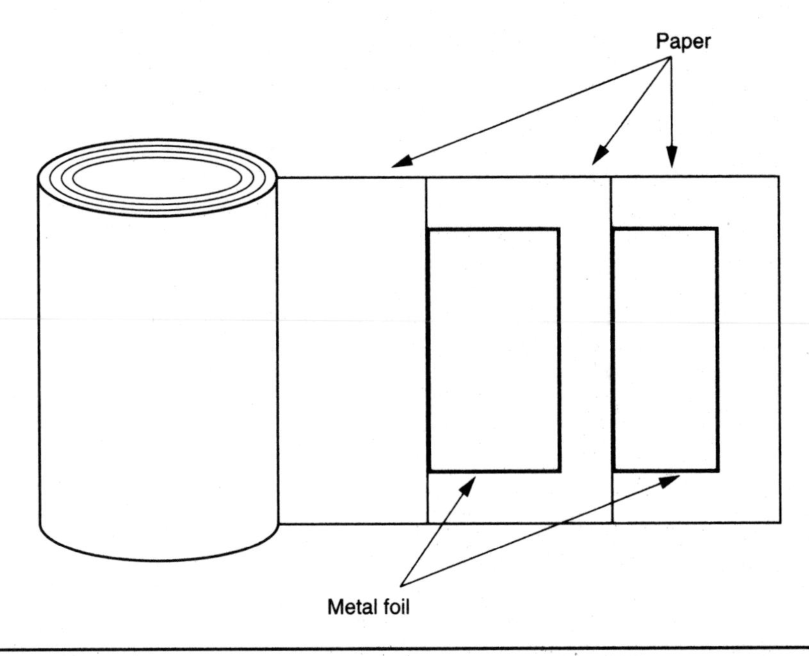

Figure 18–14 In an oil-filled paper capacitor, metal-foil plates are separated by sheets of paper soaked in dielectric oil.

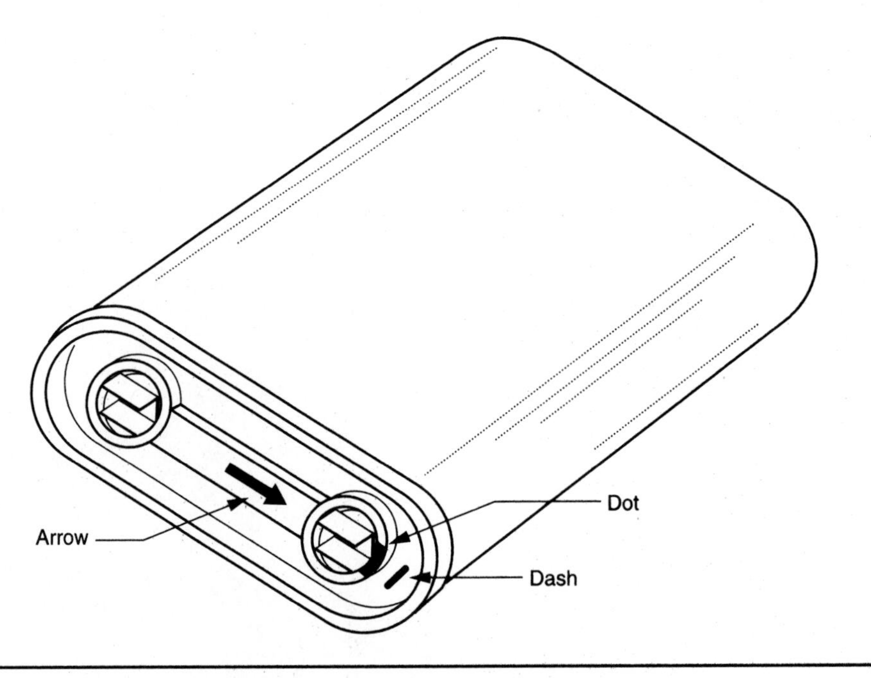

Figure 18–15 Markings indicate plate nearest capacitor case.

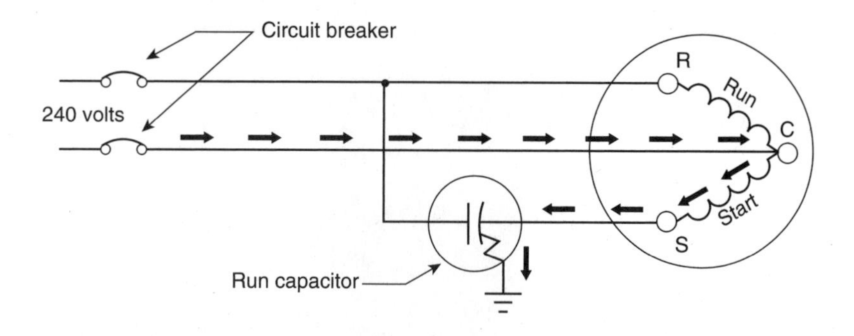

Figure 18–16 Identified capacitor terminal is incorrectly connected to motor start winding.

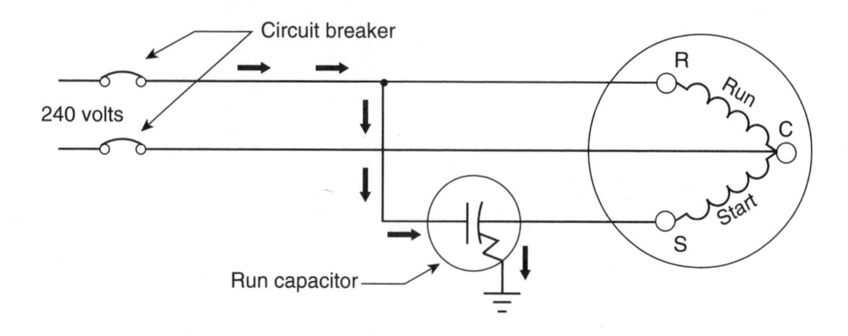

Figure 18–17 Identified capacitor terminal is correctly connected to the line.

In Figure 18–17, the run capacitor has been connected in the circuit in such a manner that the identified terminal is connected to the line side. If the capacitor should become shorted to ground, a current path exists directly to ground, bypassing the motor start winding. When the capacitor is connected in this manner, the start winding does not limit current flow, permitting the fuse or circuit breaker to open the circuit almost immediately.

Polarized Capacitors

Polarized capacitors are generally referred to as electrolytic capacitors. These capacitors are sensitive to the polarity they are connected to and will have one terminal identified as positive or negative, as in Figure 18–18. Polarized capacitors can be used in DC circuits only. If their polarity connection is reversed, the capacitor can be damaged and will sometimes explode. The advantage of electrolytic capacitors is that they can have very high capacitance in a small case size.

There are two basic types of electrolytic capacitors—the wet type and the dry type. The wet-type electrolytic, Figure 18–19, has a positive plate made of aluminum foil. The negative plate is actually an electrolyte made from a borax solution. A second piece of aluminum foil is placed in contact with the electrolyte and becomes the negative terminal. When a source of direct current is connected to the capacitor, the borax solution forms an insulating oxide film on the positive plate. This film is only a few molecules thick and acts as the insulator to separate the plates. The capacitance is very high because the distance between the plates is so small.

Figure 18–18 Polarized capacitors have very high resistance in a small case size.

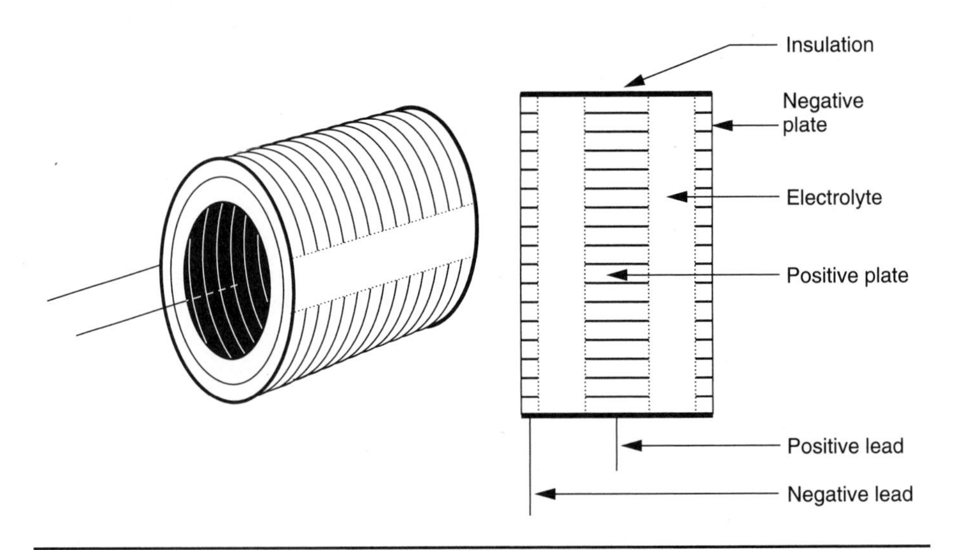

Figure 18–19 Construction of a wet type electrolytic capacitor.

If the polarity of the wet-type electrolytic capacitor should become reversed, the oxide insulating film will dissolve and the capacitor will become shorted. If the polarity connection is corrected, the film will reform and restore the capacitor.

AC Electrolytic Capacitors

This ability of the wet-type electrolytic capacitor to be shorted and then reformed is the basis for a special type of nonpolarized electrolytic capacitor called the AC electrolytic capacitor. This capacitor is used as the starting capacitor for many small single-phase motors, as the run capacitor in many ceiling-fan motors, and for low-power electronic cir-

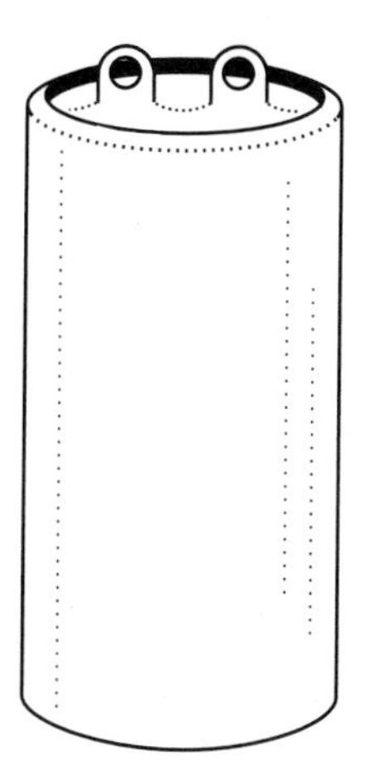

Figure 18–20 An AC electrolytic capacitor.

cuits when a nonpolarized capacitor with a high capacitance is required. The AC electrolytic capacitor is made by connecting two wet-type electrolytic capacitors together inside the same case. When alternating current is applied to the leads, one capacitor will be connected to reverse polarity and become shorted. The other capacitor will be connected to the correct polarity and will form. During the next half cycle, the polarity changes. The capacitor that was shorted forms and the other capacitor becomes shorted. An AC electrolytic capacitor is shown in Figure 18–20.

Dry-Type Electrolytic Capacitors

The dry-type electrolytic capacitor is similar to the wet type except that gauze is used to hold the borax solution. This prevents the capacitor from leaking. Although the dry-type electrolytic has the advantage of being relatively leak proof, it does have one disadvantage. If the polarity connection should become reversed and the oxide film is broken down, it will not reform when connected to the proper polarity. Reversing the polarity of a dry-type electrolytic capacitor will permanently damage the capacitor.

Variable Capacitors

Variable capacitors are constructed in such a manner that their capacitance value can be changed over a certain range. They generally contain a set of movable plates, which are connected to a shaft, and a set of stationary plates, as in Figure 18–21. The movable plates can be interleaved with the stationary plates to increase or decrease the capacitance value. Since air is used as the dielectric and the plate area is relatively small, variable capacitors are generally rated in picofarads. Another type of small variable capacitor is called the trimmer capacitor, shown in Figure 18–22. This capacitor has one movable plate and one stationary plate. The capacitance value is changed by turning an adjustment screw, which moves the movable plate closer to or farther away from the stationary plate. Figure 18–23 shows schematic symbols used to represent variable capacitors.

Ceramic Capacitors

Another capacitor that often uses color codes is the ceramic capacitor (Figure 18–24). This capacitor will generally have one band that is wider than the others. The wide band indicates the temperature coefficient and the other bands are first and second digits, multiplier, and tolerance.

Figure 18–21 A variable capacitor.

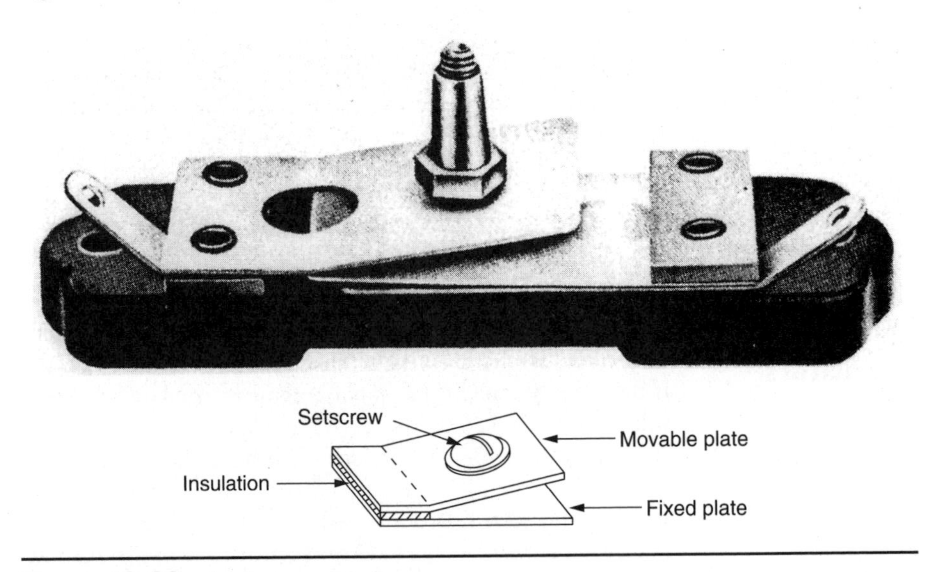

Figure 18–22 A trimmer capacitor.

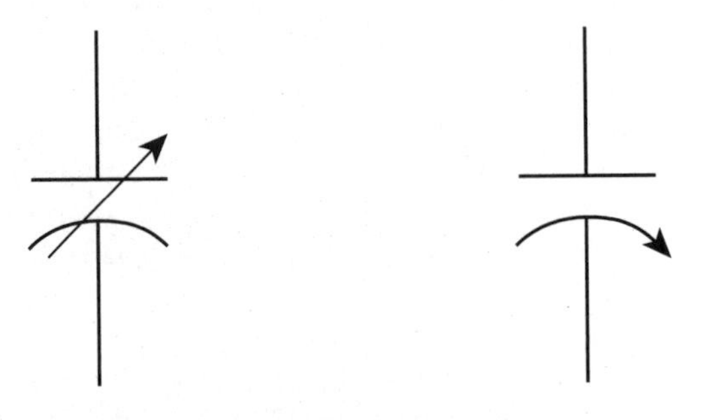

Figure 18–23 Variable capacitor symbols.

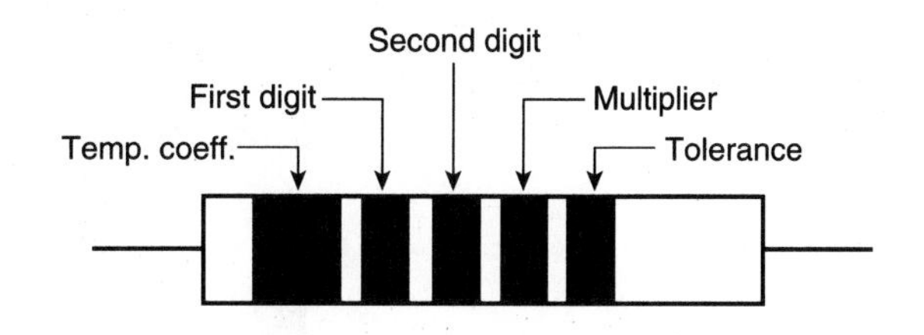

Color	Number	Multiplier	Tolerance		Temp. coeff.
			Over 10 pF	10 pF or Less	
Black	0	1	20%	2.0pF	0
Brown	1	10	1%		N30
Red	2	100	2%		N80
Orange	3	1,000			N150
Yellow	4				N220
Green	5				N330
Blue	6		5%	0.5pF	N470
Violet	7				N750
Gray	8	0.01		0.25pF	P30
White	9	0.1	10%	1.0pF	P500

Figure 18–24 Color codes for ceramic capacitors.

Dipped Tantalum Capacitors

A dipped tantalum capacitor is shown in Figure 18–25. This capacitor has the general shape of a match head but is somewhat larger in size. Color bands and dots determine the value, tolerance, and voltage. The capacitance value is given in picofarads.

Film Capacitors

Not all capacitors use color codes to indicate values. Some capacitors use numbers and letters. A film capacitor is shown in Figure 18–26. This capacitor is marked 105K. The value can be read as follows:

1. The first two numbers indicated the first two digits of the value.
2. The third number is the multiplier. Add the number of zeros to the first two numbers indicated by the multiplier. In this example, add 5 zeros to 10. The value is given in picofarads. This capacitor has a value of 1,000,000 picofarads, or 1 microfarad.
3. The K is the tolerance. In this example, K indicates a tolerance of ±10 percent.

Measuring and Testing Capacitors

Capacitor Markings

Different types of capacitors are marked in different ways. Large AC oil-filled paper capacitors generally have their capacitance and voltage values written on the capacitor. The same is true for most electrolytic and small nonpolarized capacitors. Other types of capacitors,

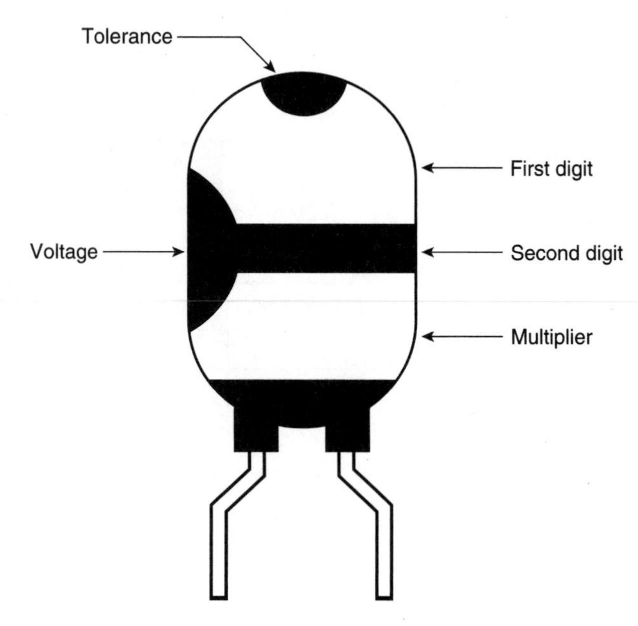

Color	Number	Multiplier	Tolerance	Voltage
			No dot 20%	
Black	0			4
Brown	1			6
Red	2			10
Orange	3			15
Yellow	4	10,000		20
Green	5	100,000		25
Blue	6	1,000,000		35
Violet	7	10,000,000		50
Gray	8			
White	9			3
Gold			5%	
Silver			10%	

Figure 18–25 Dipped tantalum capacitors.

however, depend on color codes or code numbers and letters to indicate the capacitance value, tolerance, and voltage rating. Although color coding for capacitors has been abandoned in favor of direct marking by most manufacturers, it is still used by some. Also, many older capacitors with color codes are still in use. For this reason, we will discuss color coding for several different types of capacitors. Unfortunately, there is no actual standard used by all manufacturers. The color codes presented are probably the most common. An

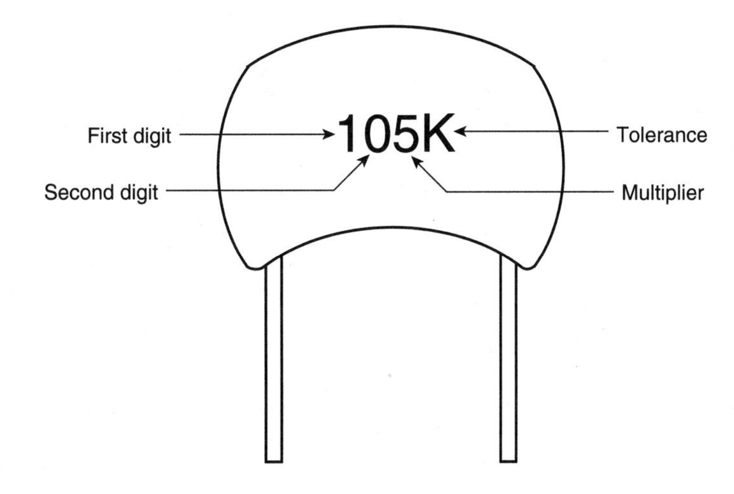

Number	Multiplier	Tolerance		
			10pF or less	Over 10pF
0	1	B	0.1pF	
1	10	C	0.25pF	
2	100	D	0.5pF	
3	1,000	F	1.0pF	1%
4	10,000	G	2.0pF	2%
5	100,000	H		3%
6		J		5%
7		K		10%
8	0.01	M		20%
9	0.1			

Figure 18–26 Film-type capacitors.

identification chart for *postage stamp* mica capacitors and tubular paper or tubular mica capacitors is shown in Figure 18–27. It should be noted that most postage stamp mica capacitors use a five-dot color code. There are six-dot color codes, however. When a six-dot color code is used, the third color dot represents a third digit and the rest of the code is the same. The capacitance values given are in picofarads. Although these markings are typical, there is no actual standard, and it may be necessary to use manufacture's literature to determine the true values.

A second method for color coding mica capacitors is called the EIA (Electronic Industries Association) standard, or the JAN (Joint Army–Navy) standard. The JAN standard is used for electronic components intended for military use. When the EIA standard is employed, the first dot will be colored white. In some instances, the first dot may be

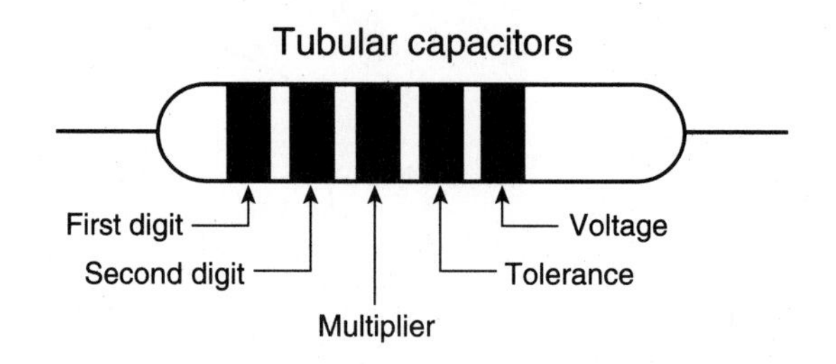

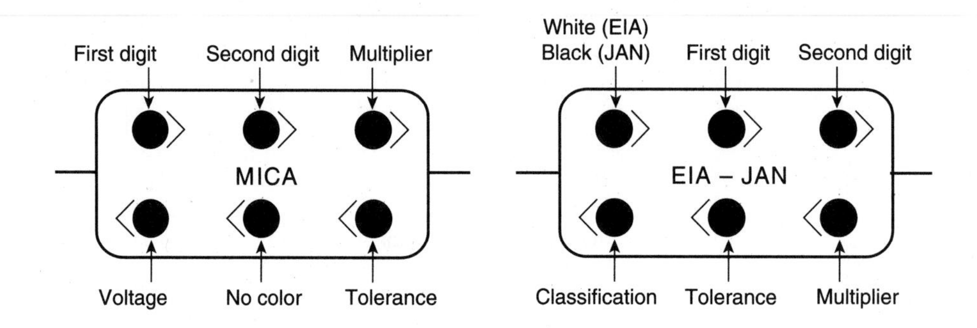

Color	Number	Multiplier	Tolerance	Voltage
No color			20%	500
Black	0	1		
Brown	1	10	1%	100
Red	2	100	2%	200
Orange	3	1,000	3%	300
Yellow	4	10,000	4%	400
Green	5	100,000	5% (EIA)	500
Blue	6	1,000,000	6%	600
Violet	7	10,000,000	7%	700
Gray	8	100,000,000	8%	800
White	9	1,000,000,000	9%	900
Gold		0.1	5% (JAN)	1,000
Silver		0.01	10%	2,000

Figure 18–27 Identification of mica and tubular capacitors.

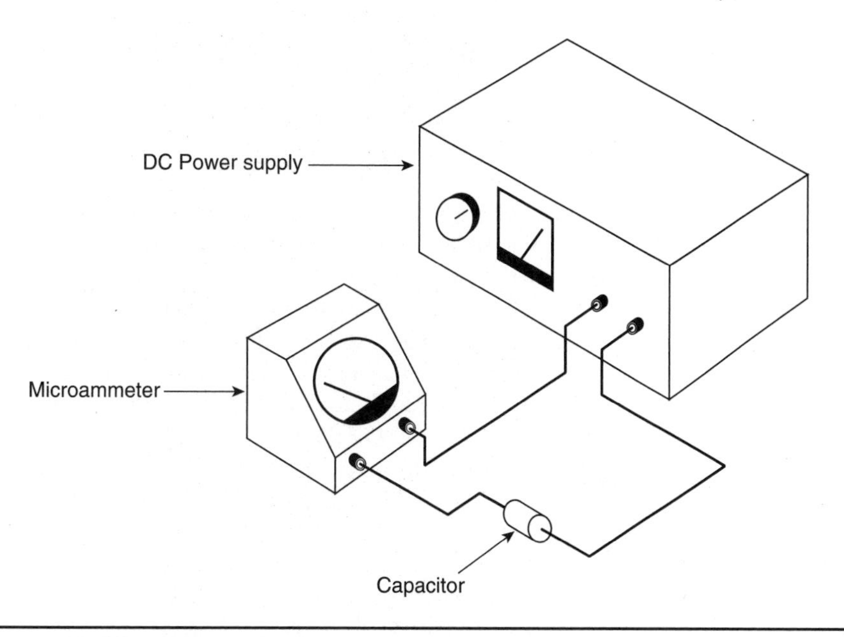

Figure 18–28 Testing a capacitor for leakage.

colored silver instead of white. This indicates that the capacitor's dielectric is paper instead of mica. When the JAN standard is used, the first dot will be colored black. The second and third dots represent digits, the fourth dot is the multiplier, the fifth dot is the tolerance, and the sixth dot indicates classes A to E of temperature and leakage coefficients.

Temperature Coefficients

The temperature coefficient indicates the amount of capacitance change with temperature. Temperature coefficients are listed in ppm (parts per million) per degree Celsius. A positive temperature coefficient indicates that the capacitor will increase its capacitance with an increase in temperature. A negative temperature coefficient indicates that the capacitance will decrease with an increase in temperature.

Testing Capacitors

Testing capacitors is difficult at best. Small electrolytic capacitors are generally tested for shorts with an ohmmeter. If the capacitor is not shorted, it should be tested for leakage using a variable DC power supply and a microammeter, as in Figure 18–28. When rated voltage is applied to the capacitor, the microammeter should indicate zero current flow.

Large AC oil-filled capacitors can be tested in a similar manner. To accurately test the capacitor, two measurements must be made. One is to measure the capacitance value of the capacitor to determine if it is the same or approximately the same as the rate value. The other is to test the strength of the dielectric.

The first test should be made with an ohmmeter. With the power disconnected, connect the terminals of an ohmmeter directly across the capacitor terminals. This test determines if the dielectric is shorted. When the ohmmeter is connected, the needle should swing up scale and return to infinity. The amount of needle swing is determined by the capacitance of the capacitor. Then reverse the ohmmeter connection and the needle should move twice as far up scale and return to the infinity setting.

If the ohmmeter test is successful, the dielectric must be tested at its rated voltage. This is called a *dielectric strength test.* To make this test, a dielectric test set must be used. This device is often referred to as a HIPOT because of its ability to produce a high voltage or high potential. The dielectric test set contains a variable voltage control, a voltmeter, and a microammeter. To use the HIPOT, connect its terminal leads to the capacitor terminals. Increase the output voltage until rated voltage is applied to the capacitor. The microammeter indicates any current flow between the plates of the dielectric. If the capacitor is good, the microammeter should indicate zero current flow.

The capacitance value must be measured to determine if there are any open plates in the capacitor. To measure the capacitance value of the capacitor, connect some value of AC voltage across the plates of the capacitor. This voltage must not be greater than the rated capacitor voltage. Then measure the amount of current flow in the circuit. Now that the voltage and current flow are known, compute the capacitive reactance of the capacitor using the formula

$$X_C = \frac{E}{I}$$

After the capacitive reactance has been determined, compute the capacitance using the formula

$$C = \frac{1}{2\pi FX_C}$$

Note: Capacitive reactance is measured in ohms and limits current flow in a manner similar to inductive reactance.

CHAPTER 19

Capacitance in Alternating Current Circuits

When a capacitor is connected to an alternating current circuit, current will appear to flow through the capacitor. The reason for this is that in an AC circuit, the current is continually changing direction and polarity. To understand this concept, consider the hydraulic circuit shown in Figure 19–1. Two tanks are connected to a common pump. Assume tank A to be full and tank B to be empty. Now assume that the pump is used to pump water from tank A to tank B. When tank B becomes full, the pump reverses and pumps the water from tank B back into tank A. Each time a tank becomes filled, the pump reverses and pumps water back into the other tank. Notice that water is continually flowing in this circuit, but there is no direct connection between the two tanks.

A similar action takes place when a capacitor is connected to an alternating current circuit, as shown in Figure 19–2. In this circuit, the AC generator or alternator charges one plate of the capacitor positive and the other plate negative. During the next half cycle, the voltage will change polarity and the capacitor will discharge and recharge to the opposite polarity. As long as the voltage continues to increase, decrease, and change polarity, current will flow from one plate of the capacitor to the other. If an ammeter were to be placed in the circuit, it would indicate a continuous flow of current, giving the appearance that current is flowing through the capacitor.

Capacitive Reactance

As the capacitor is charged, an impressed voltage develops across its plates as an electrostatic charge is built up, as in Figure 19–3. This impressed voltage opposes the applied voltage and limits the flow of current in the circuit. This counter voltage is similar to the counter voltage produced by an inductor. The counter voltage developed by the capacitor is also called reactance. Since this counter voltage is caused by capacitance, it is called capacitive reactance and is measured in ohms. The formula for finding capacitive reactance is:

$$X_C = \frac{1}{2\pi FC}$$

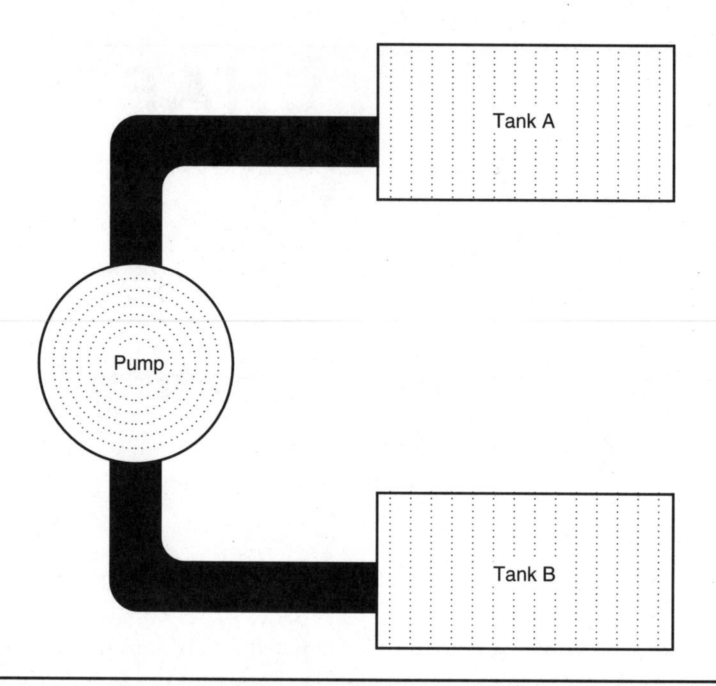

Figure 19–1 Water can flow continuously, but not between the two tanks.

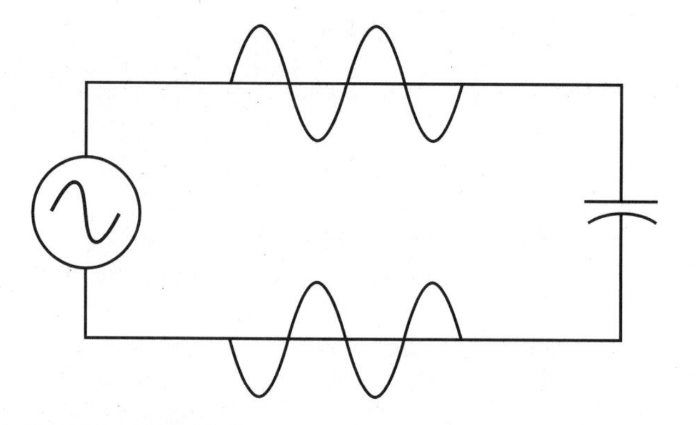

Figure 19–2 Capacitor connected to an AC circuit.

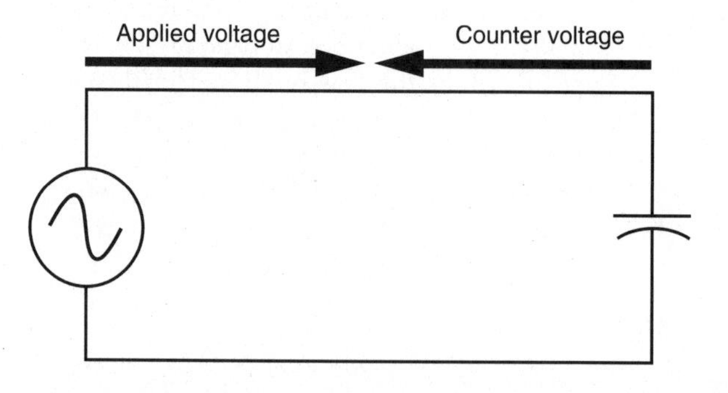

Figure 19–3 Counter voltage limits the flow of current.

where X_C = capacitive reactance
$\pi = 3.1416$
F = frequency in hertz
C = capacitance in farads

Example #1

A 35 μF capacitor is connected to a 120-volt, 60-Hz line. How much current will flow in this circuit?

Solution

The first step is to compute the capacitive reactance. Recall that the value of C in the formula is given in farads. This must be changed to the capacitive units being used. In this case, it will be microfarads.

$$X_C = \frac{1}{2 \times 3.1416 \times 60 \times 35^{\times 10^{-6}}}$$

$$X_C = 75.79\ \Omega$$

Now that the value of capacitive reactance is known, it can be used like resistance in an Ohm's law formula. Since capacitive reactance is the current limiting factor, it will replace the value of R.

$$I = \frac{E}{X_C}$$

$$I = \frac{120}{75.79}$$

$$I = 1.58 \text{ amps}$$

Capacitance (C)

If the value of capacitive reactance is known, the capacitance of the capacitor can be found using the formula:

$$C = \frac{1}{2\pi F X_C}$$

Example #2

A capacitor is connected into a 480-volt, 60-Hz circuit. An ammeter indicates a current flow of 2.6 amperes. What is the capacitance value of the capacitor?

Solution

The first step is to compute the value of capacitive reactance. Since capacitive reactance limits current flow like resistance, it can be substituted for R in an Ohm's law formula.

$$X_C = \frac{E}{I}$$

$$X_C = \frac{480}{2.6}$$

$$X_C = 184.61\ \Omega$$

Now that the capacitive reactance of the circuit is known, the value of capacitance can be found.

$$C = \frac{1}{2\pi FX_C}$$

$$C = \frac{1}{2 \times 3.1416 \times 60 \times 184.61}$$

$$C = \frac{1}{69596.49}$$

$$C = 0.00001437 \text{ farads}$$

or

$$C = 14.37\ \mu\text{F}$$

Voltage and Current Relationships in a Pure Capacitive Circuit

Earlier it was shown that the current in a pure resistive circuit is in phase with the applied voltage, and that current in a pure inductive circuit lags the applied voltage by 90 degrees. It will now be shown that in a pure capacitive circuit the current will lead the applied voltage by 90 degrees.

When a capacitor is connected to an alternating current, the capacitor will charge and discharge at the same rate and time as the applied voltage. The charge in coulombs is equal to the capacitance of the capacitor and the applied voltage ($Q = C \times V$). When the applied voltage is zero, the charge in coulombs and impressed voltage will be zero, also. When the applied voltage reaches its maximum value, positive or negative, the charge in coulombs and impressed voltage will reach maximum also, as in Figure 19–4. The impressed voltage will follow the same curve as the applied voltage.

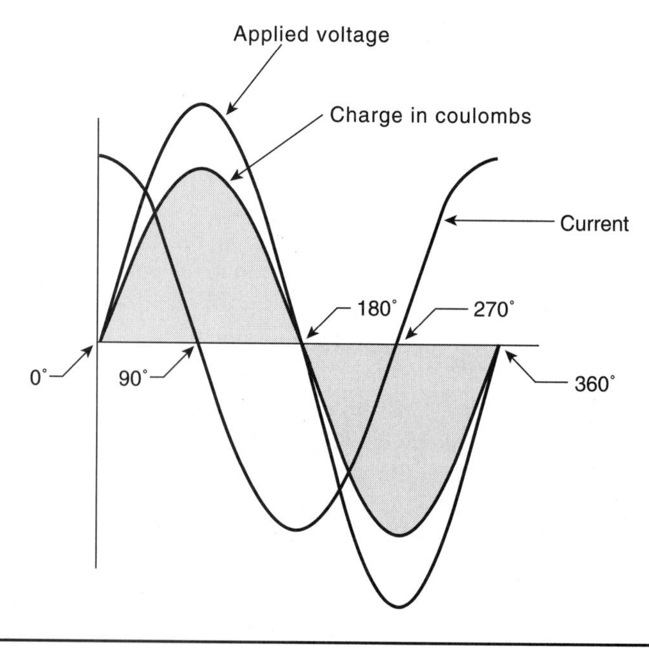

Figure 19–4 Capacitive current leads the applied voltage by 90 degrees.

In the waveform shown, voltage and charge are both shown to be zero at 0 degrees. Since there is no charge on the capacitor, there is no opposition to current flow, which is shown to be maximum. As the applied voltage increases from zero toward its positive peak at 90 degrees, the capacitor begins to charge at the same time. The charge produces an impressed voltage across the plates of the capacitor which opposes the flow of current. The impressed voltage is 180 degrees out of phase with the applied voltage, Figure 19–5. When the applied voltage reaches 90 degrees in the positive direction, the charge reaches maximum, the impressed voltage reaches peak in the negative direction, and the current flow is zero.

As the applied voltage begins to decrease, the capacitor begins to discharge causing the current to flow in the opposite or negative direction. When the applied voltage and charge reach zero at 180 degrees, the impressed voltage is zero also, and the current flow is maximum in the negative direction. As the applied voltage and charge increase in the negative direction, the increase of the impressed voltage across the capacitor again causes the current to decrease. The applied voltage and charge reach maximum negative after 270 degrees of rotation. The impressed voltage reaches maximum positive and the current has decreased to zero, as shown in Figure 19–6. As the applied voltage decreases from its maximum negative value, the capacitor again begins to discharge. This causes the current to flow in the positive direction. The current again reaches its maximum positive value when the applied voltage and charge reach zero after 360 degrees of rotation.

Power in a Pure Capacitive Circuit

Since the current flow in a pure capacitive circuit is leading the applied voltage by 90 degrees, the voltage and current will both have the same polarity for half the time during one cycle and have opposite polarities the other half of the time, as in Figure 19–7. During the period of time that both the voltage and current have the same polarity, energy is being stored in the capacitor in the form of an electrostatic field. When the voltage and current have opposite polarities, the capacitor is discharging and the energy is returned to the circuit. When these values are added, the sum will equal zero, just as it does with pure inductive circuits. Therefore, there is no true power or watts produced in a pure capacitive circuit.

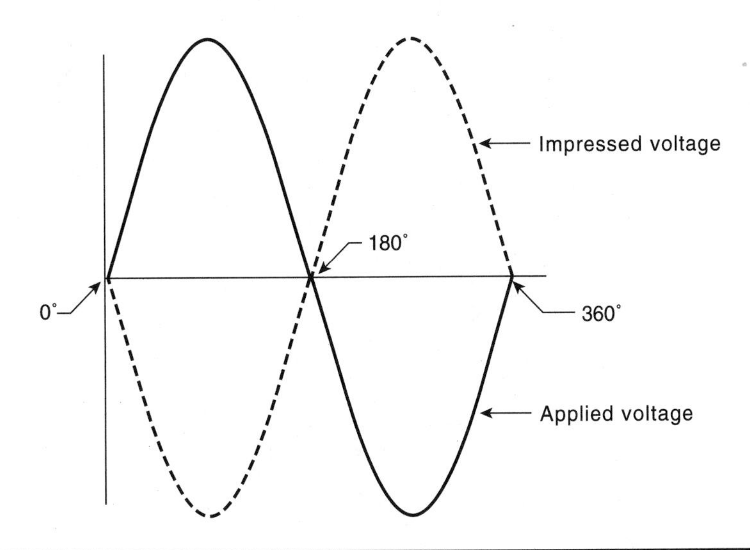

Figure 19–5 The impressed voltage is 180 degrees out of phase with the applied voltage.

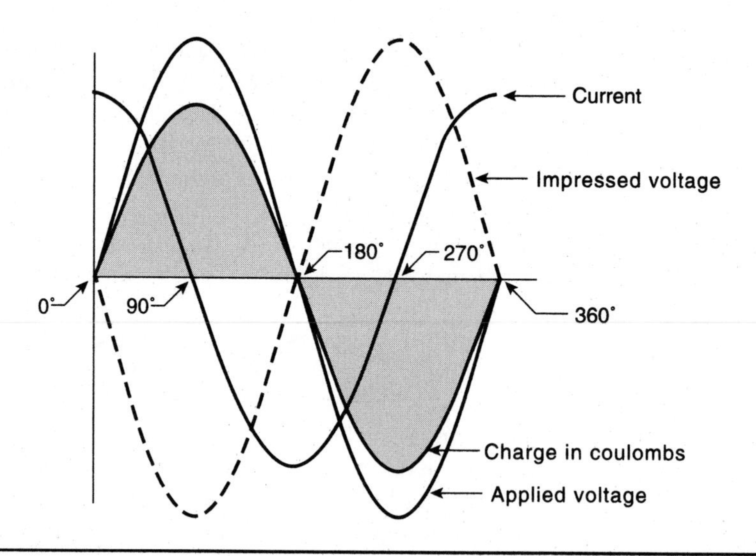

Figure 19–6 Voltage, current, and charge relationships in a capacitive circuit.

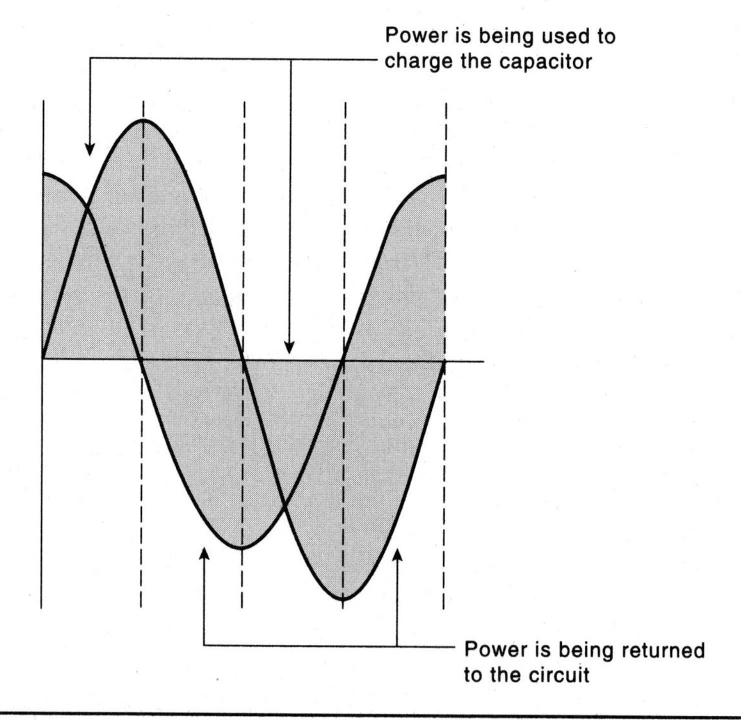

Figure 19–7 A pure capacitive circuit has no true power.

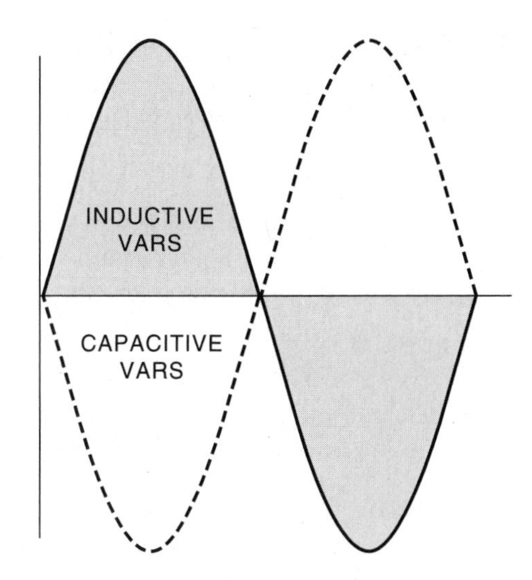

Figure 19–8 Inductive VARs and capacitive VARs are 180 degrees out of phase with each other.

The power value for a capacitor is reactive and is measured in VARs, just as it is for an inductor. Inductive VARs and capacitive VARs are 180 degrees out of phase with each other, however. This is presented in Figure 19–8, but the concept will be covered in greater detail later in this text. To distinguish between inductive and capacitive VARs, inductive VARs will be shown as $VARs_L$ and capacitive VARs will be shown as $VARs_C$.

Q of a Capacitor

The *Q* of a capacitor is generally very high. As with inductors, it is a ratio of capacitive reactance as compared to resistance.

$$Q = \frac{R}{X_C}$$

The *R* value for a capacitor is generally very high because it is the equivalent resistance of the dielectric between the plates of the capacitor. If a capacitor is leaky, however, the dielectric will appear to be a much lower resistance and the *Q* rating will decrease.

Q for a capacitor can also be found by using other formulas. One of these formulas is shown as follows:

$$Q = \frac{VARs_C}{\text{Watts}}$$

We could also set *Q* equal to the reciprocal of the power factor:

$$Q = \frac{1}{PF}$$

Capacitor Voltage Ratings

The voltage rating of a capacitor is actually the voltage rating of the dielectric. This voltage rating is extremely important concerning the life of the capacitor and should never be exceeded. Unfortunately, there are no set standards concerning how voltage ratings are marked. It is not unusual to see capacitors marked VOLTS AC, VOLTS DC, PEAK VOLTS, and WVDC (working volts DC). The voltage rating of electrolytic or polarized capacitors will always be given in DC volts. The voltage rating of nonpolarized capacitors, however, can be given as AC or DC volts.

If a nonpolarized capacitor has a voltage rating given in AC volts, the voltage indicated is the RMS value. If the voltage rating is given as PEAK or as DC volts, it indicates the peak value of AC volts. If a capacitor is to be connected to an AC circuit and the voltage rating is given as DC volts, it is necessary to compute the peak value.

Example #3

An AC oil-filled capacitor has a voltage rating of 300 WVDC. Will the voltage rating of the capacitor be exceeded if the capacitor is connected to a 240-volt, 60-Hz line?

Solution

The DC voltage rating of the capacitor indicates the peak value of voltage. To determine if the voltage rating will be exceeded, find the peak value of 240 volts by multiplying by 1.414.

$$\text{peak} = 240 \times 1.414$$

$$\text{peak} = 339.36 \text{ volts}$$

The answer is that the capacitor voltage rating will be exceeded.

Effects of Frequency in a Capacitive Circuit

One of the factors that determine the capacitive reactance of a capacitor is the frequency. Capacitive reactance is inversely proportional to frequency. As the frequency increases, the capacitive reactance decreases. The chart in Figure 19–9 shows the capacitive reactance for different values of capacitance at different frequencies. The reason that frequency has an effect on the capacitive reactance is because the capacitor charges and discharges at a faster rate at a higher frequency. Recall that current is a rate of electron flow. A current of 1 ampere is 1 coulomb per second.

$$I = \frac{C}{\tau}$$

where I = current
C = charge in coulombs
τ = time in seconds

Capacitance	Capacitive Reactance			
	30 Hz	60 Hz	400 Hz	1000 Hz
10 pF	530.5 MΩ	265.26 MΩ	39.79 MΩ	15.91 MΩ
350 pF	15.16 MΩ	7.58 MΩ	1.14 MΩ	454.73 KΩ
470 nF	112.88 KΩ	56.44 KΩ	8.47 KΩ	3.39 KΩ
750 nF	22.22 KΩ	11.11 KΩ	1.67 KΩ	666.67 Ω
1 μF	5.31 KΩ	2.65 KΩ	397.89 Ω	159.15 Ω
25 μF	212.21 Ω	106.1 Ω	159.15 Ω	6.37 Ω

Figure 19–9 Capacitive reactance is inversely proportional to frequency.

Assume that a capacitor is connected to a 30-Hz line, and 1 coulomb of charge flows each second. If the frequency is doubled to 60 Hz, 1 coulomb of charge will flow in 0.5 second because the capacitor is being charged and discharged twice as fast. This means that in a period of 1 second, 2 coulombs of charge will flow. Since the capacitor is being charged and discharged at a faster rate, the opposition to current flow is decreased.

Series Capacitors

In the first example circuit, Figure 19–10, three capacitors with values of 10 μF, 30 μF, and 15 μF are connected in series to a 480-volt, 60-Hz line. The following circuit values will be found.

X_{C1} = capacitive reactance of the first capacitor
X_{C2} = capacitive reactance of the second capacitor
X_{C3} = capacitive reactance of the third capacitor
X_{CT} = total capacitive reactance for the circuit
C_T = total capacitance for the circuit
I_T = total circuit current
E_{C1} = voltage drop across the first capacitor
$VARs_{C1}$ = reactive power of the first capacitor
E_{C2} = voltage drop across the second capacitor
$VARs_{C2}$ = reactive power of the second capacitor
E_{C3} = voltage drop across the third capacitor
$VARs_{C3}$ = reactive power of the third capacitor
$VARs_{CT}$ = total reactive power for the circuit

Capacitance Resistance (X_{C1}, X_{C2}, and X_{C3})

Since the frequency and the capacitance of each capacitor is known, the capacitive reactance for each capacitor can be found using the formula:

$$X_C = \frac{1}{2\pi FC}$$

Recall that the value for C in the formula is in farads, and the capacitors in this problem are rated in microfarads. Also recall that in a 60-Hz circuit, the value of $2\pi F$ is 377.

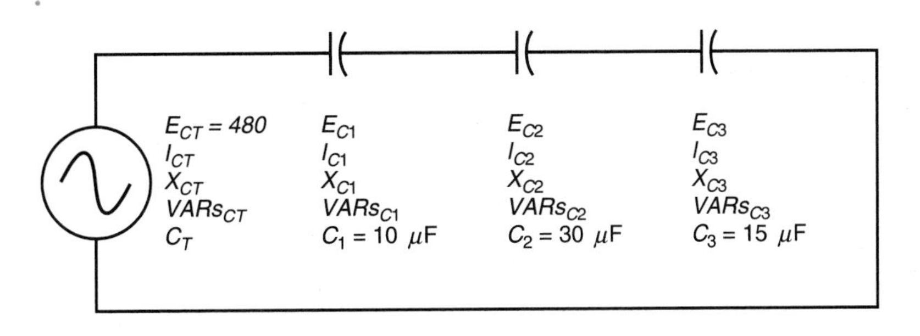

Figure 19–10 Capacitors connected in series.

$$X_{C1} = \frac{1}{377 \times 0.000010}$$

$$X_{C1} = 265.25\ \Omega$$

$$X_{C2} = \frac{1}{377 \times 0.000030}$$

$$X_{C2} = 88.417\ \Omega$$

$$X_{C3} = \frac{1}{377 \times 0.000015}$$

$$X_{C3} = 176.83\ \Omega$$

Total Capacitance Resistance (X_{CT})

Since there is no phase angle shift between any of the three capacitive reactances, the total capacitive reactance will be the sum of the three reactances.

$$X_{CT} = X_{C1} + X_{C2} + X_{C3}$$

$$X_{CT} = 265.25 + 88.417 + 176.83$$

$$X_{CT} = 530.497\ \Omega$$

Total Capacitance (C_T)

The total capacitance of a series circuit can be computed in a similar manner to computing parallel resistor. Total capacitance in this circuit will be computed using the formula:

$$C_T = \frac{1}{\frac{1}{C_1} + \frac{1}{C_2} + \frac{1}{C_3} + \frac{1}{C_N}}$$

$$C_T = \frac{1}{\frac{1}{10} + \frac{1}{30} + \frac{1}{15}}$$

$$C_T = \frac{1}{0.2}$$

$$C_T = 5\ \mu\text{F}$$

Total Current (I_T)

The total current can be found by using the total inductive reactance to substitute for R in an Ohm's law formula.

$$I_T = \frac{E_{CT}}{X_{CT}}$$

$$I_T = \frac{480}{530.497}$$

$$I_T = 0.905\ \text{amp}$$

Voltage Drop (E_{C1}, E_{C2}, E_{C3})

Since the current is the same at any point in a series circuit, the voltage drop across each capacitor can now be computed using the inductive reactance of each capacitor and the current flowing through it.

$$E_{C1} = I_{C1} \times X_{C1}$$

$$E_{C1} = 0.905 \times 265.25$$

$$E_{C1} = 240.051 \text{ volts}$$

$$E_{C2} = I_{C2} \times X_{C2}$$

$$E_{C2} = 0.905 \times 88.417$$

$$E_{C2} = 80.017 \text{ volts}$$

$$E_{C3} = I_{C3} \times X_{C3}$$

$$E_{C3} = 0.905 \times 176.83$$

$$E_{C3} = 160.031$$

Reactive Power ($VARs_{C1}$, $VARs_{C2}$, $VARs_{C3}$)

$$VARs_{C1} = E_{C1} \times I_{C1}$$

$$VARs_{C1} = 240.051 \times 0.905$$

$$VARs_{C1} = 217.246$$

$$VARs_{C2} = E_{C2} \times I_{C2}$$

$$VARs_{C2} = 80.017 \times 0.905$$

$$VARs_{C2} = 72.415$$

$$VARs_{C3} = E_{C3} \times I_{C3}$$

$$VARs_{C3} = 160.031 \times 0.905$$

$$VARs_{C3} = 144.828$$

Total Reactive Power ($VARs_{CT}$)

Power—whether true, apparent, or reactive power—will add in any type of circuit. The total reactive power in this circuit can be found by taking the sum of all the VARs for each capacitor, or by using total values of voltage and current and Ohm's law.

$$VARs_{CT} = VARs_{C1} + VARs_{C2} + VARs_{C3}$$

$$VARs_{CT} = 217.246 + 72.415 + 144.828$$

$$VARs_{CT} = 434.489$$

Parallel Capacitors

In the second example circuit, three capacitors having values of 50 μF, 75 μF, and 20 μF are connected in parallel to a 60-Hz line. The circuit has a total reactive power of 787.08 VARs (see Figure 19–11). The following unknown values will be found:

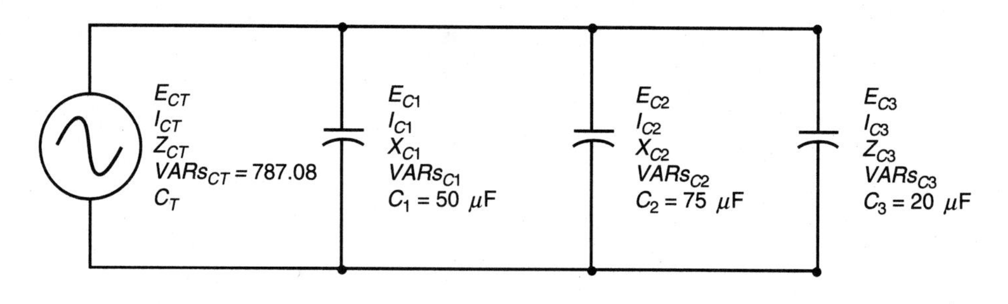

Figure 19–11 Capacitors connected in parallel.

X_{C1} = capacitive reactance of the first capacitor
X_{C2} = capacitive reactance of the second capacitor
X_{C3} = capacitive reactance of the third capacitor
X_{CT} = total capacitive reactance for the circuit
C_{CT} = total capacitance for the circuit
E_T = total circuit current
I_{C1} = voltage drop across the first capacitor
$VARs_{C1}$ = reactive power of the first capacitor
I_{C2} = voltage drop across the second capacitor
$VARs_{C2}$ = reactive power of the second capacitor
I_{C3} = voltage drop across the third capacitor
$VARs_{C3}$ = reactive power of the third capacitor

Capacitance Reactance (X_{C1}, X_{C2}, and X_{C3})

Since the frequency of the circuit and the capacitance of each capacitor is known, the capacitive reactance of each capacitor can be computed using the following formula:

$$X_C = \frac{1}{2\pi FC}$$

$$X_{C1} = \frac{1}{377 \times 0.000050}$$

$$X_{C1} = 53.05\ \Omega$$

$$X_{C2} = \frac{1}{377 \times 0.000075}$$

$$X_{C2} = 35.367\ \Omega$$

$$X_{C3} = \frac{1}{377 \times 0.000020}$$

$$X_{C3} = 132.63\ \Omega$$

Total Capacitance Reactance (X_{CT})

The total capacitive reactance can be found in a manner similar to finding the resistance of parallel resistors.

$$X_{CT} = \frac{1}{\frac{1}{X_{C1}} + \frac{1}{X_{C2}} + \frac{1}{X_{C3}} + \frac{1}{X_{CN}}}$$

$$X_{CT} = \frac{1}{\frac{1}{53.05} + \frac{1}{35.367} + \frac{1}{132.63}}$$

$$X_{CT} = \frac{1}{0.05466}$$

$$X_{CT} = 18.295\ \Omega$$

Total Circuit Current (E_T)

Now that the total capacitive reactance of the circuit is known and the total reactive power is known, we can find the voltage applied to the circuit using the following formula:

$$E_T = \sqrt{VARs_{CT} \times X_{CT}}$$

$$E_T = \sqrt{787.08 \times 18.295}$$

$$E_T = 120 \text{ volts}$$

In a parallel circuit the voltage must be the same across each branch of the circuit. Therefore, 120 volts is applied across each capacitor.

Voltage Drop (I_{CT}, I_{C1}, I_{C2}, and I_{C3})

Now that the circuit voltage is known, we can find the amount of total current for the circuit and the amount of current in each branch using Ohm's law.

$$I_{CT} = \frac{E_{CT}}{X_{CT}}$$

$$I_{CT} = \frac{120}{18.295}$$

$$I_{CT} = 6.559 \text{ amps}$$

$$I_{C1} = \frac{E_{C1}}{X_{C1}}$$

$$I_{C1} = \frac{120}{53.05}$$

$$I_{C1} = 2.262 \text{ amps}$$

$$I_{C2} = \frac{E_{C2}}{X_{C2}}$$

$$I_{C2} = \frac{120}{35.367}$$

$$I_{C2} = 3.393 \text{ amps}$$

$$I_{C3} = \frac{E_{C3}}{X_{C3}}$$

$$I_{C3} = \frac{120}{132.63}$$

$$I_{C3} = 0.905 \text{ amp}$$

Reactive Power ($VARs_{C1}$, $VARs_{C2}$, and $VARs_{C3}$)

The amount of reactive power for each capacitor can now be computed using Ohm's law.

$$VARs_{C1} = E_{C1} \times I_{C1}$$

$$VARs_{C1} = 120 \times 2.262$$

$$VARs_{C1} = 271.442$$

$$VARs_{C2} = E_{C2} \times I_{C2}$$

$$VARs_{C2} = 120 \times 3.393$$

$$VARs_{C2} = 407.159$$

$$VARs_{C3} = E_{C3} \times I_{C3}$$

$$VARs_{C3} = 120 \times 0.905$$

$$VARs_{C3} = 108.573$$

To make a quick check of the circuit values, add the VARs for each capacitor and see if they equal the total circuit VARs.

$$VARs_{CT} = VARs_{C1} + VARs_{C2} = VARs_{C3}$$

$$VARs_{CT} = 271.442 + 407.159 + 108.573$$

$$VARs_{CT} = 787.174$$

The slight difference in answers is caused by rounding off of values.

CHAPTER 20

Resistive–Capacitive Series Circuits

When a circuit containing both resistance and capacitance is connected to an alternating current circuit, the voltage and current will be out of phase with each other by some amount between 0 degrees and 90 degrees. The exact amount of phase angle difference is determined by the ratio of resistance to capacitance. Resistive–capacitive series circuit are very similar to resistive–inductive series circuits. Other than changing a few formulas, the procedure for solving circuit values will be the same.

Example Circuit #1

In the following example, a series circuit containing 12 ohms of resistance and 16 ohms of capacitive reactance is connected to a 240-volt, 60-Hz line, shown in Figure 20–1. The following unknown values will be computed:

Z = total circuit impedance
I = current flow
E_R = voltage drop across the resistor
P = watts (true power)
C = capacitance
E_C = voltage drop across the capacitor
E_T = total voltage
$VARs_C$ = volt-amperes-reactive (reactive power)
VA = volt-amperes (apparent power)
PF = power factor
$\angle\varnothing$ = angle theta (indicates the angle the voltage and current are out of phase with each other)

Impedance (Z)

The impedance (Z) is the total current limiting element in the circuit. It is a combination of both resistance and capacitive reactance. Since this is a series circuit, the current-limiting elements must be added. Resistance and capacitive reactance are 90 degrees out of

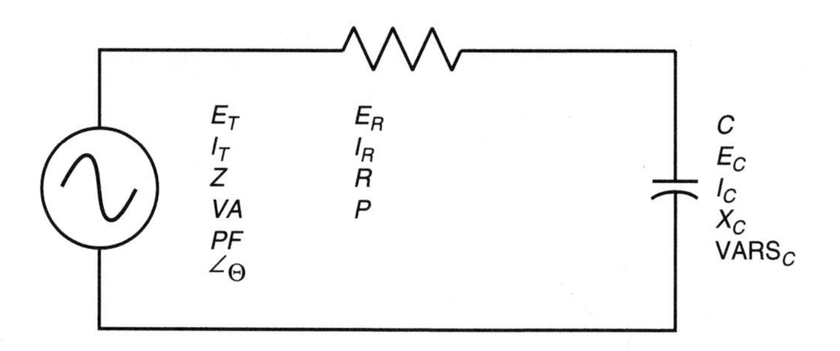

Figure 20–1 Resistive–capacitive series circuit.

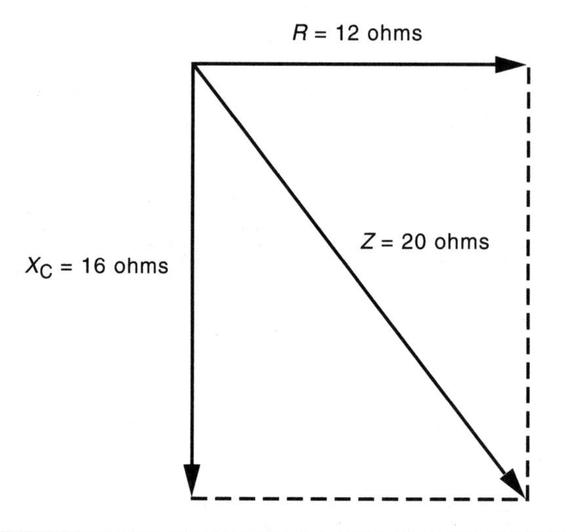

Figure 20–2 Impedance vector for example circuit #1.

phase with each other, forming a right triangle with impedance being the hypotenuse. A vector diagram illustrating this relationship is shown in Figure 20–2. Impedance can be computed using the following formula:

$$Z = \sqrt{R^2 + X_C{}^2}$$

$$Z = \sqrt{12^2 + 16^2}$$

$$Z = \sqrt{144 + 256}$$

$$Z = \sqrt{400}$$

$$Z = 20\ \Omega$$

Total Current (I_T)

Now that the total impedance of the circuit is known, the current can be computed using the formula:

$$I_T = \frac{E}{Z}$$

$$I_T = \frac{240}{20}$$

$$I_T = 12 \text{ amps}$$

Voltage Drop Across the Resistor (E_R)

In a series circuit, the current is the same at any point in the circuit. Therefore, 12 amps flow through both the resistor and the capacitor. The amount of voltage dropped across the resistor can be computed by using the formula:

$$E_R = I \times R$$

$$E_R = 12 \times 12$$

$$E_R = 144 \text{ volts}$$

True Power (P)

True power for the circuit can be computed by using any of the watts formulas, as long as values that apply only to the resistive part of the circuit are used. Recall that current and voltage must be in phase with each other for true power to be produced. The formula used in this example will be:

$$P = E_R \times I$$

$$P = 144 \times 12$$

$$P = 1728 \text{ watts}$$

Capacitance (C)

The amount of capacitance can be computed using the formula:

$$C = \frac{1}{2\pi FX_C}$$

$$C = \frac{1}{6031.858}$$

$$C = 0.0001658 \text{ farads}$$

or

$$C = 165.8\ \mu\text{F}$$

Voltage Drop Across the Capacitor (E_C)

The amount of voltage dropped across the capacitor can be computed using the formula:

$$E_C = I \times X_C$$

$$E_C = 12 \times 16$$

$$E_C = 192 \text{ volts}$$

Total Voltage (E_T)

Although the amount of total voltage applied to the circuit is given as 240 volts in this circuit, it is possible to compute the total voltage if it is not known by adding the voltage drop across the resistor and the voltage drop across the capacitor together. In a series circuit, the voltage drops across the resistor and capacitor are 90 degrees out of phase with each other, and vector addition must be used. These two voltage drops form the legs of a right triangle and the total voltage forms the hypotenuse. The total voltage can be computed using the following formula:

$$E_T = \sqrt{E_R^2 + E_C^2}$$

$$E_T = \sqrt{144^2 + 192^2}$$

$$E_T = 240 \text{ volts}$$

Reactive Power ($VARs_C$)

VARs is the amount of reactive power in the circuit. It can be computed in a similar manner as watts, except that reactive values of voltage and current are used instead of resistive values. In this example, the following formula will be used:

$$VARs_C = E_C \times I$$

$$VARs_C = 192 \times 12$$

$$VARs_C = 2304$$

Apparent Power (*VA*)

Volt-amperes is the apparent power of the circuit. It can be computed in a similar manner as watts or VARs, except that total values of voltage and current are used. In this example, the following formula will be used:

$$VARs_C = E_C \times I$$

$$VARs_C = 192 \times 12$$

$$VARs_C = 2304$$

The apparent power can also be determined by vector addition of the true power and reactive power (see Figure 20–3).

$$VA = \sqrt{P^2 + VARs_C^2}$$

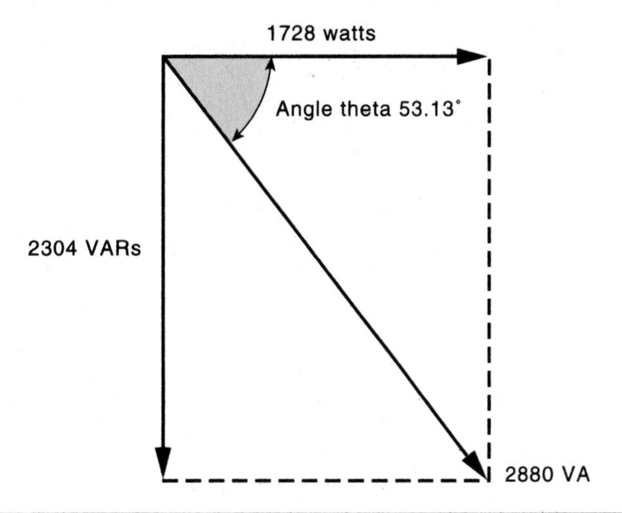

Figure 20–3 Power vector for circuit #1.

Power Factor (*PF*)

Power factor is a ratio of the true power as compared to the apparent power. It can be computed by dividing any resistive value by its like total value. In this circuit the following formula will be used:

$$PF = \frac{P}{VA}$$

$$PF = \frac{1728}{2880}$$

$$PF = .6 \times 100, \text{ or } 60\%$$

Angle Theta

The power factor of a circuit is the cosine of the phase angle. Since the power factor of this circuit is 0.6, angle theta will be:

$$\cos \angle\varnothing = PF$$

$$\cos \angle\varnothing = 0.60$$

$$\angle\varnothing = 53.13° \text{ (from a trigonometric table)}$$

In this circuit, the current leads the applied voltage by 53.13 degree (see Figure 20–4).

Example Circuit #2

In the next example, a series circuit containing one resistor and two capacitors are connected in series to a 42.5-volt, 60-Hz line, as in Figure 20–5. Capacitor #1 has a reactive power of 3.75 VARs, the resistor has a true power of 5 watts, and the second capacitor has a reactive power of 5.625 VARs. The following values will be found:

VA = apparent power
I_T = total circuit current

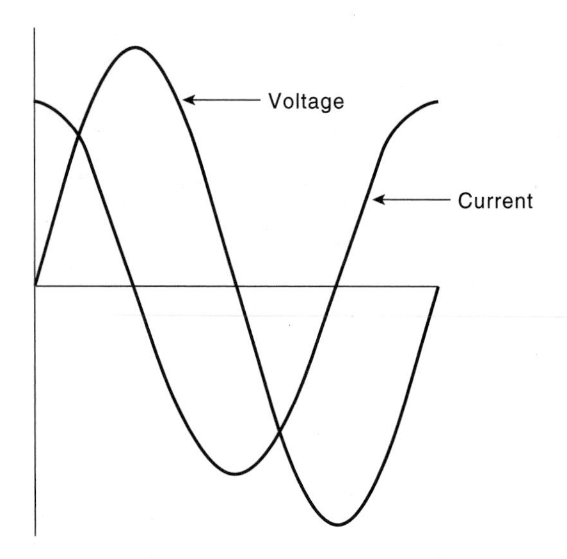

Figure 20–4 The current leads the voltage by 53.13 degrees.

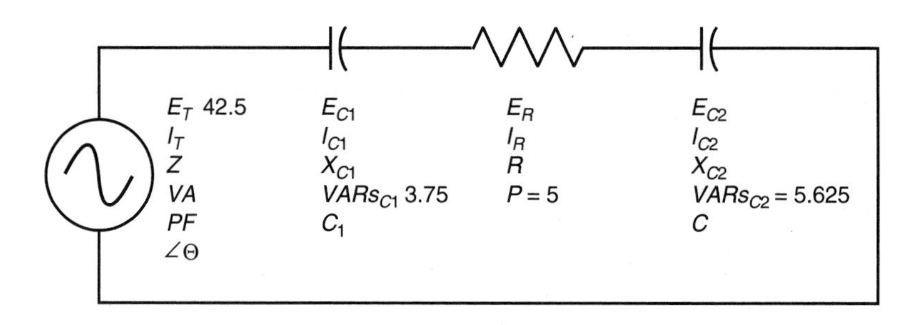

Figure 20–5 Example circuit #2.

Z = impedance of the circuit
PF = power factor
$\angle\varnothing$ = angle theta
E_{C1} = voltage drop across the first capacitor
X_{C1} = capacitive reactance of the first capacitor
C_1 = capacitance of the first capacitor
E_R = voltage drop across the resistor
R = resistance of the resistor
E_{C2} = voltage drop across the second capacitor
X_{C2} = capacitive reactance of the second capacitor
C_2 = capacitance of the second capacitor

Volt-Amperes (*VA*)

Since the reactive power of the two capacitors is known, and the true power of the resistor is known, the apparent power can be found using the formula:

$$VA = \sqrt{P^2 + VARs_C^{\ 2}}$$

In this circuit, $VARs_C$ will be sum of the reactive VARs of both capacitors.

$$VA = \sqrt{5^2 + (3.75 + 5.625)^2}$$

$$VA = \sqrt{25 + 87.891}$$

$$VA = 10.625$$

Total Circuit Current (I_T)

Now that the apparent power and applied voltage are known, the total circuit current can be computed using the formula:

$$I_T = \frac{VA}{E_T}$$

$$I_T = \frac{10.625}{42.5}$$

$$I_T = 0.25 \text{ amp}$$

In a series circuit, the current must be the same at any point in the circuit. Therefore, I_{C1}, I_R, and I_{C2} will all have a value of 0.25 amp.

Impedance (Z)

The impedance of the circuit can now be computed using the formula:

$$Z = \frac{E_T}{I_T}$$

$$Z = \frac{42.5}{0.25}$$

$$Z = 170\ \Omega$$

Power Factor (*PF*)

The power factor will be computed using the formula:

$$PF = \frac{P}{VA}$$

$$PF = \frac{5}{10.625}$$

$$PF = 0.4706, \text{ or } 47.06\%$$

Angle Theta

The cosine of angle theta is the power factor.

$$\cos \angle\varnothing = 0.4706$$

$$\angle\varnothing = 61.93° \text{ (from a trigonometric table)}$$

A vector diagram is shown in Figure 20–6 illustrating the relationship of angle theta to the reactive power, true power, and apparent power.

Voltage Drop Across the Capacitors and Resistors (EC1, ER, and EC2)

Now that the current through each circuit element is known, and the power of each element is known, the voltage drop across each element can be computed.

$$E_{C1} = \frac{VARs_{C1}}{I_{C1}}$$

$$E_{C1} = \frac{3.75}{0.25}$$

$$E_{C1} = 15 \text{ volts}$$

$$E_R = \frac{P}{I_R}$$

$$E_R = \frac{5}{0.25}$$

$$E_R = 20 \text{ volts}$$

$$E_{C2} = \frac{VARs_{C2}}{I_{C2}}$$

$$E_{C2} = \frac{5.625}{0.25}$$

$$E_{C2} = 22.5 \text{ volts}$$

Capacitive Reactance and Capacitance (X_{C1}, X_{C2}, C_1, C_2)

For the first capacitor, we can compute the capacitive reactance and capacitance using the following formula:

$$X_{C1} = \frac{E_{C1}}{I_{C1}}$$

$$X_{C1} = \frac{15}{0.25}$$

$$X_{C1} = 60\ \Omega$$

$$C_1 = \frac{1}{2\pi F X_{C1}}$$

$$C_1 = \frac{1}{377 \times 60}$$

$$C_1 = 0.0000442 \text{ or } 44.2\ \mu\text{F}$$

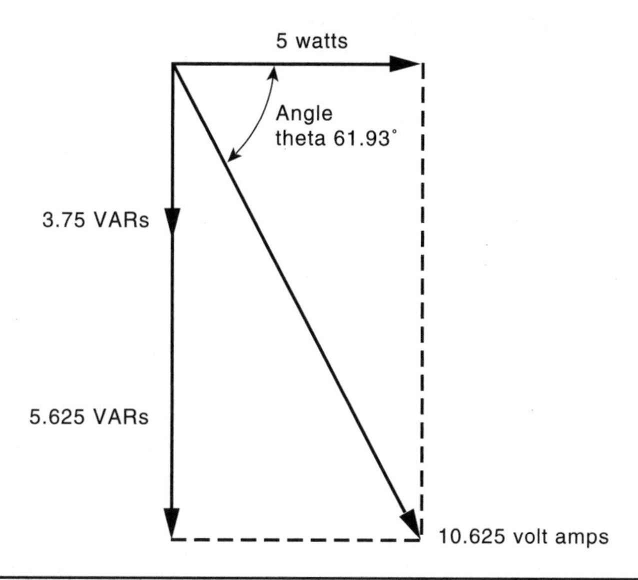

Figure 20–6 Vector relationship of reactive, true, and apparent power.

For the second capacitor, we can compute the capacitive reactance and capacitance with the formula:

$$X_{C2} = \frac{E_{C2}}{I_{C2}}$$

$$X_{C2} = \frac{22.5}{0.25}$$

$$X_{C2} = 90\ \Omega$$

$$C_2 = \frac{1}{2\pi F X_{C2}}$$

$$C_2 = \frac{1}{377 \times 90}$$

$$C_2 = 0.0000295 \text{ farads, or } 29.5\ \mu\text{F}$$

Resistance (*R*)

Formulas for finding values in circuits containing resistance and capacitance connected in series are shown in Figure 20–7. We can compute the resistance of the resistor using the formula:

$$R = \frac{E_R}{I_R}$$

$$R = \frac{20}{0.25}$$

$$R = 80\ \Omega$$

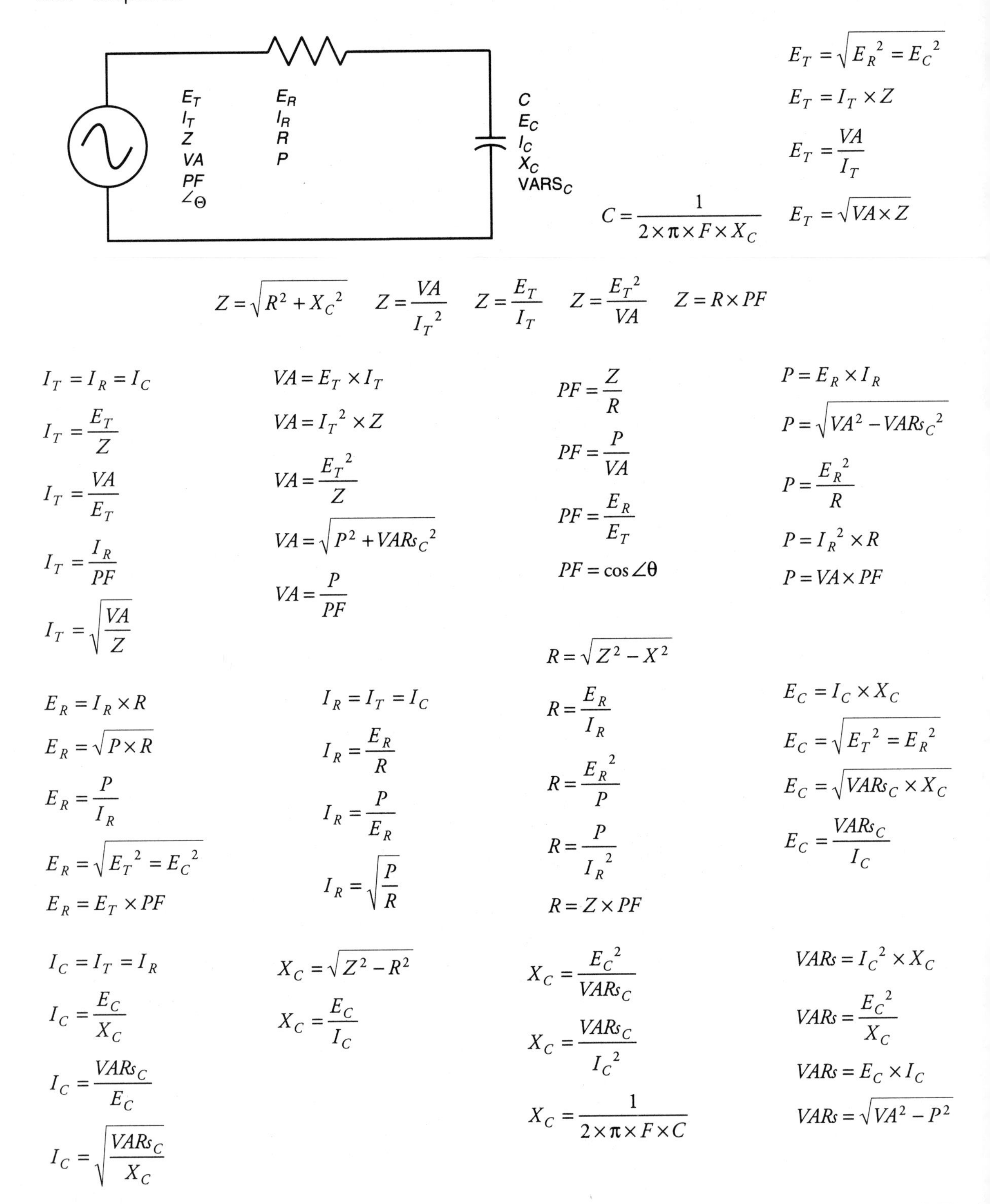

Figure 20–7 Formulas for resistive–capacitive series circuit.

CHAPTER 21

Resistive–Capacitive Parallel Circuits

When resistance and capacitance are connected in parallel, the voltage across each device will be in phase and have the same value. The current flow through the capacitor, however, will be 90 degrees out of phase with the current flow through the resistor. The amount of phase angle shift between the total circuit current and voltage is determined by the ratio of the amount of resistance as compared to the amount of capacitance. The circuit power factor is still determined by the ratio of resistance and capacitance.

Example Circuit #1

An RC parallel circuit is shown in Figure 21–1. In this example circuit, assume a resistance of 30 ohms is connected in parallel with a capacitive reactance of 20 ohms. The circuit is connected to a voltage of 240 volts AC and a frequency of 60 Hz. In this example problem, the following circuit values will be computed:

I_R = current flow through the resistor
P = watts (true power)
I_C = current flow through the capacitor
VARs = reactive power
C = capacitance of the capacitor

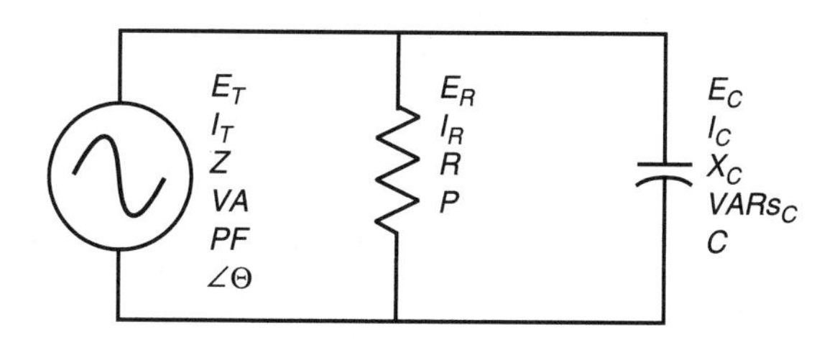

Figure 21–1 Resistive–capacitive parallel circuit.

I_T = total circuit current
Z = total impedance of the circuit
VA = volt-amperes (apparent power)
PF = power factor
$\angle\varnothing$ = angle theta

Resistive Current (I_R)

The amount of current flow through the resistor can be computed by using the following formula:

$$I_R = \frac{E}{R}$$

$$I_R = \frac{240}{30}$$

$$I_R = 8 \text{ amps}$$

True Power (P)

The amount of total power in the circuit can be determined by using any of the values associated with the pure resistive part of the circuit. In this example, true power will be found using the formula:

$$P = E \times I_R$$

$$P = 240 \times 8$$

$$P = 1920 \text{ watts}$$

Capacitive Current (I_C)

The amount of current flow through the capacitor will be computed using the formula:

$$I_C = \frac{E}{X_C}$$

$$I_C = \frac{240}{20}$$

$$I_C = 12 \text{ amps}$$

Reactive Power (VARs)

The amount of reactive power, VARs, can be found using any of the total capacitive values. In this example, VARs will be computed using the following formula:

$$VARs = E \times I_C$$

$$VARs = 240 \times 12$$

$$VARs = 2880$$

Capacitance (C)

The capacitance of the capacitor can be computed using the formula:

$$C = \frac{1}{2\pi FX_C}$$

$$C = \frac{1}{377 \times 20}$$

$$C = 0.0001326 \text{ farads, or } 132.6\ \mu\text{F}$$

Total Current (I_T)

The voltage is the same across all legs of a parallel circuit. The current flow through the resistor is in phase with the voltage and the current flow through the capacitor is leading the voltage by 90 degrees (Figure 21–2). Since these two currents are connected in parallel, vector addition can be used to find the total current flow in the circuit. The total current flow through the circuit can be computed by using the formula:

$$I_T = \sqrt{I_R^2 + I_C^2}$$

$$I_T = \sqrt{8^2 + 12^2}$$

$$I_T = \sqrt{208}$$

$$I_T = 14.42 \text{ amps}$$

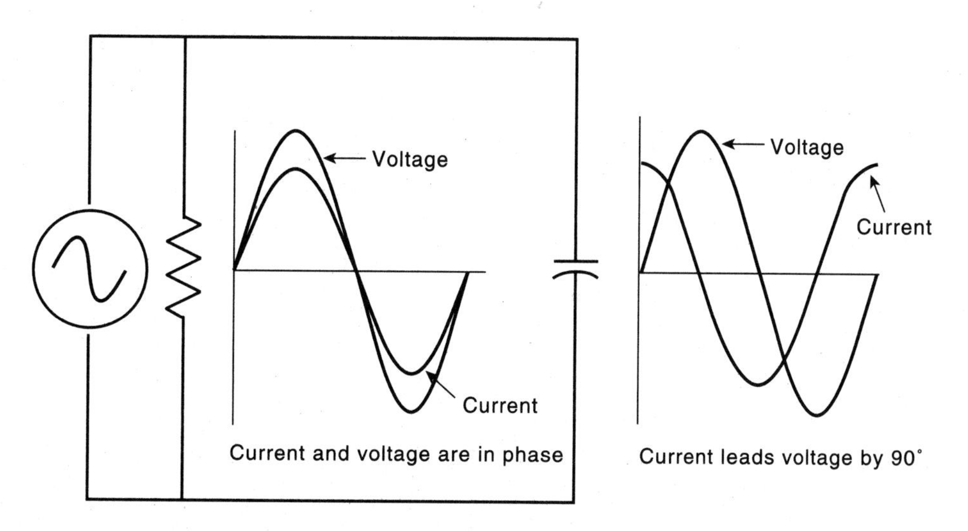

Figure 21–2 Phase relationship of current and voltage in an RC parallel circuit.

Impedance (*Z*)

The total impedance of the circuit can be found by using any of the total values and substituting *Z* for *R* in an Ohm's law formula. The total impedance of this circuit will be computed using the formula:

$$Z = \frac{E}{I_T}$$

$$Z = \frac{240}{14.42}$$

$$Z = 16.64\ \Omega$$

The impedance can also be found by adding the reciprocals of the resistance and capacitive reactance. Since the resistance and capacitive reactance are 90 degrees out of phase with each other, vector addition must be used.

$$Z = \frac{1}{\sqrt{\left(\frac{1}{R}\right)^2 + \left(\frac{1}{X_C}\right)^2}}$$

Apparent Power (*VA*)

The apparent power can be computed by multiplying the circuit voltage by the total current flow.

$$VA = E \times I_T$$

$$VA = 240 \times 14.42$$

$$VA = 3460.8$$

Power Factor (*PF*)

The power factor is the ratio of true power as compared to apparent power. The circuit power factor can be computed using the formula:

$$PF = \frac{P}{VA}$$

$$PF = \frac{1920}{3460.8}$$

$$PF = 0.5548,\ \text{or } 55.48\%$$

Angle Theta

The cosine of angle theta is equal to the power factor.

$$\cos \angle\varnothing = 0.5548$$

$$\angle\varnothing = 56.3°$$

A vector diagram of apparent, true, and reactive power is shown in Figure 21–3. Angle theta is the angle developed between the apparent and true power.

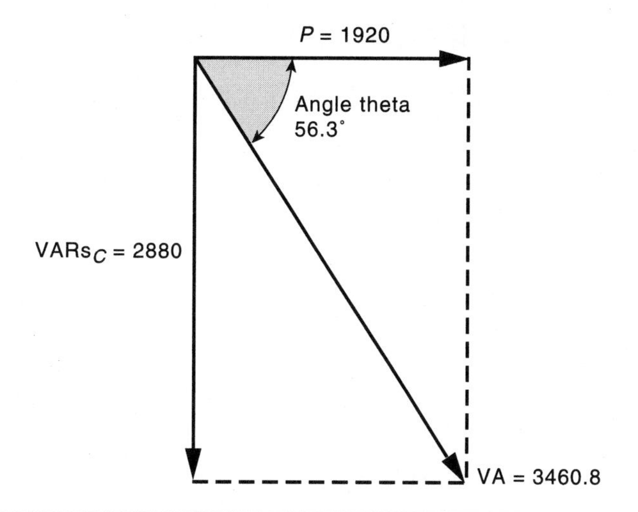

Figure 21–3 Vector relationship of apparent, true, and reactive power.

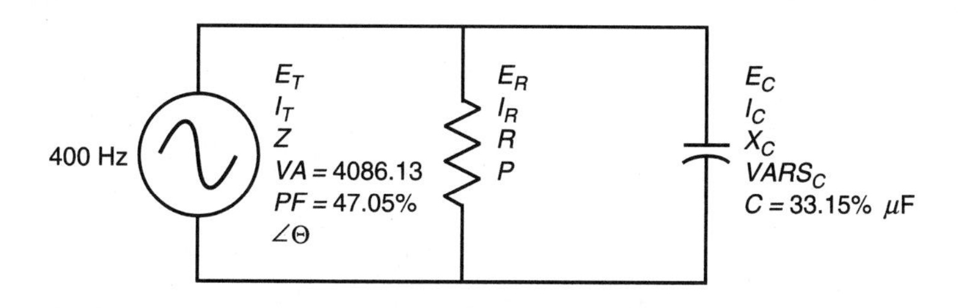

Figure 21–4 Example circuit #2.

Example Circuit #2

In this circuit a resistor and capacitor are connected in parallel to a 400-Hz line. The power factor is 47.05 percent, the apparent power is 4086.13 volt amps, and the capacitance of the capacitor is 33.15 μF, Figure 21–4. The following unknown values will be found:

$\angle\varnothing$ = angle theta
P = true power
$VARs_C$ = capacitive VARs
X_C = capacitive reactance
E_C = voltage drop across the capacitor
I_C = capacitive current
E_R = voltage drop across the resistor
I_R = resistive current
R = resistance of the resistor
E_T = applied voltage
I_T = total circuit current
Z = impedance of the circuit

Angle Theta

The power factor is the cosine of angle theta. To find angle theta, change the power factor from a percentage into a decimal fraction by dividing by 100.

$$PF = \frac{47.05\%}{100}$$

$$PF = 0.4705$$

$$\cos \angle\varnothing = 0.4705$$

$$\angle\varnothing = 61.93^\circ \text{ (from a trigonometric table)}$$

True Power (*P*)

The power factor is determined by the amount of true power, as compared to apparent power.

$$PF = \frac{P}{VA}$$

This formula can be changed to compute the true power when the power factor and apparent power are known.

$$P = VA \times PF$$

$$P = 4086.13 \times 0.4705$$

$$P = 1922.53 \text{ watts}$$

Reactive Power (VARs)

The apparent power, true power, and reactive power form a right triangle. Since these powers form a right triangle, the Pythagorean theorem can be used to find the leg of the triangle represented by the reactive power.

$$VARs = \sqrt{VA^2 - P^2}$$

$$VARs = \sqrt{4086.13^2 - 1922.53^2}$$

$$VARs = \sqrt{13000337}$$

$$VARs = 3605.60$$

Figure 21–5 presents a vector diagram showing the relationship of apparent power, true power, reactive power, and angle theta.

Capacitive Reactance (X_C)

Since the capacitance of the capacitor and the frequency are known, the capacitive reactance can be found using the following formula:

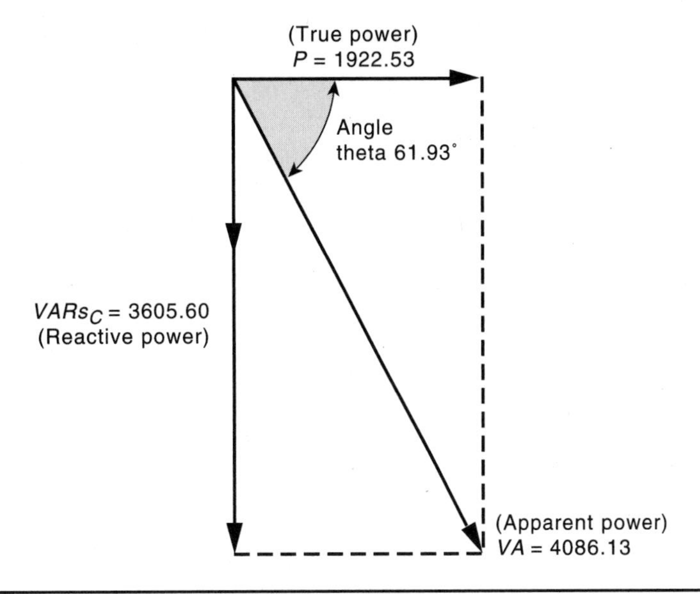

Figure 21–5 Vector diagram of apparent, true, reactive power, and angle theta.

$$X_C = \frac{1}{2\pi FC}$$

$$X_C = \frac{1}{2 \times 3.1416 \times 400 \times 0.00003315}$$

$$X_C = 12\ \Omega$$

Note: Be sure to convert the capacitance value to farads.

Voltage Drop Across the Capacitor (E_C)

$$E_C = \sqrt{VARs \times X_C}$$

$$E_C = \sqrt{3605.6 \times 12}$$

$$E_C = \sqrt{43267.2}$$

$$E_C = 208 \text{ volts}$$

Applied Voltage and Voltage Drop (E_T and E_R)

The voltage must be the same across all branches of a parallel circuit. Therefore, if 208 volts are applied across the capacitive branch, 208 volts must be the total voltage of the circuit as well as the voltage applied across the resistive branch.

$$E_T = 208 \text{ volts} \qquad E_R = 208 \text{ volts}$$

Capacitive Current (I_C)

The amount of current flowing in the capacitive branch can be computed using the formula:

$$I_C = \frac{E}{X_C}$$

$$I_C = \frac{208}{12}$$

$$I_C = 17.33 \text{ amps}$$

Resistive Current (I_R)

The amount of current flowing through the resistor can be computed using the following formula:

$$I_R = \frac{P}{E_R}$$

$$I_R = \frac{1922.53}{208}$$

$$I_R = 9.243 \text{ amps}$$

Resistance (R)

The amount of resistance can be computed using the formula:

$$R = \frac{E_R}{I_R}$$

$$R = \frac{208}{9.243}$$

$$R = 22.5\ \Omega$$

Total Current (I_T)

The total current can be computed using Ohm's law or by vector addition, since both the resistive and capacitive currents are known. Vector addition will be used in this example.

$$I_T = \sqrt{I_R^2 + I_C^2}$$

$$I_T = \sqrt{9.243^2 + 17.33^2}$$

$$I_T = \sqrt{385.726}$$

$$I_T = 19.641 \text{ amps}$$

Impedance (Z)

The impedance of the circuit will be computed using the formula

$$Z = \frac{1}{\sqrt{\frac{1^2}{R} + \frac{1^2}{X_C}}}$$

$$Z = \frac{1}{\sqrt{\frac{1^2}{22.5} + \frac{1^2}{12}}}$$

$$Z = \frac{1}{\sqrt{0.00892}}$$

$$Z = 10.588\ \Omega$$

Formulas for circuits containing resistance and capacitance connected in parallel are shown in Figure 21–6.

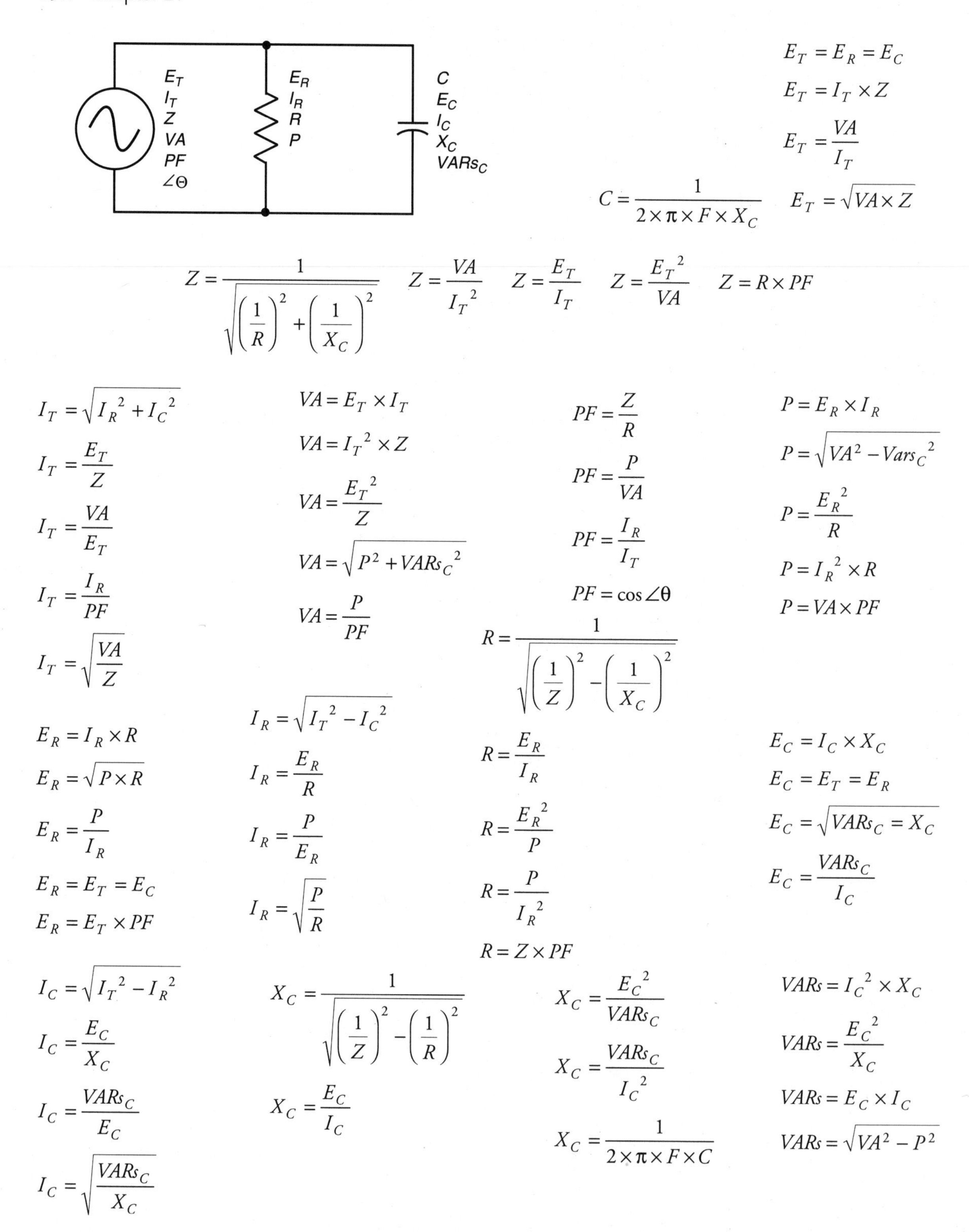

Figure 21–6 Formulas for resistive–capacitive parallel circuits.

CHAPTER 22

Resistive–Inductive–Capacitive Series Circuits

When an alternating current circuit contains elements of resistance, inductance, and capacitance connected in series, the current will be the same through all components, but the voltages dropped across each element will be out of phase with each other. The voltage dropped across the resistance will be in phase with the current, the voltage dropped across the inductor will lead the current by 90 degrees, and the voltage dropped across the capacitor will lag the current by 90 degrees, as shown in Figure 22–1. An RLC series circuit is shown in Figure 22–2. The ratio of resistance, inductance, and capacitance will determine how much the applied voltage will lead or lag the circuit current. If the circuit contains more inductance VARs than capacitive VARs, the current will lag the applied voltage and the power factor will be a lagging power factor. If there are more capacitive VARs than inductive VARs, the current will lead the voltage and the power factor will be a leading power factor.

Since inductive reactance and capacitive reactance are 180 degrees out of phase with each other, they cancel each other in an AC circuit. This can permit the impedance of the circuit to become less than either or both of the reactances, producing a high amount of current flow through the circuit. When Ohm's law is applied to the circuit values, it will be seen that the voltage drops developed across these components can be higher than the applied voltage.

Example Circuit #1

It is assumed that the circuit shown in Figure 22–2 has an applied voltage of 240 volts at 60 Hz and the resistor has a value of 12 ohms, the inductor has an inductive reactance of 24 ohms, and capacitor has a capacitive reactance of 8 ohms. The following unknown values will be found.

Z = impedance of the circuit
I = circuit current
E_R = voltage drop across the resistor
P = true power (watts)
L = inductance of the inductor
E_L = voltage drop across the inductor

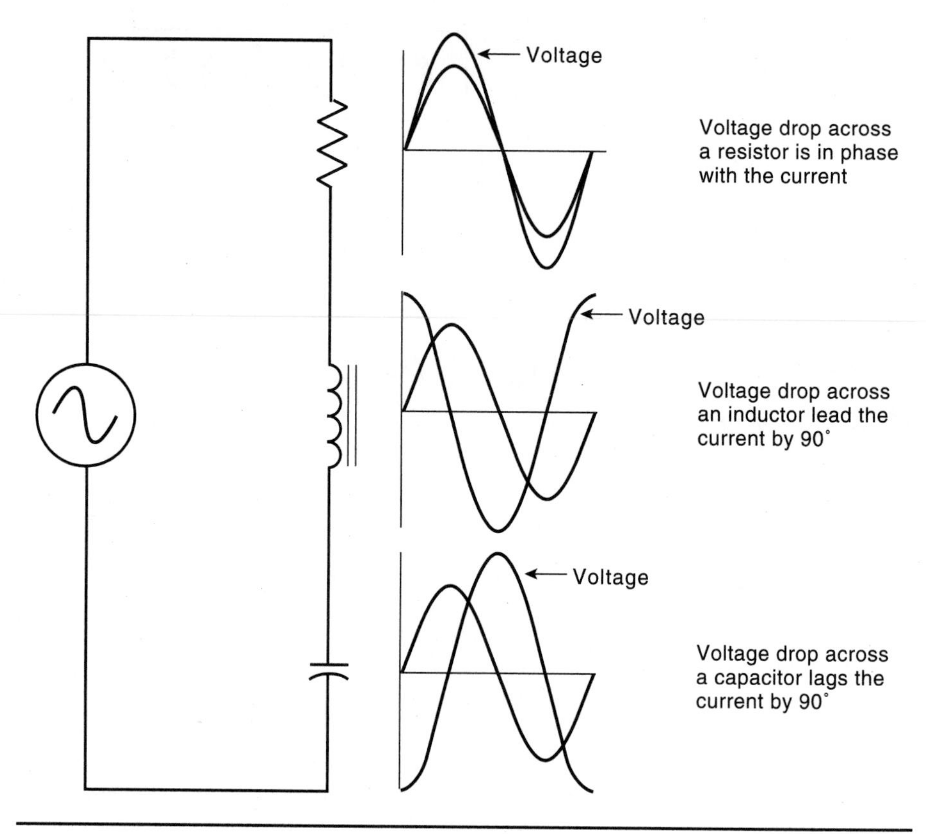

Figure 22–1 Voltage and current relationship in an RLC series circuit.

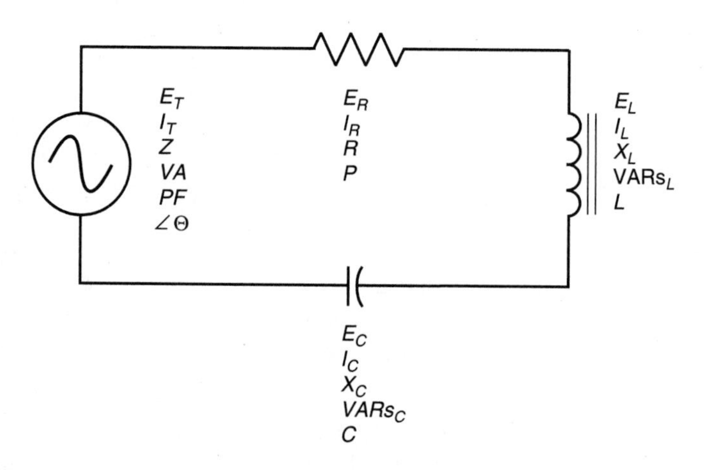

Figure 22–2 Resistive–inductive–capacitive series circuit.

$VARs_L$ = reactive power of the inductor
C = capacitance
E_C = voltage drop across the capacitor
$VARs_C$ = reactive power of the capacitor
VA = volt-amperes (apparent power)
PF = power factor
$\angle\varnothing$ = angle theta

Total Impedance (Z)

The impedance of the circuit is the sum of resistance, inductive reactance, and capacitive reactance. Since inductive reactance and capacitive reactance are 180 degrees out of phase with each other, vector addition must be used to find their sum. This results in the smaller of the two reactive values being subtracted from the larger, as in Figure 22–3. When this is done, the smaller value is eliminated and the larger value is reduced by the amount of the smaller value. The total impedance will be the hypotenuse formed by the resulting right triangle (see Figure 22–4). The impedance will be computed by using the formula:

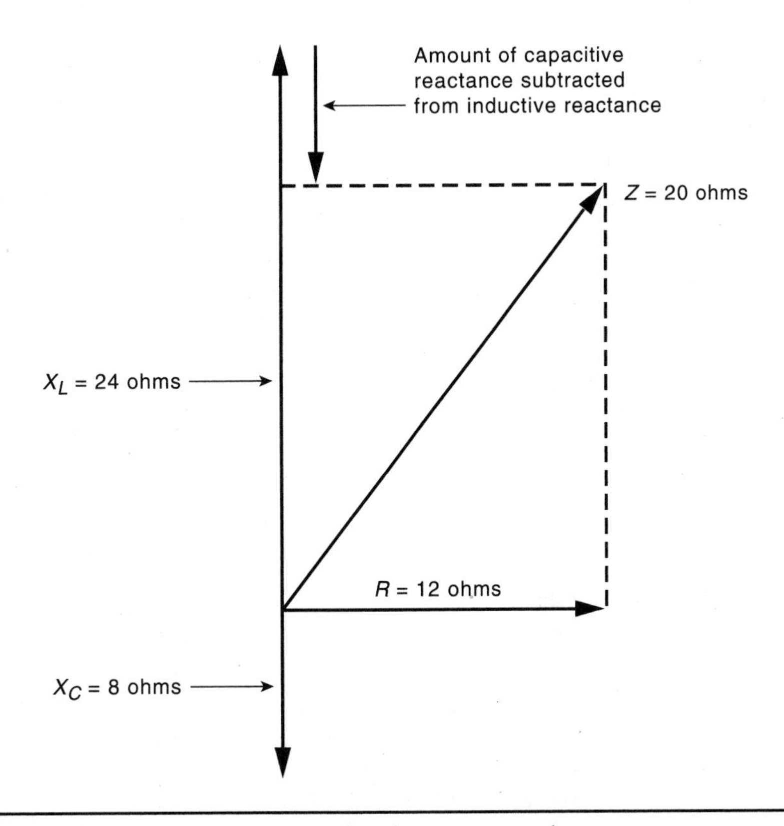

Figure 22–3 Vector addition is used to determine impedance.

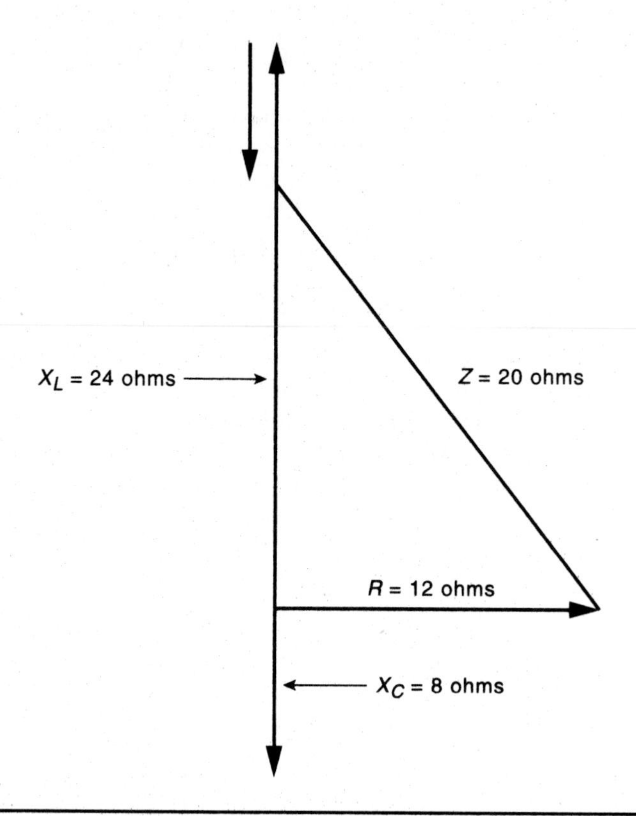

Figure 22–4 Right triangle formed by circuit impedance.

$$Z = \sqrt{R^2 + (X_L - X_C)^2}$$

$$Z = \sqrt{12^2 + (24 - 8)^2}$$

$$Z = \sqrt{12^2 + 16^2}$$

$$Z = \sqrt{144 + 256}$$

$$Z = \sqrt{400}$$

$$Z = 20\ \Omega$$

In this formula, the capacitive reactance is subtracted from the inductive reactance and then the difference is squared. If the capacitive reactance should be a larger value than inductive reactance, the difference will be a negative number. This will have no effect on the answer, however, because the square of a negative or positive number will always be positive.

Total Current (*I*)

The total current flow through the circuit can now be computed using the formula:

$$I = \frac{E}{Z}$$

$$I = \frac{240}{20}$$

$$I = 12 \text{ amps}$$

In a series circuit, the current flow is the same at any point in the circuit. Therefore, 12 amps flow through each of the circuit components.

Resistive Voltage Drop (E_R)

The voltage drop across the resistor can be computed using the formula:

$$E_R = I \times R$$

$$E_R = 12 \times 12$$

$$E_R = 144 \text{ volts}$$

Watts (*P*)

The true power of the circuit can be computed using any of the pure resistive values. In this example, true power will be found using the formula:

$$P = E_R \times I$$

$$P = 144 \times 12$$

$$P = 1728 \text{ watts}$$

Inductance (*L*)

The amount of inductance in the circuit can be computed using the formula:

$$L = \frac{X_L}{2\pi F}$$

$$L = \frac{24}{377}$$

$$L = 0.0637 \text{ henry}$$

Voltage Drop Across the Inductor (E_L)

The amount of voltage drop across the inductor can be computed using the formula:

$$E_L = I \times X_L$$

$$E_L = 12 \times 24$$

$$E_L = 288 \text{ volts}$$

Notice that the voltage drop across the inductor is greater than the applied voltage.

Inductive VARs ($VARs_L$)

The amount of reactive power of the inductor can be computed by using inductive values.

$$VARs_L = E_L \times I$$

$$VARs_L = 288 \times 12$$

$$VARs_L = 3456$$

Capacitance (C)

The amount of capacitance in the circuit can be computed by using the formula:

$$C = \frac{1}{2\pi FX_C}$$

$$C = \frac{1}{377 \times 8}$$

$$C = \frac{1}{3016}$$

$$C = 0.0003316 \text{ farads, or } 331.6\ \mu\text{F}$$

Voltage Drop Across the Capacitor (E_C)

The voltage dropped across the capacitor can be computed using the formula:

$$E_C = I \times X_C$$

$$E_C = 12 \times 8$$

$$E_C = 96 \text{ volts}$$

Capacitive VARs ($VARs_C$)

The amount of capacitive VARs can be computed using the formula:

$$VARs_C = E_C \times I$$

$$VARs_C = 96 \times 12$$

$$VARs_C = 1152$$

Apparent Power (VA)

The volt-amperes (apparent power) can be computed by multiplying the applied voltage and the circuit current.

$$VA = E_T \times I$$

$$VA = 240 \times 12$$

$$VA = 2880$$

The apparent power can also be found by vector addition of true power, inductive VARs and capacitive VARs, as in Figure 22–5. As with the addition of resistance, inductive reactance, and capacitive reactance, inductive VARs ($VARs_L$) and capacitive VARs ($VARs_C$) are 180 degrees out of phase with each other. This will result in the elimination of the smaller and a reduction of the larger. The following formula can be used to determine apparent power.

$$VA = \sqrt{P^2 + (VARs_L - VARs_C)^2}$$

$$VA = \sqrt{1728^2 + (3456 - 1152)^2}$$

$$VA = \sqrt{1728^2 + 2304^2}$$

$$VA = \sqrt{8294400}$$

$$VA = 2880$$

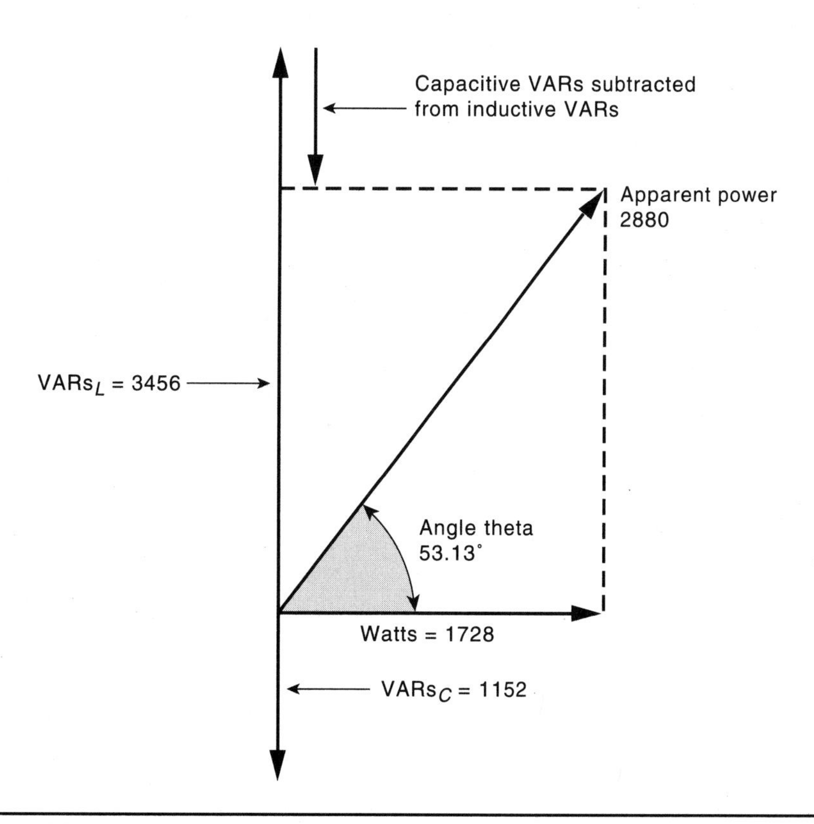

Figure 22–5 Vector addition of apparent, true, and reactive power.

Power Factor (*PF*)

The power factor can be computed by dividing the true power of the circuit by the apparent power. The answer is multiplied by 100 to change the decimal into a percent.

$$PF = \frac{P}{VA} \times 100$$

$$PF = \frac{1728}{2880} \times 100$$

$$PF = 0.60 \times 100$$

$$PF = 60\%$$

Angle Theta

The power factor is the cosine of angle theta.

$$\cos \angle\varnothing = 0.60$$

$$\angle\varnothing = 53.13°$$

Example Circuit #2

An RLC series circuit contains a capacitor with a capacitance of 66.3 µF, an inductor with an inductance of 0.0663 henrys, and a resistor with a value of 8 ohms connected to a 120-volt, 60-Hz line (Figure 22–6). How much current will flow in this circuit?

Capacitance and Inductive Reactance (X_C and X_L)

The first step in solving this problem is to find the values of capacitive and inductive reactance.

$$X_C = \frac{1}{2\pi FC}$$

$$X_C = \frac{1}{377 \times 0.0000663}$$

$$X_C = 40\ \Omega$$

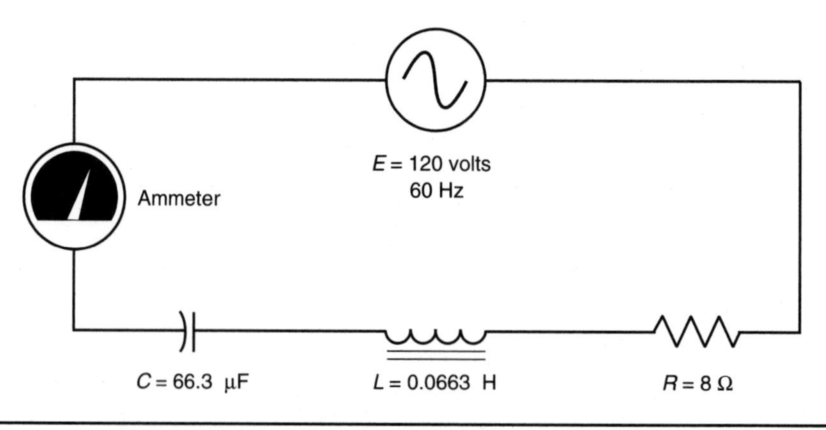

Figure 22–6 Example circuit #2.

$$X_L = 2\pi FL$$

$$X_L = 377 \times 0.0663$$

$$X_L = 24.995\Omega$$

Impedance (*Z*)

Now that the capacitive and inductive reactance values are known, the circuit impedance can be found using the following formula:

$$Z = \sqrt{R^2 + (X_L - X_C)^2}$$

$$Z = \sqrt{8^2 + (24.995 - 40)^2}$$

$$Z = \sqrt{62 + 225.15}$$

$$Z = 17\ \Omega$$

Now that the circuit impedance is known, the current flow can be found using Ohm's law.

$$I_T = \frac{E_T}{Z}$$

$$I_T = \frac{120}{17}$$

$$I_T = 7.059 \text{ amps}$$

Series Resonance (F_R)

When an inductor and capacitor are connected in series, Figure 22–7, there will be one frequency at which the inductive reactance and capacitive reactance will become equal. The reason for this is that as frequency increases, inductive reactance increases and capacitive

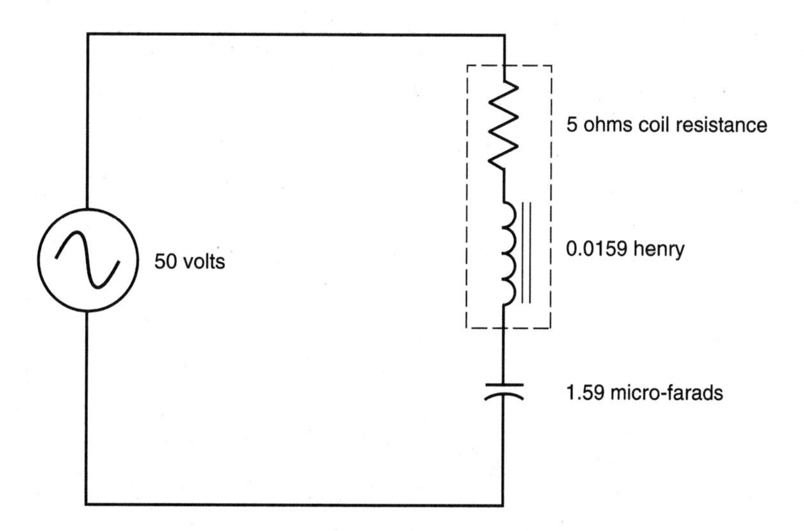

Figure 22–7 LC series circuit.

reactance decreases. The point at which both reactances become equal is called resonance. Resonant circuits are used to provide great increases of current and voltage at the resonant frequency. The following formula can be used to determine the resonant frequency when the values of L and C are known.

$$F_R = \frac{1}{2\pi\sqrt{LC}}$$

where F_R = frequency at resonance
L = inductance in henrys
C = capacitance in farads

In the circuit shown in Figure 22–7, an inductor has an inductance of 0.0159 henry and a wire resistance in the coil of 5 ohms. The capacitor connected in series with the inductor has a capacitance of 1.59 μF. This circuit will reach resonance at 1000 Hz when both the inductor and capacitor produce reactances of 100 ohms. At this point, the two reactances are equal and opposite in direction and the only current limiting factor in the circuit is the 5 ohms of wire resistance in the coil, Figure 22–8.

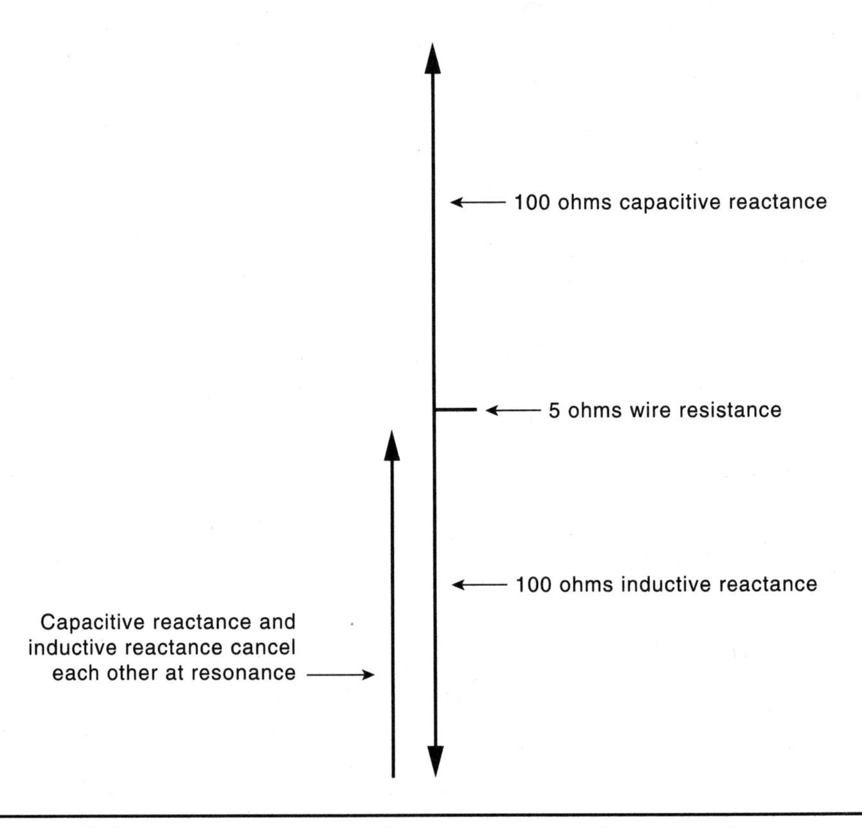

Figure 22–8 Inductive reactance and capacitive reactance become equal at resonance.

During the period of time that the circuit is not at resonance, current flow is limited by the combination of inductive reactance and capacitive reactance. At 600 Hz the inductive reactance will be 59.94 ohms and the capacitive reactance will be 166.83 ohms. The total circuit impedance will be

$$Z = \sqrt{R^2 + (X_L - X_C)^2}$$

$$Z = \sqrt{5^2 + (59.94 - 166.83)^2}$$

$$Z = 107\ \Omega$$

If 50 volts are applied to the circuit, the current flow will be 0.467 amps (50 / 107).

If the frequency is greater than 1000 Hz the inductive reactance will increase and the capacitive reactance will decrease. At a frequency of 1400 Hz, for example, the inductive reactance has become 139.86 ohms and the capacitive reactance has become 71.5 ohms. The total impedance of the circuit at this point will be 68.54 ohms. The circuit current will be 0.729 amps (50 / 68.54).

When the circuit reaches resonance, the current will suddenly increase to 10 amps due to the fact that the only current limiting factor is the 5 ohms of wire resistance (50 volts / 5 Ω = 10 amps). A graph illustrating the effect of current in a resonant circuit is shown in Figure 22–9.

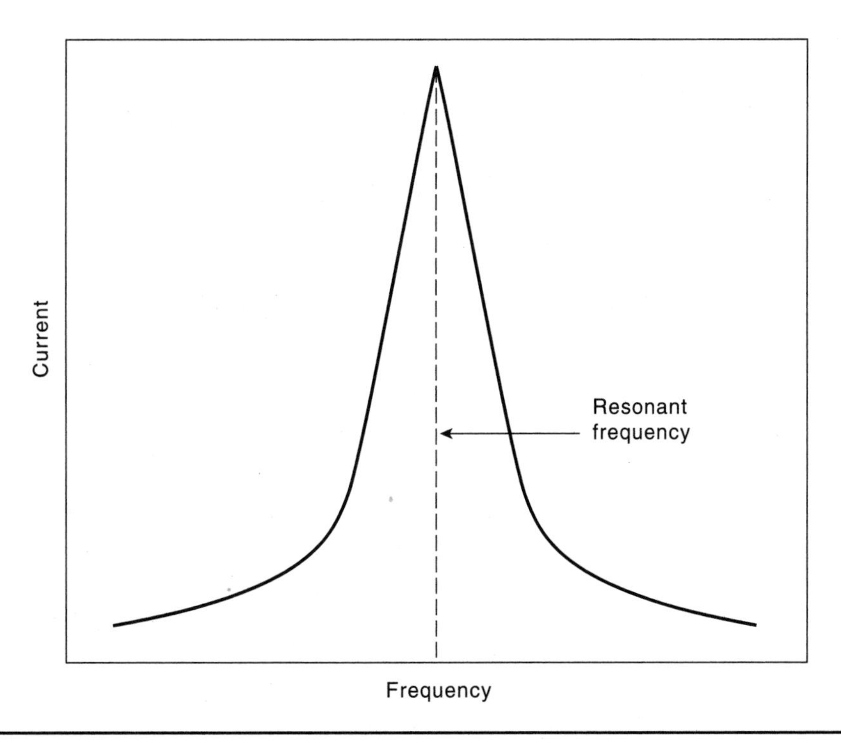

Figure 22–9 The current increases sharply at the resonant frequency.

Although inductive and capacitive reactance cancel each other at resonance, each is still a real value. In this example, both the inductive reactance and capacitive reactance have an ohmic value of 100 ohms at the resonant frequency. The voltage drop across each component will be proportional to the reactance and the amount of current flow. If voltmeters were to be connected across each component, a voltage of 1000 volts would be seen (10 amps × 100 Ω = 1000 volts).

Bandwidth

The rate of current increase and decrease is proportional to the Q of the components in the circuit.

$$B = \frac{F_R}{Q}$$

where B = bandwidth
Q = q of circuit components
F_R = frequency at resonance

High Q components result in a sharp increase of current, as illustrated by the curve of Figure 22–9. Not all series resonant circuits produce as sharp an increase or decrease of current, as illustrated in Figure 22–9. The term used to describe this rate of increase or decrease is *bandwidth.* Bandwidth is a frequency measurement. It is the difference between the two frequencies at which the current is at a value of 0.707 of the maximum current value, Figure 22–10.

$$B = F_2 - F_1$$

Assume the circuit producing the curve in Figure 22–10 reaches resonance at a frequency of 1000 Hz. Also assume that the circuit reaches a maximum value of 1 amp at resonance. The bandwidth of this circuit can be determined by finding the lower and upper frequencies on either side of 1000 Hz at which the current reaches a value of 0.707 of the maximum value. In this illustration, that will be 0.707 amp (1 × 0.707 = 0.707). Assume the lower frequency value to be 995 Hz and the upper value to be 1005 Hz. This circuit has a bandwidth of 10 Hz (1005 – 995 = 10). When resonant circuit are constructed with components that have a relatively high Q, the difference between the two frequencies is small. These circuits are said to have a narrow bandwidth.

If a resonant circuit using components with a lower Q rating is constructed, the current will not increase as sharply as shown in Figure 22–11. In this circuit, it is assumed that resonance is reached at a frequency of 1200 Hz, and the maximum current flow at resonance is 0.8 amp. The bandwidth is determined by the difference between the two frequencies at which the current is at a value of 0.566 amp (0.8 × 0.707 = 0.566). Assume the lower frequency to be 1150 Hz and the upper frequency to be 1250 Hz. This circuit has a bandwidth of 100 Hz (1250 – 1150 = 100). This circuit is said to have a wide bandwidth.

Formulas for finding values in series circuits containing resistance, inductance, and capacitance are shown in Figure 22–12.

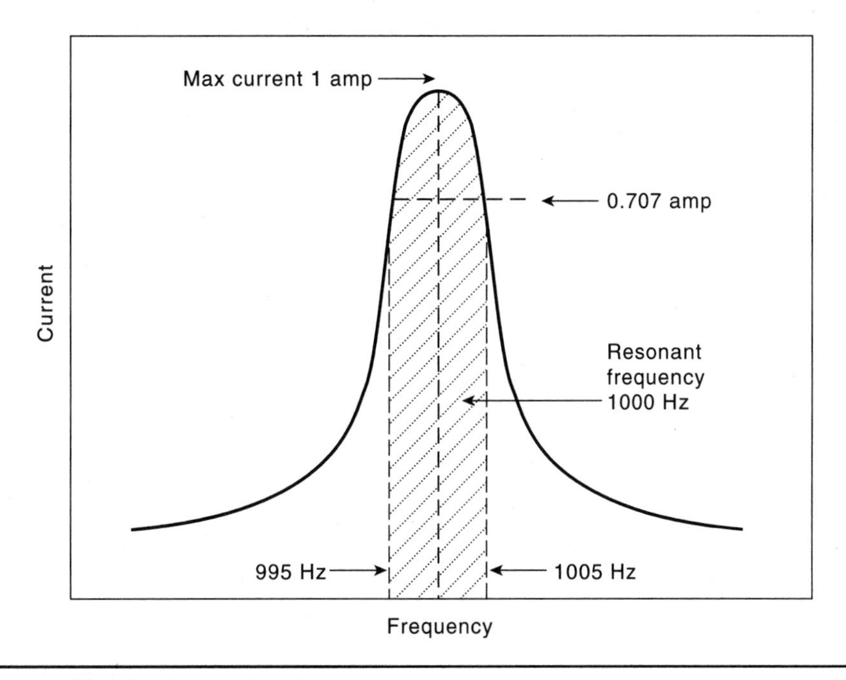

Figure 22–10 Narrow-band resonant circuit.

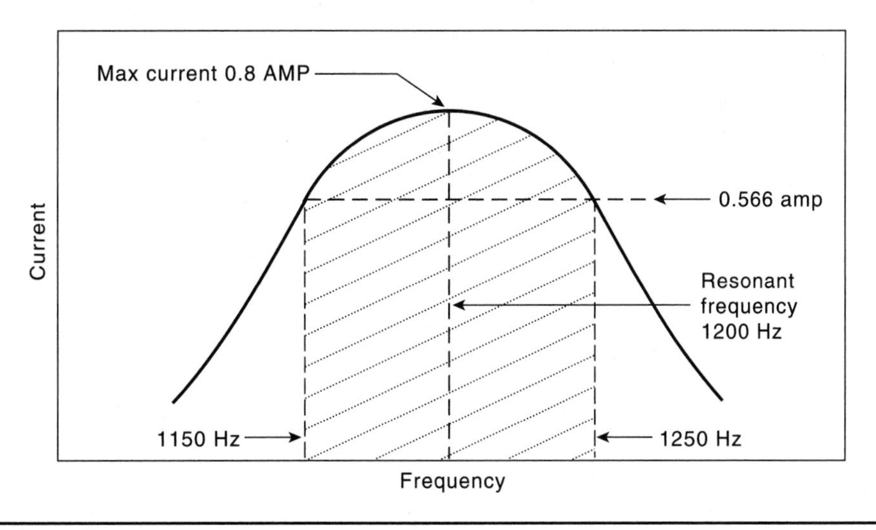

Figure 22–11 Wide-band resonant circuit.

$$E_T = \sqrt{E_{R^2}\left(E_L - E_C\right)^2} \qquad E_T = \frac{VA}{I_T}$$

$$E_T = I_T \times Z \qquad E_T = \frac{E_R}{PF}$$

$$Z = \sqrt{R^2 + \left(X_L - X_C\right)^2} \qquad Z = \frac{VA}{I_{T^2}}$$

$$Z = \frac{E_T}{I_T} \qquad Z = \frac{R}{PF}$$

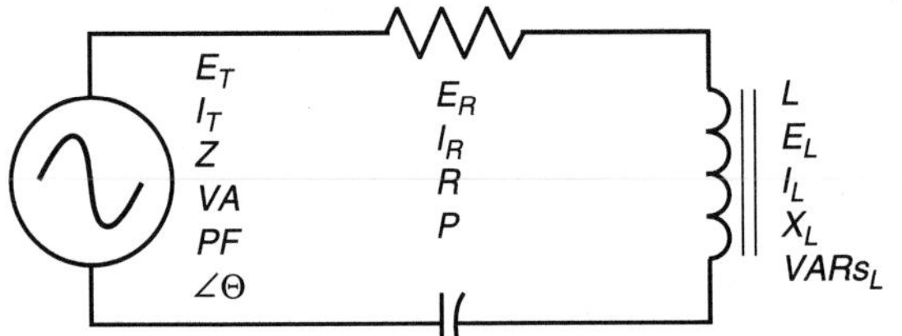

E_T
I_T
Z
VA
PF
$\angle\Theta$

E_R
I_R
R
P

L
E_L
I_L
X_L
$VARs_L$

C
E_C
I_C
X_C
$VARs_C$

$$I_T = I_R = I_L = I_C \qquad I_T = \frac{VA}{E_T}$$

$$I_T = \frac{E_T}{Z} \qquad I_T = \sqrt{\frac{VA}{Z}}$$

$$VA = E_T \times I_R \qquad VA = \frac{P}{PF}$$

$$VA = I_T^{\,2} \times Z \qquad VA = \frac{E_T^{\,2}}{Z}$$

$$VA = \sqrt{P^2 + \left(VARs_L - VARs_C\right)^2}$$

$$I_L = \frac{E_L}{X_L}$$

$$I_L = \frac{VARs_L}{E_L}$$

$$I_L = \sqrt{\frac{VARs_L}{X_L}}$$

$$I_L = I_R = I_T = I_C$$

$$PF = \frac{R}{Z}$$

$$PF = \frac{P}{VA}$$

$$PF = \frac{E_R}{E_T}$$

$$PF = \cos\angle\theta$$

$$P = E_R \times I_R \qquad P = VA \times PF$$

$$P = \sqrt{VA^2 - \left(VARs_L - VARs_C\right)^2}$$

$$P = \frac{E_R^{\,2}}{R} \qquad E_R = \sqrt{P \times R}$$

$$P = I_R^{\,2} \times R$$

$$E_R = I_R \times R$$

$$E_R = \frac{P}{I_R}$$

$$E_R = \sqrt{E_T^{\,2} - \left(E_L - E_C\right)^2}$$

$$E_R = E_T \times PF$$

$$E_L = I_L \times X_L$$

$$E_L = \sqrt{VARs_L \times X_L}$$

$$E_L = \frac{VARs_L}{I_L}$$

$$I_R = I_T = I_C = I_L$$

$$I_R = \frac{E_R}{R}$$

$$I_R = \frac{P}{E_R}$$

$$I_R = \sqrt{\frac{P}{R}}$$

$$R = \sqrt{Z^2 - \left(X_L - X_C\right)^2}$$

$$R = \frac{E_R}{I_R} \qquad R = Z \times PF$$

$$R = \frac{E_{R^2}}{P} \qquad R = \frac{P}{I_{R^2}}$$

$$E_C = I_C \times X_C$$

$$E_C = \sqrt{VARs_C \times X_C}$$

$$E_C = \frac{VARs_C}{I_C}$$

$$C = \frac{1}{2 \times \pi \times F \times X_C}$$

$$VARs_L = E_L \times I_L$$

$$VARs_L = \frac{E_L^{\,2}}{X_L}$$

$$VARs_L = I_L^{\,2} \times X_L$$

$$I_C = I_R = I_T = I_L \qquad X_C = \frac{1}{2 \times \pi \times F \times V}$$

$$I_C = \frac{E_C}{X_C}$$

$$I_C = \frac{VARs_C}{E_C}$$

$$I_C = \sqrt{\frac{VARs_C}{X_C}}$$

$$X_C = \frac{E_C^{\,2}}{VARs_C} \qquad X_C = \frac{VARs_C}{I_C^{\,2}}$$

$$X_C = \frac{E_C}{I_C}$$

$$VARs_C = E_C \times I_C$$

$$VARs_C = I_C^{\,2} \times X_C$$

$$VARs_C = \frac{E_C^{\,2}}{X_C}$$

$$L = \frac{X_L}{2 \times \pi \times F} \qquad X_L = 2\pi \times F \times L$$

$$X_L = \frac{E_L}{I_L} \qquad X_L = \frac{VARs_L}{I_L^{\,2}}$$

$$X_L = \frac{E_L^{\,2}}{VARs_L}$$

Figure 22–12 RLC series circuit formulas.

CHAPTER 23

Resistive–Inductive–Capacitive Parallel Circuits

When an alternating current circuit contains elements of resistance, inductance, and capacitance connected in parallel, the voltage dropped across each element will be the same and in phase. The current flowing through each branch, however, will be out of phase with each other, as shown in Figure 23–1. The current flowing through a pure resistive element will be in phase with the applied voltage. The current flowing through a pure inductive element lags the applied voltage by 90 electrical degrees, and the current flowing through a pure capacitive element will lead the voltage by 90 electrical degrees. The phase angle difference between the applied voltage and the total current is determined by the ratio of resistance, inductance, and capacitance connected in parallel. As with RLC series circuit, if the inductive VARs are greater than the capacitive VARs, the current will lag the voltage and the power factor will be lagging. If the capacitive VARs are greater, the current will lead the voltage and the power factor will be leading.

Example Circuit #1

An RLC parallel circuit is shown in Figure 23–2. It is assumed that the circuit is connected to a 240-volt, 60-Hz line. The resistor has a resistance of 12 ohms, the inductor has an inductive reactance of 8 ohms and the capacitor has a capacitive reactance of 16 ohms. The following unknown value will be computed.

Z = impedance of the circuit
I_R = current flow through the resistor
P = true power (watts)
L = inductance of the inductor
I_L = current flow through the inductor
$VARs_L$ = reactive power of the inductor
C = capacitance
I_C = current flow through the capacitor
$VARs_C$ = reactive power of the capacitor
I_T = total circuit current
VA = volt-amperes (apparent power)
PF = power factor
$\angle\varnothing$ = angle theta

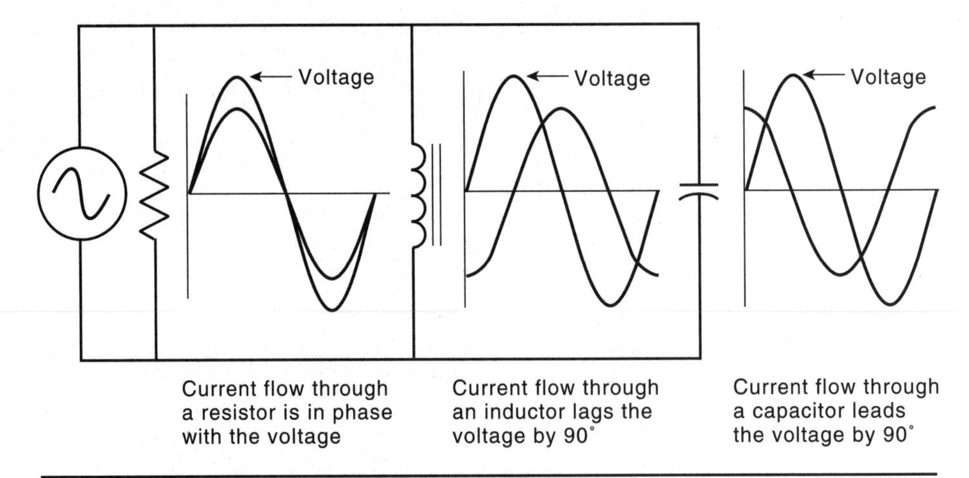

Figure 23–1 Voltage and current relationship in an RLC parallel circuit.

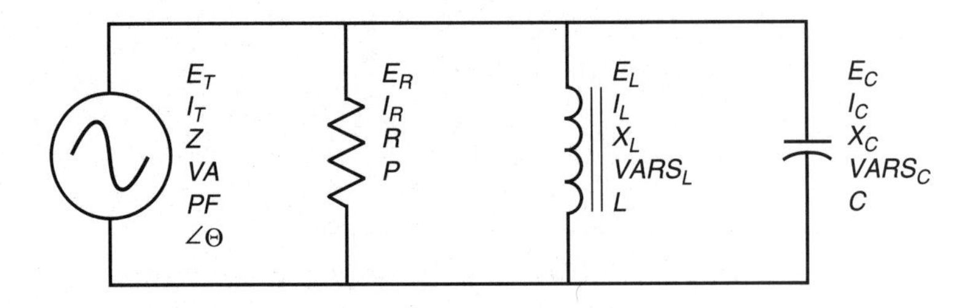

Figure 23–2 RLC parallel circuit.

Impedance (Z)

The impedance of the circuit is the reciprocal of the sum of the reciprocals of each leg. Since these values are out of phase with each other, vector addition must be used.

$$Z = \frac{1}{\sqrt{\left(\frac{1}{R}\right)^2 + \left(\frac{1}{X_L} - \frac{1}{X_C}\right)^2}}$$

$$Z = \frac{1}{\sqrt{\left(\frac{1}{12}\right)^2 + \left(\frac{1}{8} - \frac{1}{16}\right)^2}}$$

$$Z = \frac{1}{\sqrt{0.006944 + 0.003906}}$$

$$Z = \frac{1}{\sqrt{0.01085}}$$

$$Z = \frac{1}{0.10416}$$

$$Z = 9.6\ \Omega$$

Resistive Current (I_R)

The first unknown value to be found will be the current flow through the resistor. This can be computed by using the formula:

$$I_R = \frac{E}{R}$$

$$I_R = \frac{240}{12}$$

$$I_R = 20 \text{ amps}$$

True Power (*P*)

The true power or watts can be computed using the following formula:

$$P = E \times I_R$$

$$P = 240 \times 20$$

$$P = 4800 \text{ watts}$$

Inductive Reactance (I_L)

The amount of current flow through the inductor can be computed using the following formula:

$$I_L = \frac{E}{X_L}$$

$$I_L = \frac{240}{8}$$

$$I_L = 30 \text{ amps}$$

Inductive VARs ($VARs_L$)

The amount of reactive power or VARs produced by the inductor can be computed using the formula:

$$VARs_L = E \times I_L$$

$$VARs_L = 240 \times 30$$

$$VARs_L = 7200$$

Inductance (L)

The amount of inductance in the circuit can be computed using the formula:

$$L = \frac{X_L}{2\pi F}$$

$$L = \frac{30}{2 \times 3.1416 \times 60}$$

$$L = \frac{30}{377}$$

$$L = 0.0212 \text{ henry}$$

Capacitive Current (I_C)

The current flow through the capacitor can be computed using the formula:

$$I_C = \frac{E}{X_C}$$

$$I_C = \frac{240}{16}$$

$$I_C = 15 \text{ amps}$$

Capacitance (C)

The amount of circuit capacitance can be computed using the formula:

$$C = \frac{1}{2\pi F X_C}$$

$$C = 0.0001658 \text{ farads, or } 165.8\ \mu\text{F}$$

Capacitive VARs ($VARs_C$)

The capacitive VARs can be computed using the formula:

$$VARs_C = E \times I_C$$

$$VARs_C = 240 \times 15$$

$$VARs_C = 3600$$

Total Circuit Current (I_T)

The amount of total current flow in the circuit can be computed by vector addition of the currents flowing through each leg of the circuit, Figure 23–3. The inductive current is 180 degrees out of phase with the capacitive current. These two currents will tend to cancel each other, resulting in the elimination of the smaller and reduction of the larger. The total circuit is the hypotenuse of the resulting right triangle, shown in Figure 23–4. The following formula can be used to find total circuit current.

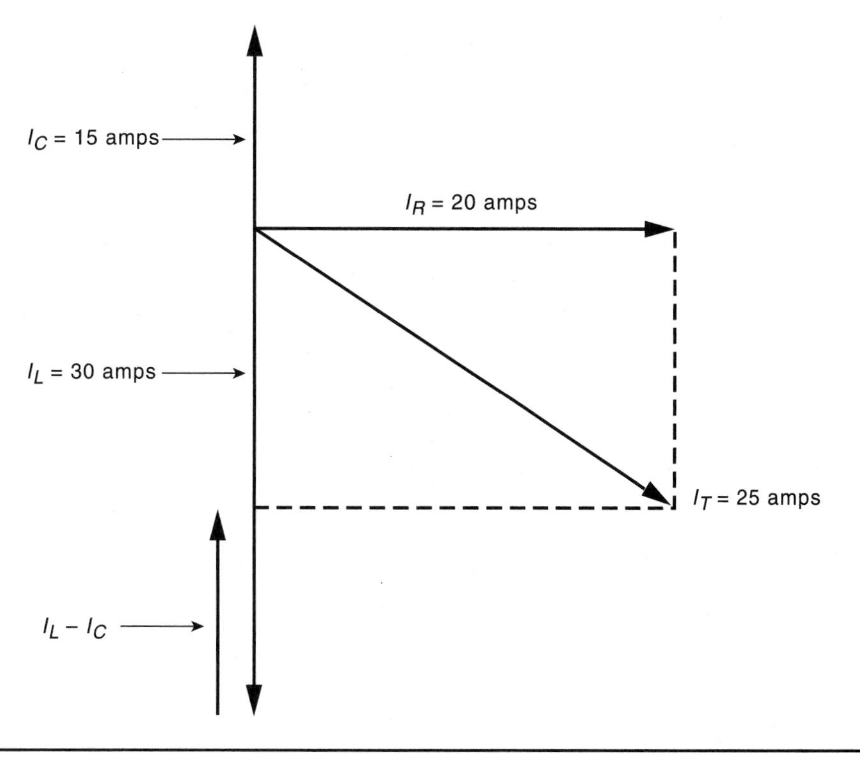

Figure 23–3 Vector diagram of currents in example circuit #1.

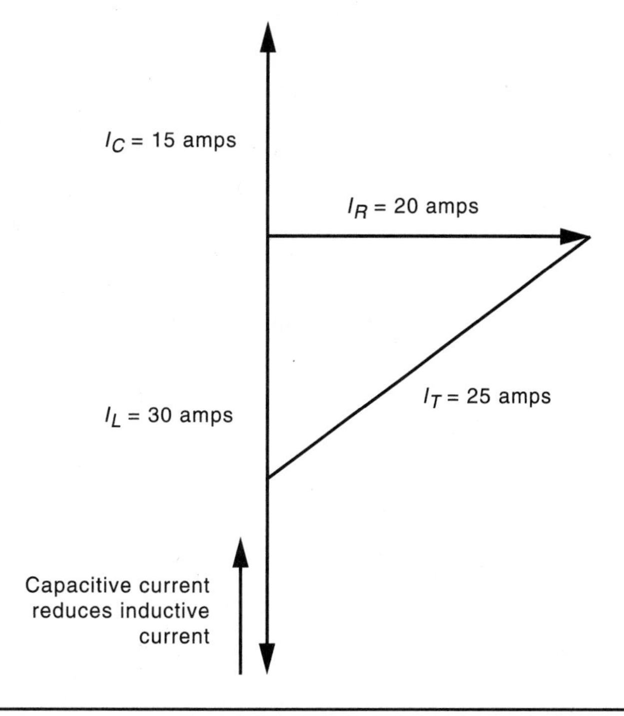

Figure 23–4 Inductive and capacitive currents cancel each other.

$$I_T = \sqrt{I_R^2 + (I_L - I_C)^2}$$

$$I_T = \sqrt{20^2 + (30 - 15)^2}$$

$$I_T = \sqrt{20^2 + 15^2}$$

$$I_T = \sqrt{400 + 225}$$

$$I_T = \sqrt{625}$$

$$I_T = 25 \text{ amps}$$

The total current could also be computed by using the value of impedance found earlier in the problem.

$$I_T = \frac{E}{Z}$$

$$I_T = \frac{240}{9.6}$$

$$I_T = 25 \text{ amps}$$

Apparent Power (*VA*)

Now that the total circuit current has been computed, the apparent power or volt–amps can be found using the formula

$$VA = E \times I_T$$

$$VA = 240 \times 25$$

$$VA = 6000$$

The apparent power can also be found by vector addition of the true power and reactive power.

$$VA = \sqrt{P^2 + (VARs_L - VARs_C)^2}$$

Power Factor (*PF*)

The power factor can now be computed using the formula

$$PF = \frac{P}{VA} \times 100$$

$$PF = \frac{4800}{6000} \times 100$$

$$PF = 0.80 \times 100$$

$$PF = 80\%$$

Angle Theta

The power factor is the cosine of angle theta. Angle theta is, therefore,

$$\cos \angle \varnothing = 0.80$$

$$\angle \varnothing = 36.87°$$

Example Circuit #2

In the circuit shown in Figure 23–5, a resistor, inductor, and capacitor are connected to a 1200-Hz power source. The resistor has a resistance of 18 ohms, the inductor has an inductance of 9.76 mH (0.00976 H), and the capacitor has a capacitance of 5.5 μF. What is the impedance of this circuit?

Solution

The first step in finding the impedance of this circuit is to find the values of inductive and capacitive reactance.

$$X_L = 2\pi FL$$

$$X_L = 2 \times 3.1416 \times 1200 \times 0.00976$$

$$X_L = 73.589\ \Omega$$

$$X_C = \frac{1}{2\pi FC}$$

$$X_C = \frac{1}{2 \times 3.1416 \times 1200 \times 0.0000055}$$

$$X_C = 24.114\ \Omega$$

The impedance of the circuit can now be computed using the formula:

$$Z = \frac{1}{\sqrt{\left(\frac{1}{R}\right)^2 + \left(\frac{1}{X_L} - \frac{1}{X_C}\right)^2}}$$

$$Z = \frac{1}{\sqrt{\left(\frac{1}{18}\right)^2 + \left(\frac{1}{74.589} - \frac{1}{24.114}\right)^2}}$$

$$Z = \frac{1}{\sqrt{0.0555^2 + (0.0134 - 0.0415)^2}}$$

$$Z = \frac{1}{0.00387}$$

$$Z = 16.075\ \Omega$$

Parallel Resonant Circuits

When values of inductive reactance and capacitive reactance become equal, they are said to be resonant. In a parallel circuit, inductive current and capacitive current cancel each other because they are 180 degrees out of phase with each other. This produces minimum line current at the point of resonance. An LC parallel circuit is shown in Figure 23–6. LC parallel circuits are often referred to as tank circuits. In the example circuit the inductor has an inductance of 0.0398 Henry and a wire resistance of 10 ohms. The capacitor has a

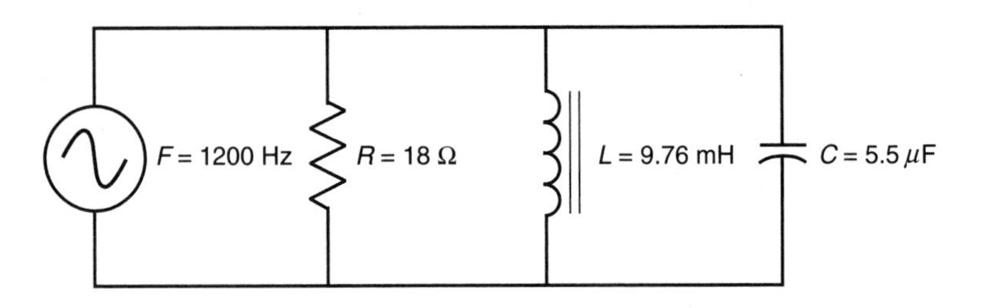

Figure 23–5 Example circuit #2.

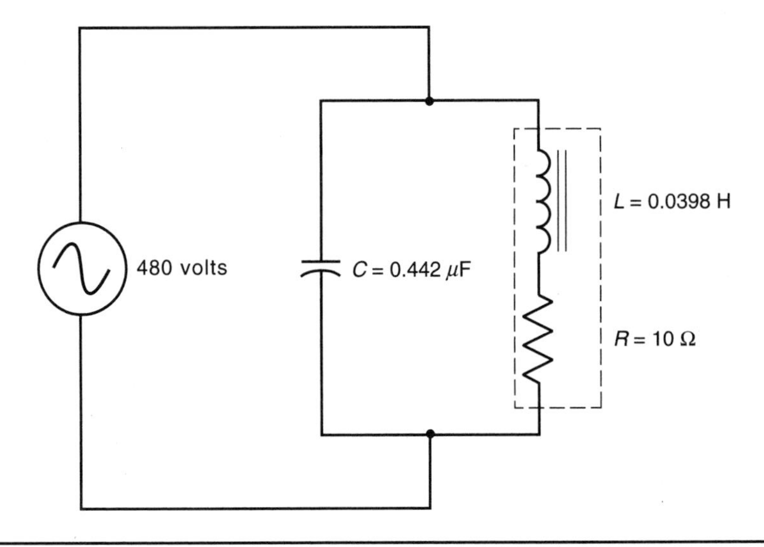

Figure 23–6 Parallel resonant circuit.

capacitance of 0.442 μF. This circuit will reach resonance at 1200 Hz when both the capacitor and inductor exhibit reactances of 300 ohms each.

Computing the values for a parallel resonant circuit is a bit more involved than computing the values for a series resonance circuit. In theory, when a parallel circuit reaches resonance, the total circuit current should reach zero and total circuit impedance should become infinite because the capacitive current and inductive currents cancel each other. In practice, the *Q* of the circuit components determines total circuit current and, therefore, total circuit impedance. Since capacitors generally have an extremely high *Q* by their very nature, the *Q* of the inductor is the determining factor (see Figure 23–7).

$$Q = \frac{I_{\text{Tank}}}{I_{\text{Line}}}$$

or

$$Q = \frac{X_L}{R}$$

In the example shown in Figure 23–6, the inductor has an inductive reactance of 300 ohms at the resonant frequency and a wire resistance of 10 ohms. The *Q* of this inductor

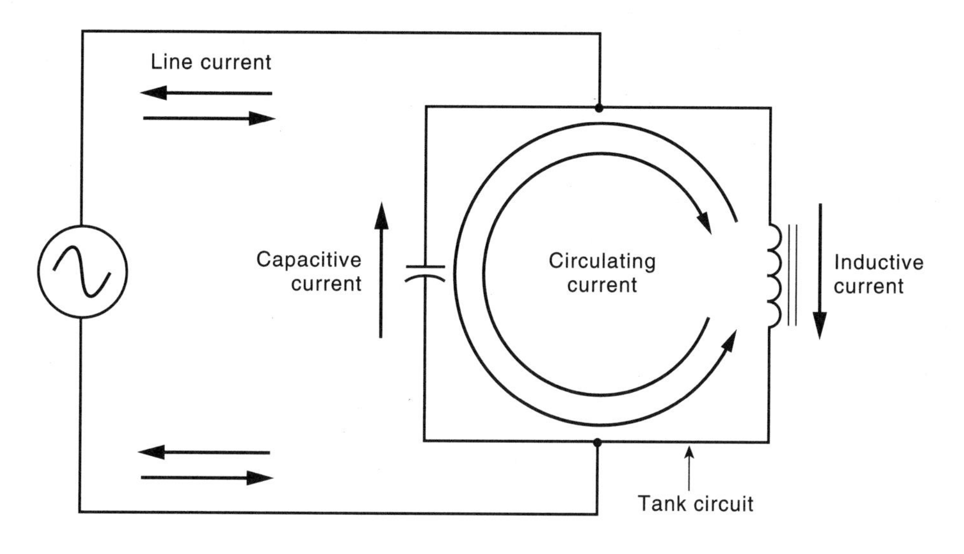

Figure 23–7 Current circulates inside the tank.

is 30 at the resonant frequency ($Q = X_L / R$). To determine the total circuit current at resonance, it is first necessary to determine the amount of current flow through each of the components at the resonant frequency. Since this is a parallel circuit, the inductor and capacitor will have the alternator voltage of 480 volts applied to them. At the resonant frequency of 1200 Hz, both the inductor and capacitor will have a current flow of 1.6 amps (480 volts / 300 ohms = 1.6 amps). The total current flow in the circuit will be the in-phase current caused by the wire resistance of the coil, Figure 23–8. This value can be computed by dividing the circulating current inside the LC parallel loop by the Q of the circuit. The total current in this circuit will be 0.0533 amps (1.6 / 30 = 0.0533). Now that the total circuit current is known, we can find the total impedance at resonance using Ohm's law.

$$Z = \frac{E}{I_T}$$

$$Z = \frac{480}{0.0533}$$

$$Z = 9006.63\ \Omega$$

Another method of computing total current for a tank circuit is to determine the true power in the circuit caused by the resistance of the coil. The resistive part of a coil is considered to be in series with the reactive part, as shown in Figure 23–6. Since this is a series connection, the current flow through the inductor is the same for both the reactive and resistive elements. The coil has a resistance of 10 ohms. The true power produced by the coil can be computed by using the formula

$$P = I^2 R$$

$$P = 1.6^2 \times 10$$

$$P = 25.6 \text{ watts}$$

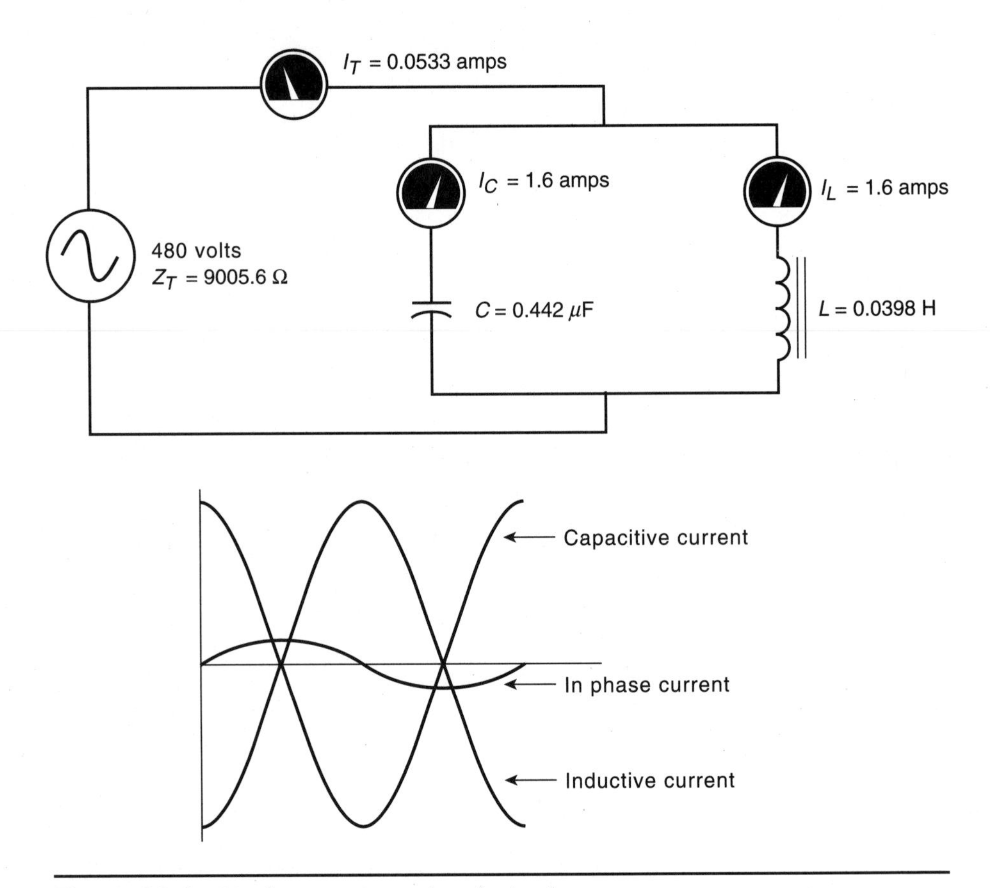

Figure 23–8 Total current is equal to the in-phase current.

Now that the true power is known, we can find the total circuit current using Ohm's law.

$$I_T = \frac{P}{E}$$

$$I_T = \frac{25.6}{480}$$

$$I_T = 0.0533 \text{ amps}$$

Charts illustrating the decrease of current and increase of impedance in a parallel resonant circuit are shown in Figure 23–9.

Bandwidth

The bandwidth for a parallel resonant circuit is determined in a similar manner as a series resonant circuit. The bandwidth of a parallel circuit is determined by computing the frequency either side of resonance at which the impedance is 0.707 of maximum. As in series resonant circuits, the Q of the parallel circuit determines the bandwidth. Circuits that have a high Q will have a narrow bandwidth and circuits with a low Q will have a wide bandwidth, as shown in Figure 23–10.

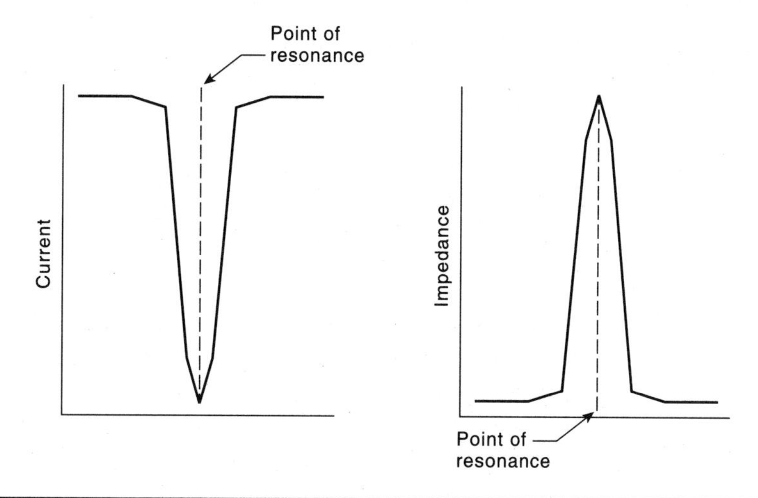

Figure 23–9 Characteristic curves of an LC parallel circuit at resonance.

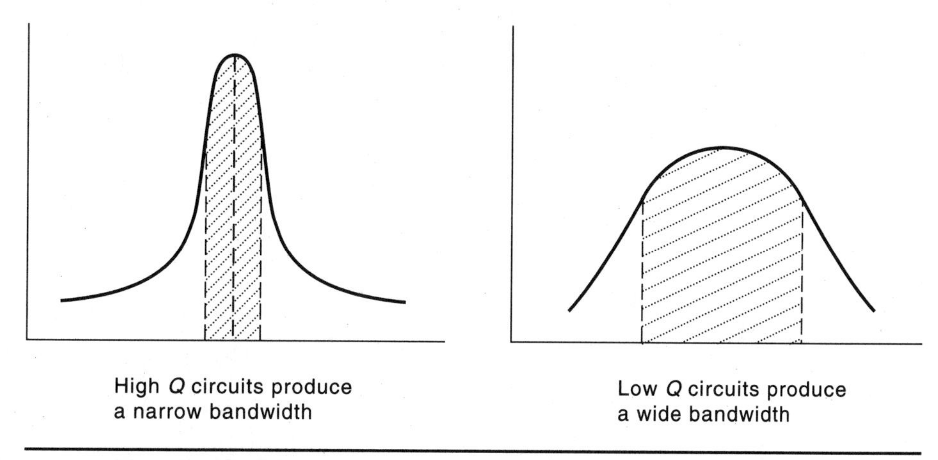

Figure 23–10 The *Q* of the circuit determines the bandwidth.

Induction Heating with Resonant Circuits

The tank circuit is often used when a large amount of current flow is needed. Recall that the formula for *Q* of a parallel resonant circuit is:

$$Q = \frac{I_{\text{Tank}}}{I_{\text{Line}}}$$

If this formula is changed, it will be seen that the current circulating inside the tank is equal to the line current times the *Q* of the circuit:

$$I_{\text{Tank}} = I_{\text{Line}} \times Q$$

A high *Q* circuit can produce an extremely high current inside the tank with very little line current. A good example of this is an induction heater used to heat pipe for tempering, as shown in Figure 23–11. In this example, the coil is the inductor and the pipe acts as the core of the inductor. The capacitor is connected in parallel with the coil to produce resonance at a desired frequency. The pipe is heated by eddy current induction. Assume that the coil has a Q of 10. If this circuit has a total current of 100 amperes, then 1000 amps of current flows in the tank. This 1000 amps is used to heat the pipe. Since the pipe acts as a core for the inductor, the inductance of the coil will change when the pipe is not in the coil. Therefore, the circuit is resonant only during the times that the pipe is in the coil.

Induction heaters of this type have another advantage over other methods that heat pipe with flames produced by oil- or gas-fired furnaces. When induction heating is used, the resonant frequency can be changed by adding or subtracting capacitance in the tank circuit. This ability to control the frequency greatly affects the tempering of metal. If the frequency is relatively low, 400 Hz or less, the metal is heated evenly. If the frequency is increased to 1000 Hz or greater, skin effect causes most of the heating effect to localize at the surface of the metal. This permits a hard coating to develop at the outer surface of the metal without greatly changing the temper of the inside of the metal, as shown in Figure 23–12.

Formulas for finding values in parallel circuits containing resistance, inductance, and capacitance are shown in Figure 23–13.

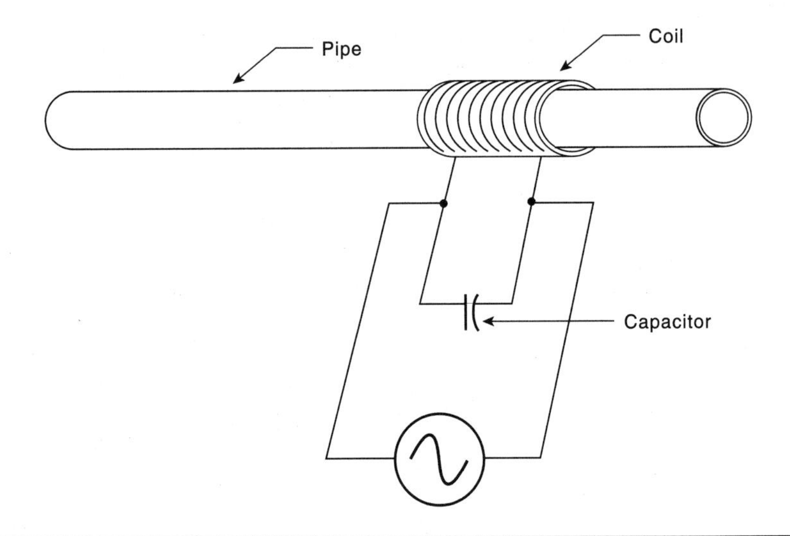

Figure 23–11 An induction heating system: The coil is the inductor and the pipe acts as the core.

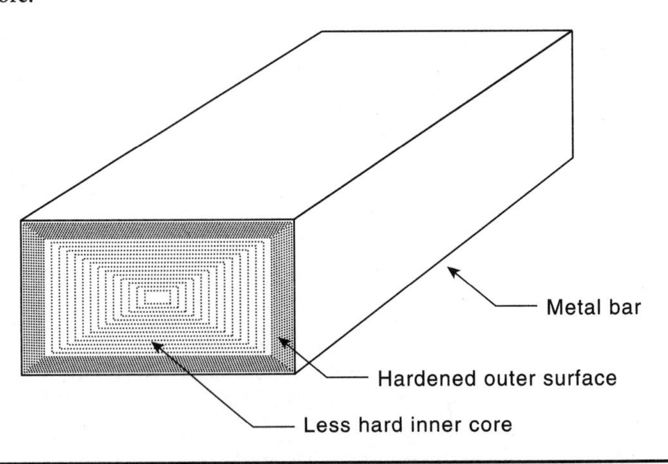

Figure 23–12 Frequency controls depth of heat penetration.

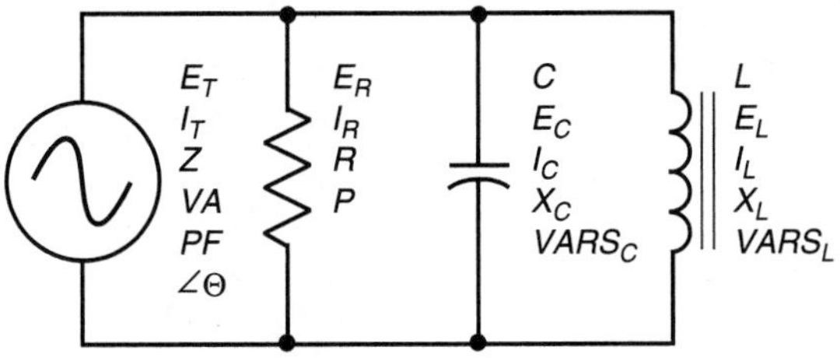

E_T I_T Z VA PF $\angle\Theta$

E_R I_R R P

C E_C I_C X_C $VARS_C$

L E_L I_L X_L $VARS_L$

$$E_T = E_R = E_L = E_C$$

$$E_T = I_T \times Z$$

$$E_T = \frac{VA}{I_T}$$

$$E_T = \sqrt{VA \times Z}$$

$$E_C = \frac{VARs_C}{I_C}$$

$$E_C = I_C \times X_C$$

$$E_C = E_T = E_R = E_L$$

$$E_C = \sqrt{VARs_C \times X_C}$$

$$Z = \frac{1}{\sqrt{\left(\frac{1}{R}\right)^2 + \left(\frac{1}{X_L} - \frac{1}{X_C}\right)^2}} \qquad Z = \frac{VA}{{I_T}^2}$$

$$Z = \frac{E_T}{I_T} \qquad Z = \frac{{E_T}^2}{VA} \qquad Z = R \times PF$$

$$I_T = \sqrt{{I_R}^2 + \left(I_L - I_C\right)^2} \qquad I_T = \frac{VA}{E_T} \qquad I_T = \sqrt{\frac{VA}{Z}}$$

$$I_T = \frac{E_T}{Z} \qquad I_T = \frac{I_R}{PF}$$

$$I_C = \frac{E_C}{X_C}$$

$$I_C = \frac{VARs_C}{E_C}$$

$$I_C = \sqrt{\frac{VARs_C}{X_C}}$$

$$VA = E_T \times I_T$$

$$VA = {I_T}^2 \times Z$$

$$VA = \frac{{E_T}^2}{Z}$$

$$PF = \frac{Z}{R}$$

$$PF = \frac{P}{VA}$$

$$PF = \frac{I_R}{I_T}$$

$$PF = \cos \angle\theta$$

$$E_L = I_L \times X_L$$

$$E_L = \frac{VARs_L}{I_L}$$

$$E_L = E_T = E_R = E_C$$

$$E_L = \sqrt{VARs_L \times X_L}$$

$$L = \frac{X_L}{2 \times \pi \times F}$$

$$VA = \sqrt{P^2 + \left(VARs_L - VARs_C\right)^2}$$

$$VA = \frac{P}{PF}$$

$$I_L = \frac{E_L}{X_L}$$

$$X_L = \frac{E_L}{I_L}$$

$$I_L = \frac{VARs_L}{E_L}$$

$$I_L = \sqrt{\frac{VARs_L}{X_L}}$$

$$X_L = 2 \times \pi \times F \times L$$

$$X_L = \frac{{E_L}^2}{VARs_L}$$

$$X_L = \frac{VARs_L}{{I_L}^2}$$

$$VARs_L = \frac{{E_L}^2}{X_L}$$

$$VARs_L = {I_L}^2 \times X_L$$

$$VARs_L = E_L \times I_L$$

$$X_C = \frac{E_C}{I_C}$$

$$X_C = \frac{{E_C}^2}{VARs_C}$$

$$X_C = \frac{VARs_C}{{I_C}^2}$$

$$E_R = I_R \times R$$

$$E_R = \sqrt{P \times R}$$

$$E_R = \frac{P}{I_R}$$

$$E_R = E_T = E_L = E_C$$

$$I_R = \sqrt{{I_T}^2 - \left(I_L - I_C\right)^2} \qquad I_R = I_T \times PF \qquad X_C = \frac{1}{2 \times \pi \times F \times C}$$

$$I_R = \frac{E_R}{R} \qquad R = \frac{E_R}{I_R}$$

$$I_R = \frac{P}{E_R}$$

$$I_R = \sqrt{\frac{P}{R}}$$

$$R = \frac{1}{\left(\frac{1}{Z}\right)^2 - \left(\frac{1}{X_L} - \frac{1}{X_C}\right)^2}$$

$$R = \frac{P}{{I_R}^2} \qquad R = \frac{Z}{PF} \qquad R = \frac{{E_R}^2}{P} \qquad P = \frac{{E_R}^2}{R}$$

$$P = \sqrt{VA^2 - \left(VARs_L - VARs_C\right)^2} \qquad P = E_R \times I_R \qquad P = {I_R}^2 \times R \qquad P = VA \times PF$$

$$VARs_C = E_C \times I_C \qquad VARs_C = {I_C}^2 \times X_C \qquad VARs_C = \frac{{E_C}^2}{X_C}$$

Figure 23–13 Formulas for RLC parallel circuits.

CHAPTER 24

Power Factor Correction for Single-Phase Circuits

Another very common application for LC parallel circuits is the correction of power factor. Assume that a motor is connected to a 240-volt single-phase line with a frequency of 60 Hz. An ammeter indicates a current flow of 10 amperes and a wattmeter indicates a true power of 1630 watts when the motor is at full load. In this problem, the existing power factor will be determined and then the amount of capacitance needed to correct the power factor will be computed.

Although an AC motor is an inductive device, when it is loaded it must produce true power to overcome the load connected to it. For this reason, the motor appears to be a resistance connected in series with an inductance, Figure 24–1. Also, the inductance of the motor remains constant regardless of the load connected to it. Recall that true power or watts can exist only when electrical energy is converted into some other form. A resistor produces true power because it converts electrical energy into heat energy. In the case of a motor, electrical energy is being converted into both heat energy and mechanical energy. When a motor is operated at a no-load condition, the current is relatively small as compared to the full load current. At no load, most of the current is used to magnetize the iron core of the stator and rotor. This current is inductive and is 90 degrees out of phase with the voltage. The only true power produced at no load is caused by motor losses, such as eddy currents being induced into the iron core, the heating effect caused by the resistance of the wire in the windings, hysteresis losses, and the small amount of mechanical energy being produced to overcome the losses of bearing friction and windage. At no load, the motor would appear to be a circuit containing a large amount of inductance and a small amount of resistance.

As load is added to the motor, more electrical energy is converted into mechanical energy to drive the load. The increased current used to produce the mechanical energy is in phase with the voltage. This causes the circuit to appear to be more resistive. By the time the motor reaches full load, the circuit appears to be more resistive than inductive. Notice that as load is added or removed, only the resistive value of the motor changes, which means that once the power factor has been corrected, it will remain constant regardless of the motor load.

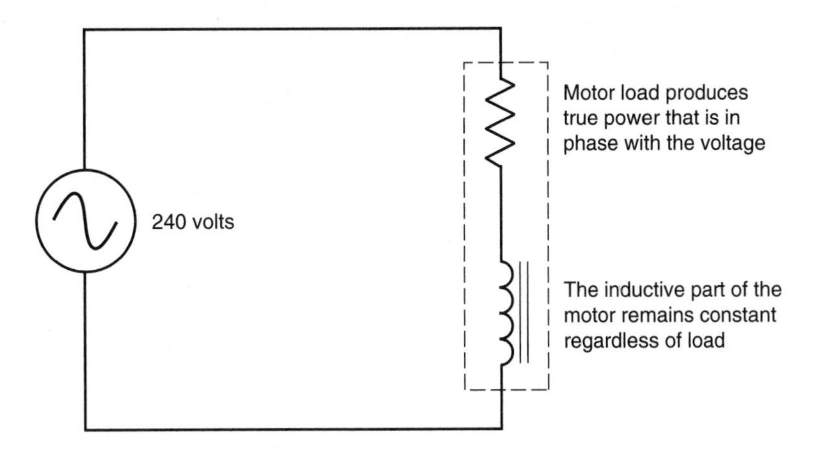

Figure 24–1 Equivalent motor circuit.

To determine the existing motor power factor, compute the apparent power by multiplying the applied voltage and total current.

$$VA = E \times I$$

$$VA = 240 \times 10$$

$$VA = 2400$$

Now that the apparent power is known, the power factor can be computed using the formula:

$$PF = \frac{P}{VA}$$

$$PF = \frac{1630}{2400}$$

$$PF = 67.9\%$$

Before the power factor can be corrected, we must determine the part of the circuit that is reactive. The reactive part of the circuit can be determined by finding the reactive power produced by the inductance:

$$VARs_L = \sqrt{VA^2 - P^2}$$

$$VARs_L = \sqrt{2400^2 - 1630^2}$$

$$VARs_L = 1761.6$$

To correct the power factor to 100 percent or unity, an equal amount of capacitive VARs must be connected in parallel with the motor. In actual practice, however, it is generally not considered practical to correct the power factor to unity or 100 percent. It is common practice to correct motor power factor to a value of about 95 percent. To correct the power factor to 95 percent, first determine the apparent power required to produce a power factor of 95 percent.

$$VA = \frac{P}{PF}$$

$$VA = \frac{1630}{0.95}$$

$$VA = 1715.789$$

The amount of inductive VARs needed to produce this amount of apparent power can be determined using the formula:

$$VARs_L = \sqrt{VA^2 - P^2}$$

$$VARs_L = \sqrt{1715.789^2 - 1630^2}$$

$$VARs_L = \sqrt{287031.89}$$

$$VARs_L = 535.754$$

Currently, the inductive VARs is 1761.6. To find the capacitive VARs needed to produce a total reactive power of 535.754 in the circuit, subtract the amount of reactive power needed from the current amount.

$$VARs_C = VARs_L - 535.754$$

$$VARs_C = 1761.6 - 535.754$$

$$VARs_C = 1225.846$$

To determine the capacitive reactance needed to produce the required reactive power at 240 volts, the following formula can be used.

$$X_C = \frac{E^2}{VARs_C}$$

$$X_C = \frac{240^2}{1225.846}$$

$$X_C = 46.988\ \Omega$$

The amount of capacitance needed to produce the required capacitive reactance at 60 Hz can be computed using the formula:

$$C = \frac{1}{2\pi\ FX_C}$$

$$C = \frac{1}{377 \times 46.988}$$

$$C = 56.5\ \mu F$$

The power factor will be corrected to 95 percent when a capacitor with a capacitance of 56.5 μF is connected in parallel with the motor.

CHAPTER 25

Three-Phase Circuits

Most of the power generated in the world today is three phase. There are several reasons why three-phase power is superior to single-phase power.

1. The horsepower rating of three-phase motors and the kVA rating of three-phase transformers is about 150 percent greater than for a single-phase motor or transformer with a similar frame size.
2. The power delivered by a single-phase system pulsates. The power falls to zero three times during each cycle. The power delivered by a three-phase circuit pulsates also, but the power never falls to zero (see Figure 25–1). In a three-phase system, the power delivered to the load is the same at any instant. This produces superior operating characteristics for three-phase motors.
3. In a balanced three-phase system, the conductors need be only about 75 percent the size of conductors for a single-phase, two-wire system of the same kVA

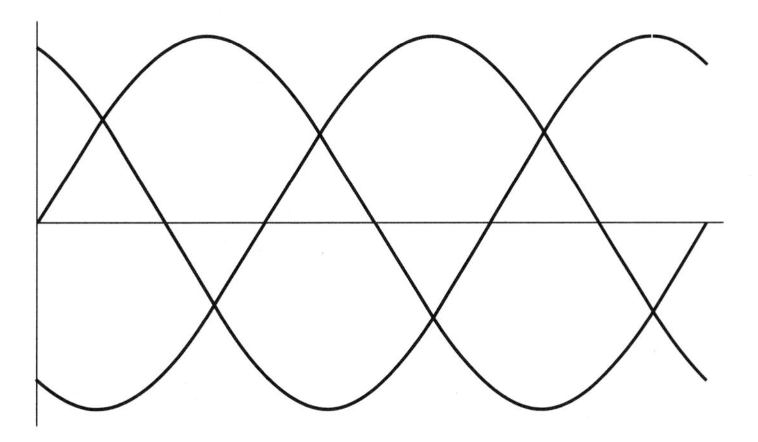

Figure 25–1 Three-phase power never falls to zero.

(kilo-volt-amp) rating. This helps offset the cost of supplying the third conductor required by three-phase systems.

A single-phase alternating voltage can be produced by rotating a magnetic field through the conductors of a stationary coil, as shown in Figure 25–2.

Since alternate polarities of the magnetic field cut through the conductors of the stationary coil, the induced voltage will change polarity at the same speed as the rotation of the magnetic field. The alternator shown in Figure 25–2 is single phase because it produces only one AC voltage.

If three separate coils are spaced 120 degrees apart, as shown in Figure 25–3, three voltages 120 degrees out of phase with each other will be produced when the magnetic field cuts through the coils. This is the manner in which a three-phase voltage is produced. There are two basic three-phase connections, the wye (or star), and the delta.

Wye (Star) Connection

The wye, or star, connection is made by connecting one end of each of the three-phase windings together. The voltage measured across a single winding or phase is known as the *phase* voltage. The voltage measured between the lines is known as the line-to-line voltage, or simply as the *line* voltage.

In Figure 25–4, ammeters have been placed in the phase winding of a wye-connected load and in the line supplying power to the load. Voltmeters have been connected across the input to the load and across the phase. A line voltage of 208 volts has been applied to the load. Notice that the voltmeter connected across the lines indicates a value of 208 volts, but the voltmeter connected across the phase indicates a value of 120 volts.

In a wye-connected system, the line voltage is higher than the phase voltage by a factor of the $\sqrt{3}$ *(1.732).* Two formulas used to compute the voltage in a wye-connected system are

$$E_{\text{Line}} = E_{\text{Phase}} \times 1.732$$

or

$$E_{\text{Phase}} = \frac{E_{\text{Line}}}{1.732}$$

Notice in Figure 25–4 that 10 amperes of current flows in both the phase and the line.

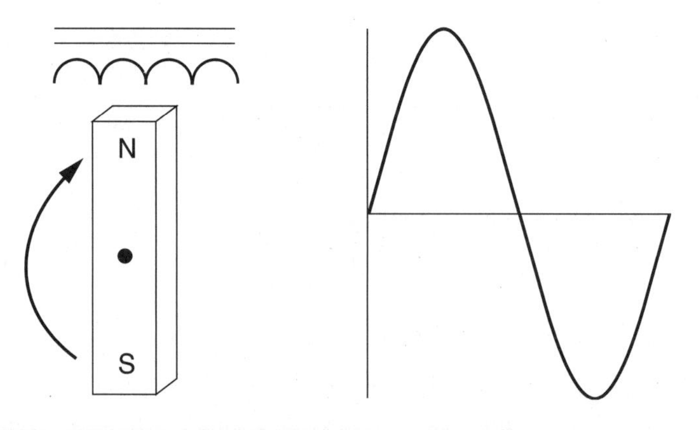

Figure 25–2 Producing a single-phase voltage.

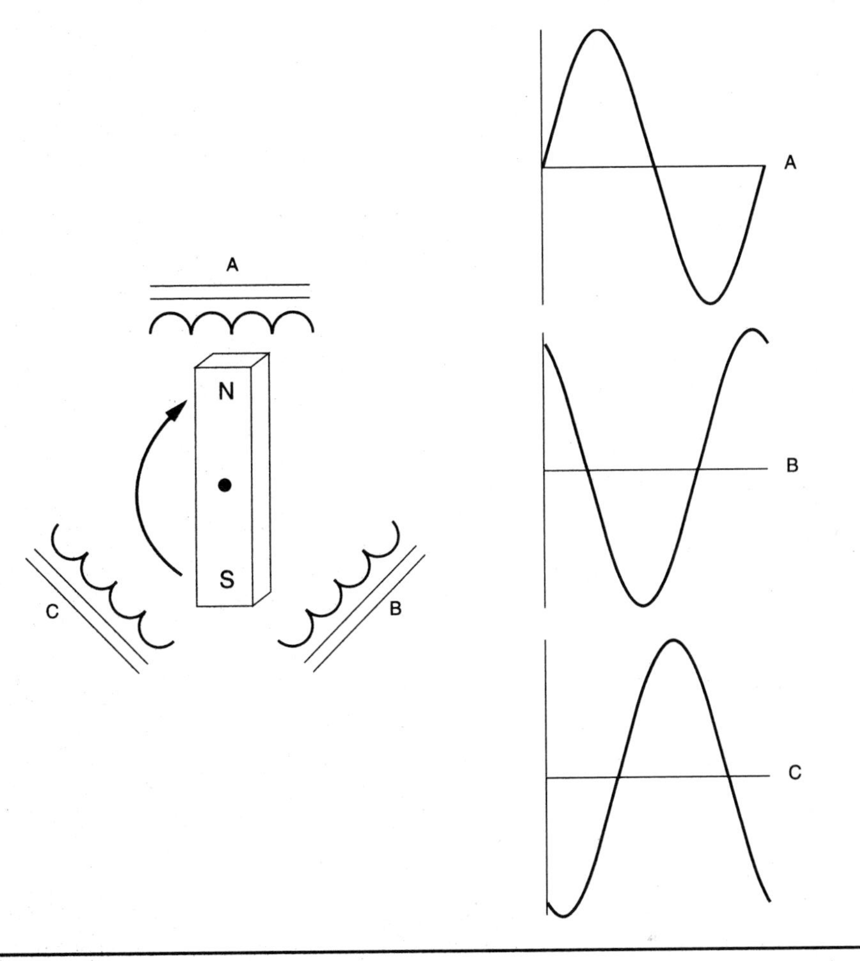

Figure 25–3 The voltages of a three-phase system are 120 degrees out of phase with each other.

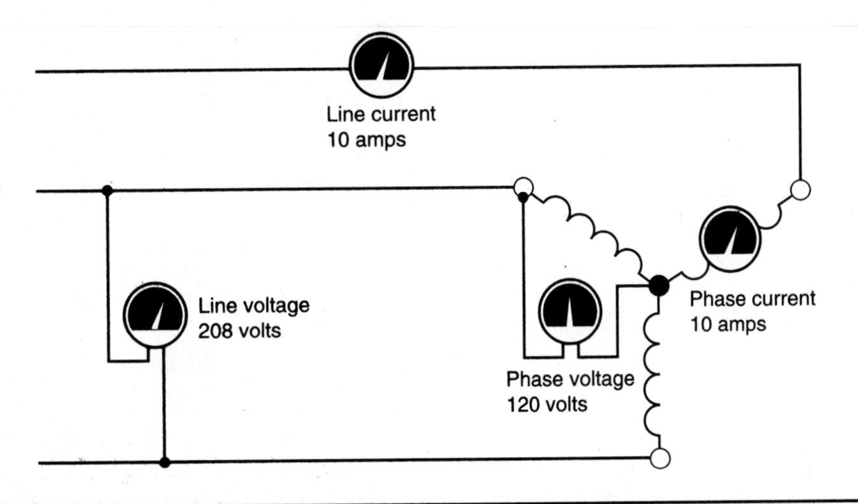

Figure 25–4 Line current and phase current are the same in a wye connection.

In a wye-connected system, phase current and line current are the same.

$$I_{\text{Line}} = I_{\text{Phase}}$$

Voltage Relationships in a Wye Connection

Many electricians have difficulty understanding why the line voltage of the wye connection used in this illustration is 208 volts instead of 240 volts. Since line voltage is measured across two phases that have a voltage of 120 volts each, it would appear that the sum of the two voltages should be 240 volts. One cause of this misconception is that many electricians are familiar with the 240/120 volt connection supplied to most homes. If voltage is measured across the two incoming lines, a voltage of 240 volts will be seen. If voltage is measured from either of the two lines to the neutral, a voltage of 120 volts will be seen. The reason is that this is a single-phase connection derived from the center tap of a transformer, shown in Figure 25–5. If the center tap is used as a common point, the two line voltages on either side of it will be in phase with each other as shown in Figure 25–6. The vector sum of these two voltages would be 240 volts.

Three-phase voltages are 120 degrees apart. The vector sum of these voltages is 208 volts, as shown in Figure 25–7. Another illustration of vector addition is shown in Figure 25–8. In this illustration, two-phase voltage vectors are added and the resultant is drawn from the starting point of one vector to the end point of the other. Figure 25–9 shows the parallelogram method of vector addition for the voltages in a wye-connected, three-phase system.

Delta Connections

In Figure 25–10, three separate inductive loads have been connected to form a delta connection. This connection receives its name from the fact that a schematic diagram of this connection resembles the Greek letter delta (Δ). In Figure 25–11, voltmeters have been

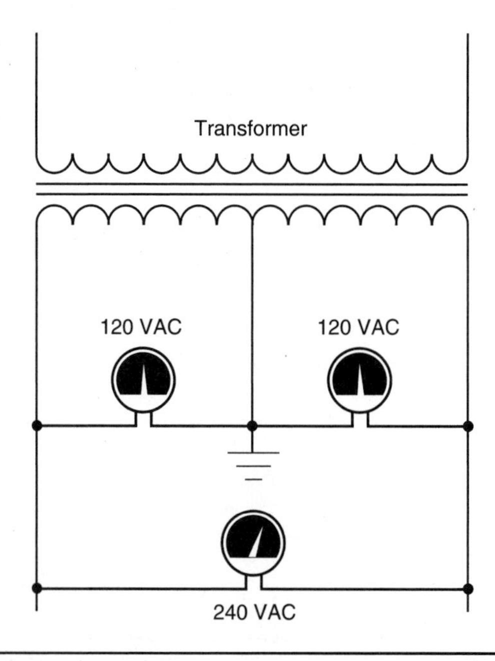

Figure 25–5 Single-phase transformer with grounded center tap.

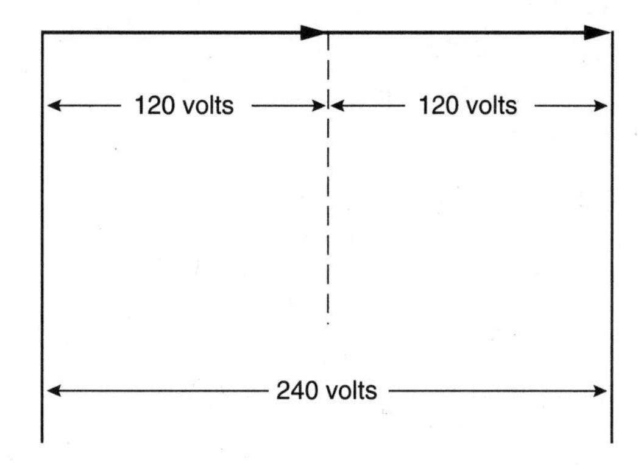

Figure 25–6 The voltages of a single-phase system are in phase with each other.

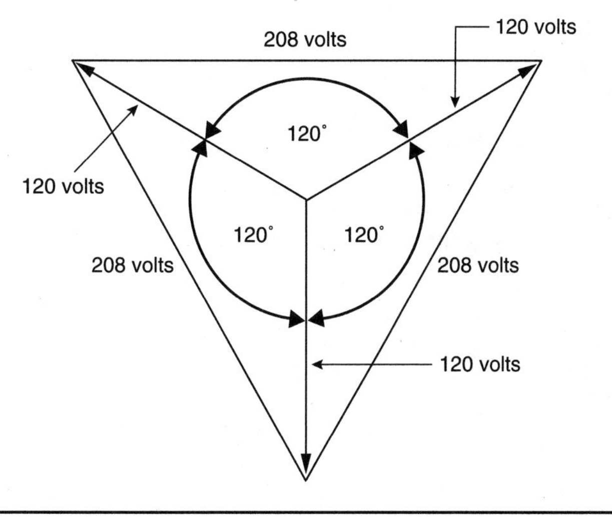

Figure 25–7 Vector sum of the voltages in a three-phase connection.

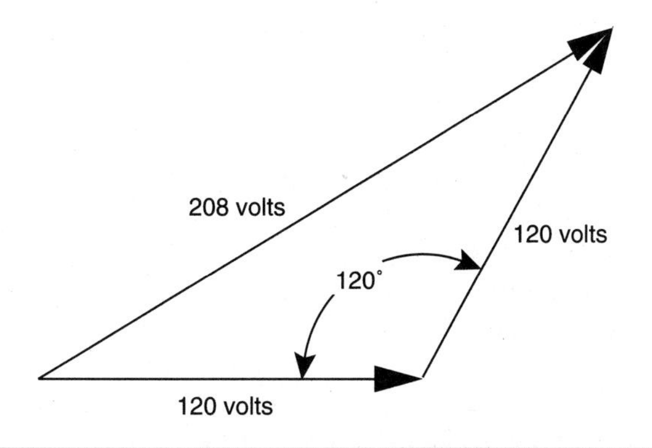

Figure 25–8 Adding voltage vectors of two-phase voltage values.

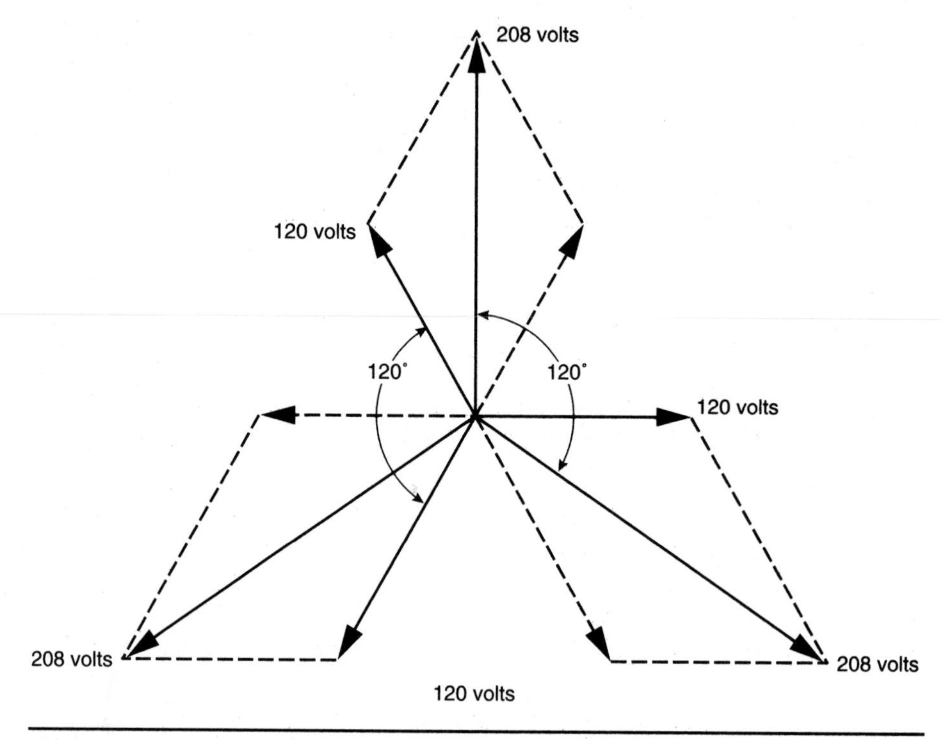

Figure 25–9 The parallelogram method of adding three-phase vectors.

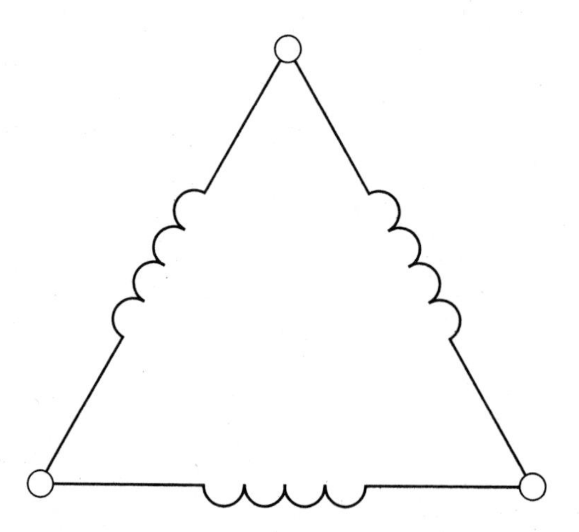

Figure 25–10 Three inductive loads are connected to form a three-phase delta connection.

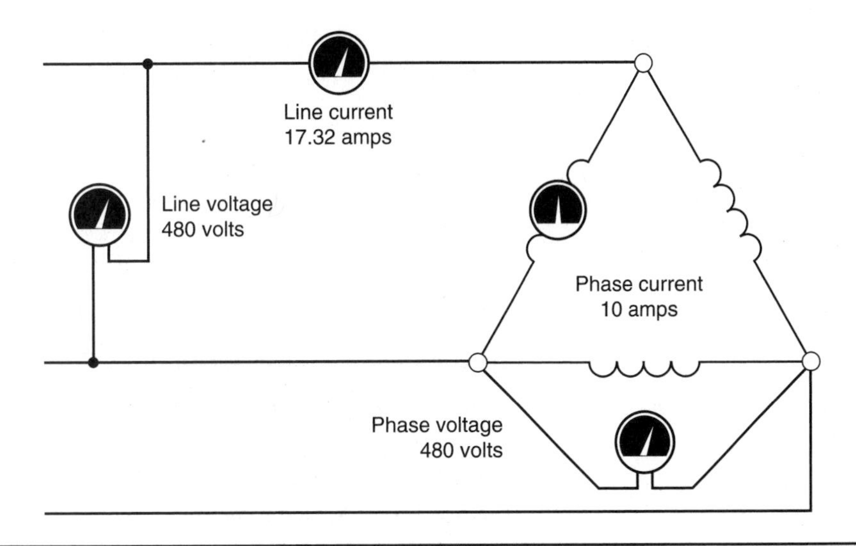

Figure 25–11 Voltage and current relationships in a delta connection.

connected across the lines and across the phase. Ammeters have been connected in the line and in the phase. *In the delta connection, line voltage and phase voltage are the same.* Notice that both voltmeters indicated a value of 480 volts.

$$E_{\text{Line}} = E_{\text{Phase}}$$

Notice that the line current and phase current are different, however. *The line current of a delta connection is higher than the phase current by a factor of the $\sqrt{3}$ (1.732).* In the example shown, it is assumed that each of the phase windings has a current flow of 10 amperes. The current in each of the lines, however, is 17.32 amperes. The reason for this difference in current is that current flows through different windings at different times in a three-phase circuit. During some periods of time, current will flow between two lines only. At other times, current will flow from two lines to the third (see Figure 25–12). The delta connection is similar to a parallel connection because there is always more than one path for current flow. Since these currents are 120 degrees out of phase with each other, vector addition must be used when finding the sum of the currents, as in Figure 25–13. Formulas for determining the current in a delta connection are

$$I_{\text{Line}} = I_{\text{Phase}} \times 1.732$$

or

$$I_{\text{Phase}} = \frac{I_{\text{Line}}}{1.732}$$

Three-Phase Apparent Power

Electricians sometimes become confused when computing values of power in three phase circuits. One reason for this confusion is because there are actually two formulas that can be used. If line values of voltage and current are known, the apparent power of the circuit can be computed using the formula:

$$VA = \sqrt{3} \times E_{\text{Line}} \times I_{\text{Line}}$$

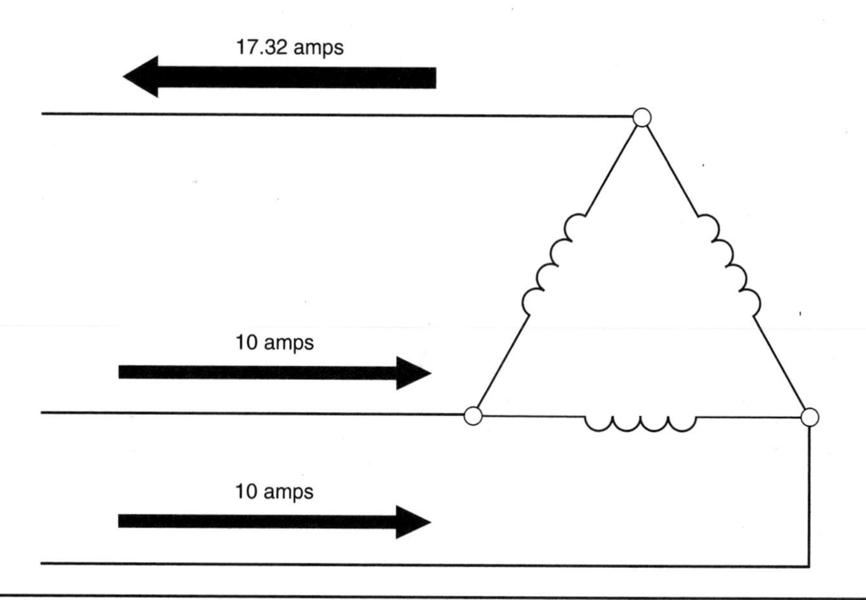

Figure 25–12 Division of currents in a delta connection.

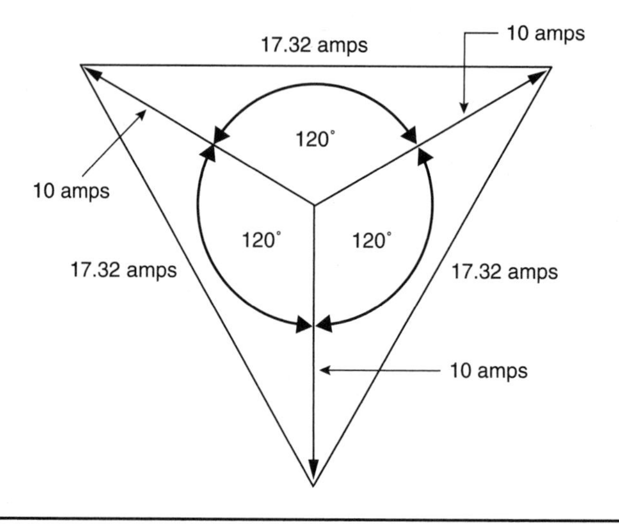

Figure 25–13 Vector addition is used to compute the sum of the currents in a delta connection.

If the phase values of voltage and current are known, the apparent power can be computed using the following formula:

$$VA = 3 \times E_{\text{Phase}} \times I_{\text{Phase}}$$

Notice in the first formula, the line values of voltage and current are multiplied by the square root of 3. In the second formula, the phase values of voltage and current are multiplied by 3. The first formula is the most used because it is generally more convenient to obtain line values of voltage and current since they can be measured with a voltmeter and clamp-on ammeter.

Three-Phase Watts and VARs

Watts and VARs can be computed in a similar manner. Watts can be computed by multiplying the apparent power by the power factor:

$$P = \sqrt{3} \times E_{\text{Line}} \times I_{\text{Line}} \times PF$$

or

$$P = 3 \times E_{\text{Phase}} \times I_{\text{Phase}} \times PF$$

Note: When computing the power of a pure resistive load, the voltage and current are in phase with each other and the power factor is 1.

VARs can be computed in a similar manner, except that voltage and current values of a pure reactive load are used. For example, assume that three capacitors are connected in wye and that the line voltage is 480 volts and the line current is 30 amperes. Capacitive VARs can be computed using the following formula:

$$VARs_C = \sqrt{3} \times E_{\text{Line(CAPACITIVE)}} \times I_{\text{Line(CAPACITIVE)}}$$

$$VARs_C = 1.732 \times 560 \times 30$$

$$VARs_C = 29{,}097.6$$

Three-Phase Circuit Calculations

In the following examples, values of line and phase voltage, line and phase current, and power will be computed for different types of three-phase connections.

Example #1

A wye-connected, three-phase alternator supplies power to a delta-connected resistive load, as shown in Figure 25–14. The alternator has a line voltage of 480 volts. Each resistor of the delta load has 8 ohms of resistance. Find the following values:

$E_{\text{Line (Load)}}$ = line voltage of the load
$E_{\text{Phase (Load)}}$ = phase voltage of the load
$I_{\text{Phase (Load)}}$ = phase current of the load
$I_{\text{Line (Load)}}$ = line current to the load
$I_{\text{Line (Alt)}}$ = line current delivered by the alternator
$E_{\text{Phase (Alt)}}$ = phase voltage of the alternator
P = true power

Solution

The load is connected directly to the alternator. Therefore, the line voltage supplied by the alternator is the line voltage of the load.

$$E_{\text{Line (Load)}} = 480 \text{ volts}$$

The three resistors of the load are connected in a delta connection. In a delta connection, the phase voltage is the same as the line voltage.

$$E_{\text{Phase (Load)}} = E_{\text{Line (Load)}}$$

$$E_{\text{Phase (Load)}} = 480 \text{ volts}$$

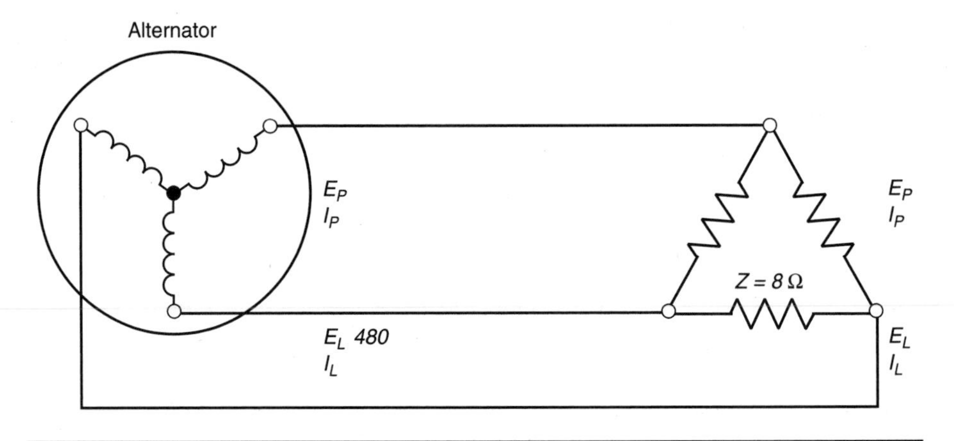

Figure 25–14 Computing three-phase values using example circuit #1.

Each of the three resistors in the load make up one phase of the load. Now that the phase voltage is known (480 volts), the amount of phase current can be computed using Ohm's law.

$$I_{\text{Phase (Load)}} = \frac{E_{P\ \text{(Load)}}}{Z}$$

$$I_{\text{Phase (Load)}} = \frac{480}{8}$$

$$I_{\text{Phase (Load)}} = 60 \text{ amps}$$

In this example, the three load resistors are connected as a delta with 60 amperes of current flow in each phase. The line current supplying a delta connection must be 1.732 times greater than the phase current.

$$I_{\text{Line (Load)}} = I_{\text{Phase}} \times 1.732$$

$$I_{\text{Line (Load)}} = 60 \times 1.732$$

$$I_{\text{Line (Load)}} = 103.92 \text{ amps}$$

The alternator must supply the line current to the load or loads to which it is connected. In this example, there is only one load connected to the alternator. Therefore, the line current of the load will be the same as the line current of the alternator.

$$I_{\text{Line (Alt)}} = 103.92 \text{ amperes}$$

The phase windings of the alternator are connected in a wye connection. In a wye connection, the phase current and line current are equal. The phase current of the alternator will, therefore, be the same as the alternator line current.

$$I_{\text{Phase (Alt)}} = 103.92 \text{ amps}$$

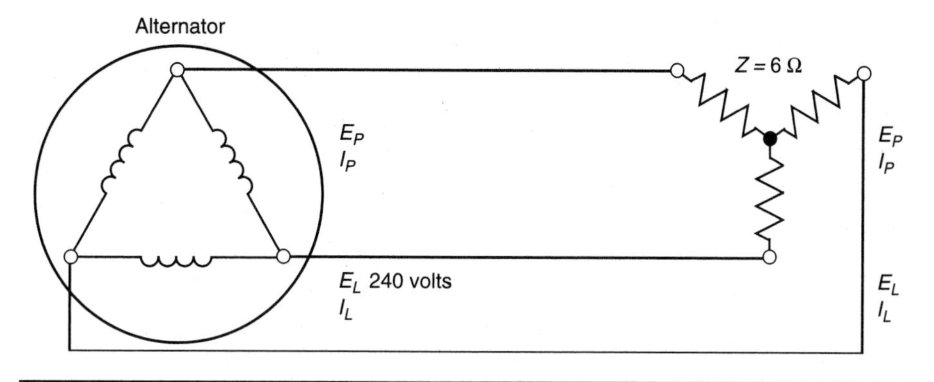

Figure 25–15 Computing three-phase values using example problem #2.

The phase voltage of a wye connection is less than the line voltage by a factor of the square root of 3. The phase voltage of the alternator will be

$$E_{\text{Phase (Alt)}} = \frac{E_{\text{Line (Alt)}}}{1.732}$$

$$E_{\text{Phase (Alt)}} = \frac{480}{1.732}$$

$$E_{\text{Phase (Alt)}} = 277.13 \text{ volts}$$

In this circuit, the load is pure resistive. The voltage and current are in phase with each other, which produces a unity power factor of 1. The true power in this circuit will be computed using the formula

$$P = 1.732 \times E_{\text{Line (Alt)}} \times I_{\text{Line (Alt)}} \times PF$$

$$P = 1.732 \times 480 \times 103.92 \times 1$$

$$P = 86{,}394.93 \text{ watts}$$

Example #2

In the next example, a delta-connected alternator is connected to a wye-connected resistive load, as in Figure 25–15. The alternator produces a line voltage of 240 volts and the resistors have a value of 6 ohms each. The following values will be found:

$E_{\text{Line (Load)}}$ = line voltage of the load
$E_{\text{Phase (Load)}}$ = phase voltage of the load
$I_{\text{Phase (Load)}}$ = phase current of the load
$I_{\text{Line (Load)}}$ = line current of the load
$I_{\text{Line (Alt)}}$ = line current delivered by the alternator
$E_{\text{Phase (Alt)}}$ = phase voltage of the alternator
P = true power

As was the case in the first example, the load is connected directly to the output of the alternator. The line voltage of the load must, therefore, be the same as the line voltage of the alternator.

$$E_{\text{Line (Load)}} = 240 \text{ volts}$$

The phase voltage of a wye connection is less than the line voltage by a factor of 1.732.

$$E_{\text{Phase (Load)}} = \frac{E_{\text{Line (Load)}}}{1.732}$$

$$E_{\text{Phase (Load)}} = \frac{240}{1.732}$$

$$E_{\text{Phase (Load)}} = 138.57 \text{ amps}$$

Each of the three 6 ohm resistors is one phase of the wye-connected load. Since the phase voltage is 138.57 volts, this voltage is applied to each of the three resistors. The amount of phase current can now be determined using Ohm's law.

$$E_{\text{Phase (Load)}} = \frac{E_{\text{Line (Load)}}}{Z}$$

$$E_{\text{Phase (Load)}} = \frac{138.57}{6}$$

$$E_{\text{Phase (Load)}} = 23.1 \text{ amps}$$

The amount of line current needed to supply a wye-connected load is the same as the phase current of the load.

$$I_{\text{Line (Load)}} = 23.1 \text{ amps}$$

In this example, there is only one load connected to the alternator. The line current supplied to the load is the same as the line current of the alternator.

$$I_{\text{Line (Alt)}} = 23.1 \text{ amps}$$

The phase windings of the alternator are connected in delta. In a delta connection, the phase current is less than the line current by a factor of 1.732.

$$I_{\text{Phase (Alt)}} = \frac{I_{\text{Line (Alt)}}}{1.732}$$

$$I_{\text{Phase (Alt)}} = \frac{23.1}{1.732}$$

$$I_{\text{Phase (Alt)}} = 13.34 \text{ amps}$$

The phase voltage of a delta is the same as the line voltage.

$$E_{\text{Phase (Alt)}} = 240 \text{ volts}$$

Since the load in this example is pure resistive, the power factor has a value of unity, or 1. Power will be computed by using the line values of voltage and current.

$$P = 1.732 \times E_{\text{Line}} \times I_{\text{Line}} \times PF$$

$$P = 1.732 \times 240 \times 23.1 \times 1$$

$$P = 9{,}602.21 \text{ watts}$$

Example #3

In the next example, the phase windings of an alternator are connected in wye. The alternator produces a line voltage of 440 volts, and supplies power to two resistive loads. One load contains resistors with a value of 4 ohms each connected in wye. The second load contains resistors with a value of 6 ohms each connected in delta, Figure 25–16. The following circuit values will be found:

$E_{\text{Line (Load 2)}}$ = line voltage of load #2
$E_{\text{Phase (Load 2)}}$ = phase voltage of load #2
$I_{\text{Phase (Load 2)}}$ = phase current of load #2
$I_{\text{Line (Load 2)}}$ = line current to load #2
$E_{\text{Line (Load 1)}}$ = line voltage of load #1
$E_{\text{Phase (Load 1)}}$ = phase voltage of load #1
$I_{\text{Phase (Load 1)}}$ = phase current of load #1
$I_{\text{Line (Load 1)}}$ = line current to load #1
$I_{\text{Line (Alt)}}$ = line current delivered by the alternator
$E_{\text{Phase (Alt)}}$ = phase voltage of the alternator
P = true power

Both loads are connected directly to the output of the alternator. The line voltage for both loads 1 and 2 will be the same as the line voltage of the alternator.

$$E_{\text{Line (Load 2)}} = 440 \text{ volts}$$

$$E_{\text{Line (Load 1)}} = 440 \text{ volts}$$

Load #2 is connected as a delta. The phase voltage will be the same as the line voltage.

$$E_{\text{Phase (Load 2)}} = 440 \text{ volts}$$

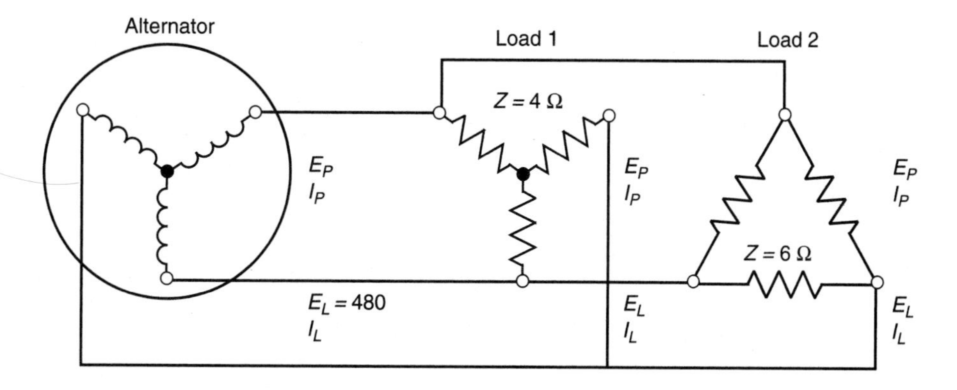

Figure 25–16 Example problem #3.

Each of the resistors that make up a phase of load #2 has a value of 6 ohms. The amount of phase current can be found using Ohm's law.

$$I_{\text{Phase (Load 2)}} = \frac{E_{\text{Phase (Load 2)}}}{Z}$$

$$I_{\text{Phase (Load 2)}} = \frac{440}{6}$$

$$I_{\text{Phase (Load 2)}} = 73.33 \text{ amps}$$

The line current supplying a delta-connected load is 1.732 times greater than the phase current. The amount of line current needed for load #2 can be computed by increasing the phase current value by 1.732.

$$I_{\text{Line (Load 2)}} = I_{\text{Phase (Load 2)}} \times 1.732$$

$$I_{\text{Line (Load 2)}} = 73.33 \times 1.732$$

$$I_{\text{Line (Load 2)}} = 127.01 \text{ amps}$$

The resistors of load #1 are connected to form a wye. The phase voltage of a wye connection is less than the line voltage by a factor of 1.732.

$$E_{\text{Phase (Load 1)}} = \frac{E_{\text{Line (Load 1)}}}{1.732}$$

$$E_{\text{Phase (Load 1)}} = \frac{440}{1.732}$$

$$E_{\text{Phase (Load 1)}} = 254.04 \text{ volts}$$

Now that the voltage applied to each of the 4-ohm resistors is known, the phase current can be computed using Ohm's law.

$$I_{\text{Phase (Load 1)}} = \frac{E_{\text{Phase (Load 1)}}}{Z}$$

$$E_{\text{Phase (Alt)}} = \frac{440}{1.732}$$

$$I_{\text{Phase (Load 1)}} = 63.51 \text{ amps}$$

The line current supplying a wye-connected load is the same as the phase current. Therefore, the amount of line current needed to supply load #1 is:

$$I_{\text{Line (Load 1)}} = 63.51 \text{ amps}$$

The alternator must supply the line current needed to operate both loads. In this example, both loads are resistive. The total line current supplied by the alternator will be the sum of the line currents of the two loads.

$$I_{\text{Line (Alt)}} = I_{\text{Line (Load 1)}} + I_{\text{Line (Load 2)}}$$

$$I_{\text{Line (Alt)}} = 63.51 \times 127.01$$

$$I_{\text{Line (Alt)}} = 190.52 \text{ amps}$$

Since the phase windings of the alternator in this example are connected in a wye, the phase current will be the same as the line current.

$$I_{\text{Phase (Alt)}} = 190.52$$

The phase voltage of the alternator will be less than the line voltage by a factor of 1.732.

$$E_{\text{Phase(Alt)}} = \frac{440}{1.732}$$

$$E_{\text{Phase(Alt)}} = 254.04 \text{ volts}$$

Both of the loads in this example are resistive and have a unity power factor of 1. The total power in this circuit can be found by using the line voltage and total line current supplied by the alternator.

$$P = 1.732 \times E_{\text{Line}} \times I_{\text{Line}} \times PF$$

$$P = 1.732 \times 440 \times 190.52 \times 1$$

$$P = 145{,}191.48 \text{ watts}$$

Example #4

In this example, a wye-connected, three-phase alternator with a line voltage of 560 volts supplies power to three different loads, Figure 25–17. The first load is formed by three resistors with a value of 6 ohms, each connected in a wye. The second load comprises three inductors with an inductive reactance of 10 ohms each connected in delta, and the third load comprises three capacitors with a capacitive reactance of 8 ohms each connected in wye. The following circuit values will be found:

$E_{\text{Line (Load 3)}}$ = line voltage of load #3 (capacitive)
$E_{\text{Phase (Load 3)}}$ = phase voltage of load #3 (capacitive)
$I_{\text{Phase (Load 3)}}$ = phase current of load #3 (capacitive)
$I_{\text{Line (Load 3)}}$ = line current to load #3 (capacitive)
$E_{\text{Line (Load 2)}}$ = line voltage of load #2 (inductive)

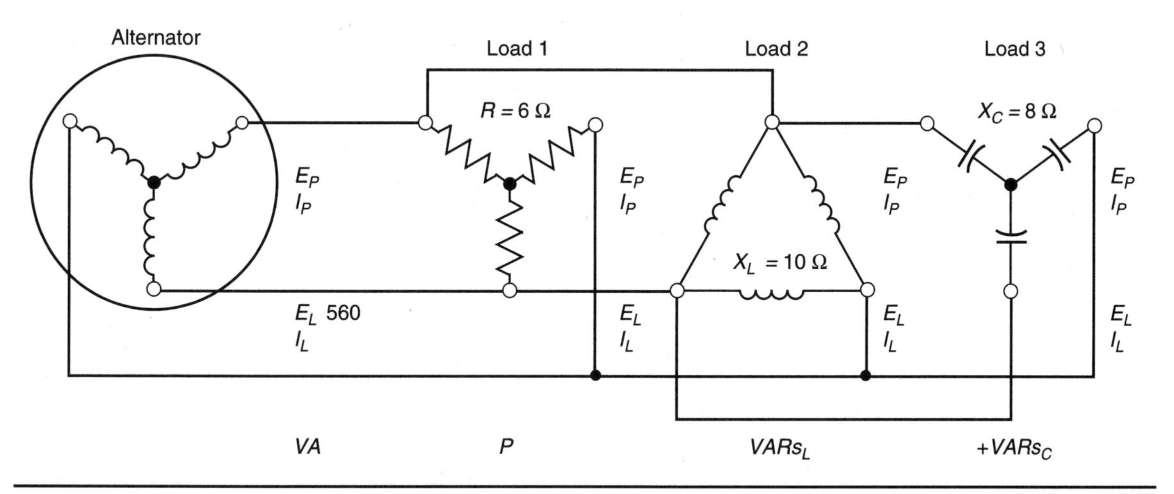

Figure 25–17 Example problem #4.

$E_{Phase\ (Load\ 2)}$ = phase voltage of load #2 (inductive)
$I_{Phase\ (Load\ 2)}$ = phase current of load #2 (inductive)
$I_{Line\ (Load\ 2)}$ = line current to load #2 (inductive)
$E_{Line\ (Load\ 1)}$ = line voltage of load #1 (resistive)
$E_{Phase\ (Load\ 1)}$ = phase voltage of load #1 (resistive)
$I_{Phase\ (Load\ 1)}$ = phase current of load #1 (resistive)
$I_{Line\ (Load\ 1)}$ = line current to load #1 (resistive)
$I_{Line\ (Alt)}$ = line current delivered by the alternator
$E_{Phase\ (Alt)}$ = phase voltage of the alternator
P = true power
$VARs_L$ = reactive power of the inductive load
$VARs_C$ = reactive power of the capacitive load
VA = apparent power
PF = power factor

All three of the loads are connected to the output of the alternator. The line voltage connected to each load is the same as the line voltage of the alternator.

$$E_{Line\ (Load\ 3)} = 560 \text{ volts}$$

$$E_{Line\ (Load\ 2)} = 560 \text{ volts}$$

$$E_{Line\ (Load\ 1)} = 560 \text{ volts}$$

Load #3 Calculations

Load #3 is formed from three capacitors with a capacitive reactance of 8 ohms each connected in a wye. Since this load is wye connected, the phase voltage will be less than the line voltage by a factor of 1.732.

$$E_{Phase\ (Load\ 3)} = \frac{E_{Line\ (Load\ 3)}}{1.732}$$

$$E_{Phase\ (Load\ 3)} = \frac{560}{1.732}$$

$$E_{Phase\ (Load\ 3)} = 323.33 \text{ volts}$$

Now that the voltage applied to each capacitor is known, the phase current can be computed using Ohm's law.

$$I_{Phase\ (Load\ 3)} = \frac{E_{Phase\ (Load\ 3)}}{X_C}$$

$$I_{Phase\ (Load\ 3)} = \frac{323.33}{8}$$

$$I_{Phase\ (Load\ 3)} = 40.42 \text{ amps}$$

The line current required to supply a wye-connected load is the same as the phase current.

$$I_{Line\ (Load\ 3)} = 40.42 \text{ amps}$$

The reactive power of load #3 can be found using a formula similar to the formula for computing apparent power. Since load #3 is pure capacitive, the current and voltage and 90 degrees out of phase with each other and the power factor is zero.

$$VARs_C = 1.732 \times E_{\text{Line (Load 3)}} \times I_{\text{Line (Load 3)}}$$

$$VARs_C = 1.732 \times 560 \times 40.42$$

$$VARs_C = 39{,}204.17$$

Load #2 Calculations

Load #2 comprises three inductors connected in a delta with an inductive reactance of 10 ohms each. Since the load is connected in delta, the phase voltage will be same as the line voltage.

$$E_{\text{Line (Load 2)}} = 560 \text{ volts}$$

The phase current can be computed by using Ohm's law:

$$I_{\text{Phase (Load 2)}} = \frac{E_{\text{Phase (Load 2)}}}{X_L}$$

$$I_{\text{Phase (Load 2)}} = \frac{560}{10}$$

$$I_{\text{Phase (Load 2)}} = 56 \text{ amps}$$

The amount of line current needed to supply a delta-connected load is 1.732 times greater than the phase current of the load.

$$I_{\text{Line (Load 2)}} = I_{\text{Phase (Load 2)}} \times 1.732$$

$$I_{\text{Line (Load 2)}} = 56 \times 1.732$$

$$I_{\text{Line (Load 2)}} = 96.99 \text{ amps}$$

Since load #2 comprises inductors, the reactive power can be computed using the line values of voltage and current supplied to the load.

$$VARs_L = 1.732 \times E_{\text{Line (Load 2)}} \times I_{\text{Line (Load 2)}}$$

$$VARs_L = 1.732 \times 560 \times 96.99$$

$$VARs_L = 94{,}072.54$$

Load #1 Calculations

Load #1 comprises three resistors with a resistance of 6 ohms each connected in wye. In a wye connection the phase voltage is less than the line voltage by a factor of 1.732. The phase voltage for load #1 will be the same as the phase voltage for load #3.

$$E_{\text{Phase (Load 1)}} = 323.33 \text{ volts}$$

The amount of phase current can now be computed using the phase voltage and the resistance of each phase.

$$I_{\text{Phase (Load 1)}} = \frac{E_{\text{Phase (Load 1)}}}{R}$$

$$I_{\text{Phase (Load 1)}} = \frac{323.33}{6}$$

$$I_{\text{Phase (Load 1)}} = 53.89 \text{ amps}$$

Since the resistors of load #1 are connected in a wye, the line current will be the same as the phase current.

$$I_{\text{Line (Load 1)}} = 53.89 \text{ amps}$$

Since load #1 is pure resistive and the power factor has a value of 1, true power can be computed using the line and phase current values.

$$P = 1.732 \times E_{\text{Line (Load 1)}} \times I_{\text{Line (Load 1)}}$$

$$P = 1.732 \times 560 \times 53.89$$

$$P = 52{,}267 \text{ watts}$$

Alternator Calculations

The alternator must supply the line current for each of the loads. In this problem, however, the line currents are out of phase with each other. To find the total line current delivered by the alternator, vector addition must be used. The current flow in load #1 is resistive and in phase with the line voltage. The current flow in load #2 is inductive and lags the line voltage by 90 degrees. The current flow in load #3 is capacitive and leads the line voltage by 90 degrees. A formula similar to the formula used to find total current flow in an RLC parallel circuit can be employed to the find the total current delivered by the alternator.

$$I_{\text{Line(Alt)}} = \sqrt{{I_{\text{Line (Load 1)}}}^2 + (I_{\text{Line (Load 2)}} - I_{\text{Line (Load 3)}})^2}$$

$$I_{\text{Line(Alt)}} = \sqrt{53.89^2 + (96.99 - 40.42)^2}$$

$$I_{\text{Line(Alt)}} = 78.13 \text{ amps}$$

We can now find the apparent power using the line voltage and current values of the alternator.

$$VA = 1.732 \times E_{\text{Line (Alt)}} \times I_{\text{Line (Alt)}}$$

$$VA = 1.732 \times 560 \times 78.13$$

$$VA = 75{,}779.85$$

The circuit power factor is the ratio of apparent power and true power.

$$PF = \frac{P}{VA}$$

$$PF = \frac{52{,}267}{75{,}779.85}$$

$$PF = 0.69, \text{ or } 69\%$$

CHAPTER 26

Power Factor Correction for a Three-Phase Circuit

Correcting the power factor of a three-phase circuit is similar to the procedure used to correct the power factor of a single-phase circuit. In this example, it will be assumed that a three-phase motor is connected to a 480-volt, 60-Hz line, as in Figure 26–1. A clamp-on ammeter indicates a running current of 68 amperes at full load and a three-phase wattmeter indicates a true power of 40,277 watts. In this example, the motor power factor will first be computed. The amount of capacitance needed to correct the power factor to 95 percent will then be found. It is to be assumed that the capacitors used for power factor correction are to be connected in wye and the capacitor bank is then to be connected in parallel with the motor.

The first step will be to find the amount of apparent power in the circuit.

$$VA = 1.732 \times E_{\text{Line}} \times I_{\text{Line}}$$

$$VA = 1.732 \times 480 \times 68$$

$$VA = 56{,}532.48$$

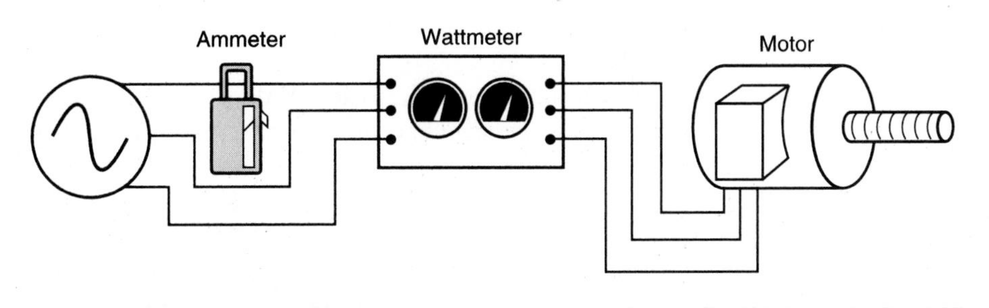

Figure 26–1 Determining apparent and true power for a three-phase motor.

The motor power factor can be computed by dividing the true power by the apparent power.

$$PF = \frac{P}{VA}$$

$$PF = \frac{40,277}{56,532.48}$$

$$PF = 0.712, \text{ or } 71.2\%$$

We can compute the inductive VARs in the circuit using the formula

$$VARs_L = \sqrt{VA^2 - P^2}$$

$$VARs_L = \sqrt{56,532.48^2 - 40,277^2}$$

$$VARs_L = 39,669.69$$

If the power factor is to be corrected to 95 percent, the apparent power at 95 percent power factor must be found. This can be done using this formula:

$$VA = \frac{P}{PF}$$

$$VA = \frac{40,277}{0.95}$$

$$VA = 42396.84$$

Find the amount of inductive VARs needed to produce an apparent power of 42,396.84 volt amps by using the following formula:

$$VARs_L = \sqrt{VA^2 - P^2}$$

$$VARs_L = \sqrt{42,396.84^2 - 40,277^2}$$

$$VARs_L = 13,238.4$$

To correct the power factor to 95 percent, the inductive VARs must be reduced from 39,669.69 to 13,238.4. This can be done by connecting a bank of capacitors in the circuit that will produce a total of 26,431.29 capacitive VARs (39,669.69 – 13,238.4 = 26,431.29). This amount of capacitive VARs will reduce the inductive VARs to the desired amount (see Figure 26–2).

Now that the amount of capacitive VARs needed to correct the power factor is known, we can compute the amount of line current supplying the capacitor bank using the formula

$$I_{\text{Line}} = \frac{VARs_C}{E_{\text{Line}} \times 1.732}$$

$$I_{\text{Line}} = \frac{26,431.29}{480 \times 1.732}$$

$$I_{\text{Line}} = 31.79 \text{ amps}$$

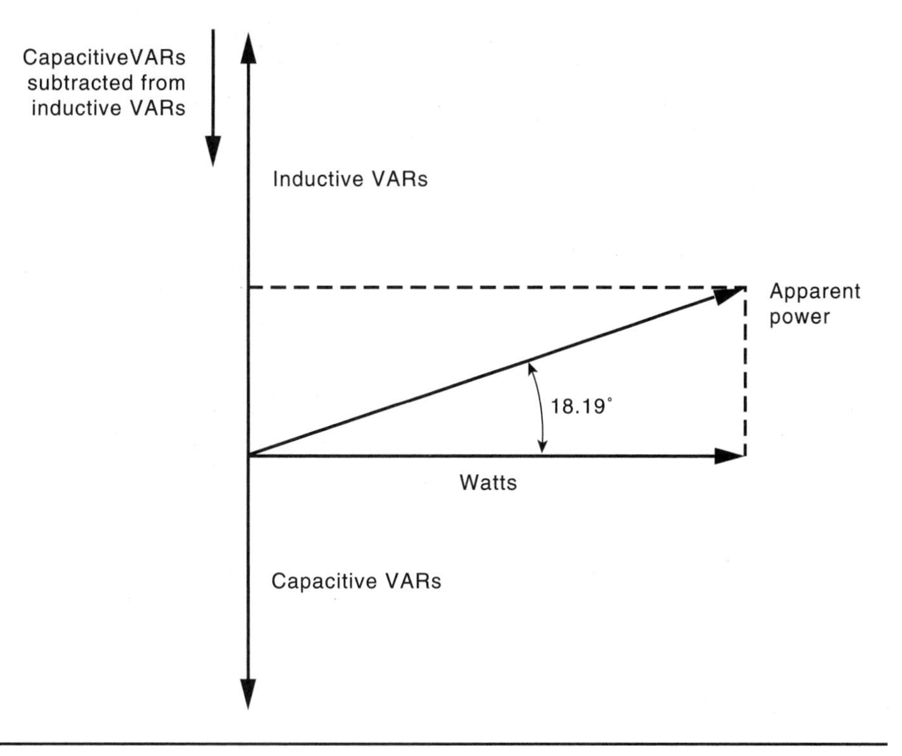

Figure 26–2 Vector relationship of powers to correct motor power factor.

The capacitive load bank is to be connected in a wye. Therefore, the phase current will be the same as the line current. The phase voltage, however, will be less than the line voltage by a factor of 1.732, or 277.14 volts (480/1.732). Ohm's law can be used to find the amount of capacitive reactance needed to produce a phase current of 31.79 amperes with an applied voltage of 277.14 volts.

$$X_C = \frac{E_{\text{Phase}}}{I_{\text{Phase}}}$$

$$X_C = \frac{277.14}{31.79}$$

$$X_C = 8.72\ \Omega$$

The amount of capacitance needed to produce a capacitive reactance of 8.72 ohms can now be computed.

$$C = \frac{1}{2\pi F X_C}$$

$$C = \frac{1}{377 \times 8.72}$$

$$C = 304.2\ \mu\text{F}$$

When a bank of wye-connected capacitors with a value of 304.2 μF each are connected in parallel with the motor, the power factor will be corrected to 95 percent, as shown in Figure 26–3.

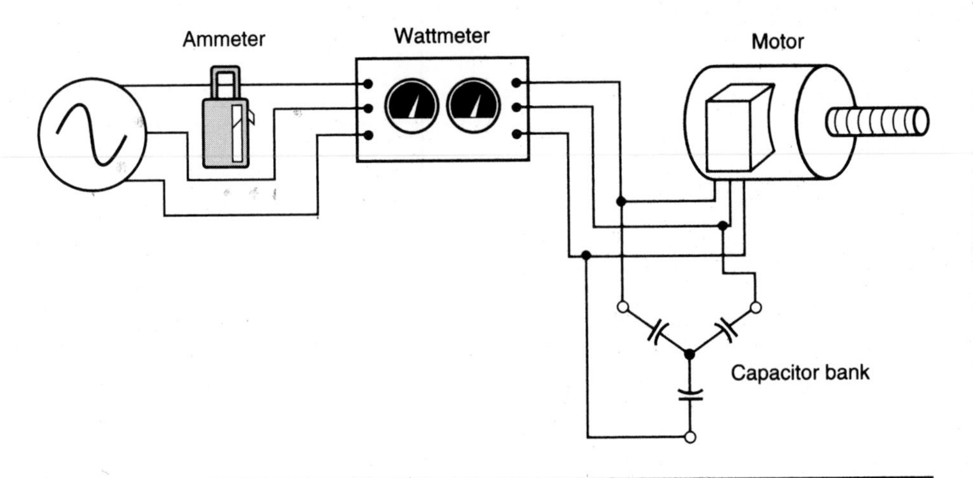

Figure 26–3 A wye-connected bank of capacitors is used to correct motor power factor.

CHAPTER 27

Batteries

The first practical battery was invented in 1800 by Alessandro Volta (see Figure 27–1). Volta's battery was constructed using zinc and silver discs separated by a piece of cardboard soaked in brine or salt water. Volta called his battery a *voltaic pile* because it was a series of individual cells connected together. Each cell produced a certain amount of voltage, depending on the materials used to make the cell. A battery is actually several cells connected together, although the word *battery* is often used in reference to a single cell. The schematic symbols for an individual cell and for a battery are shown in Figure 27–2.

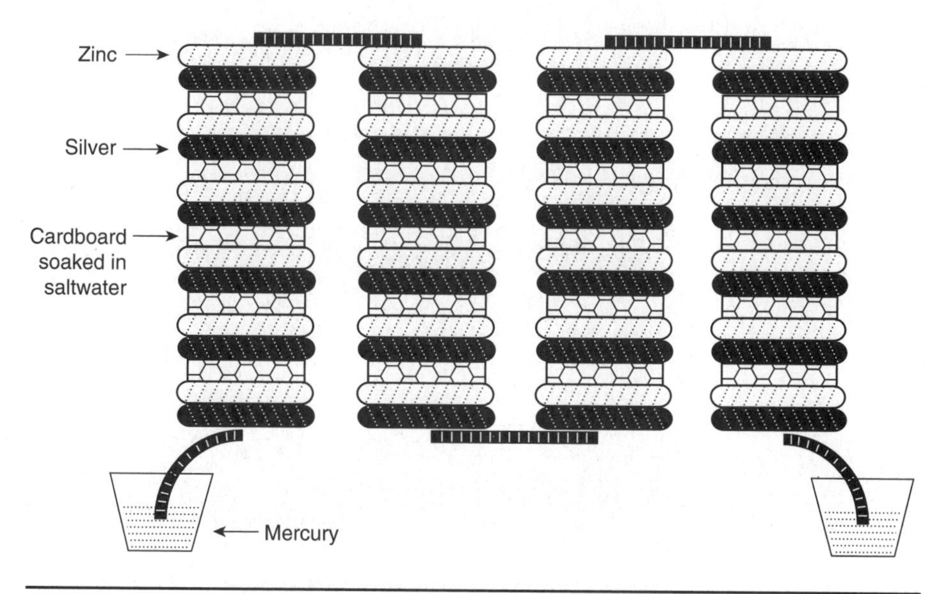

Figure 27–1 A voltaic pile.

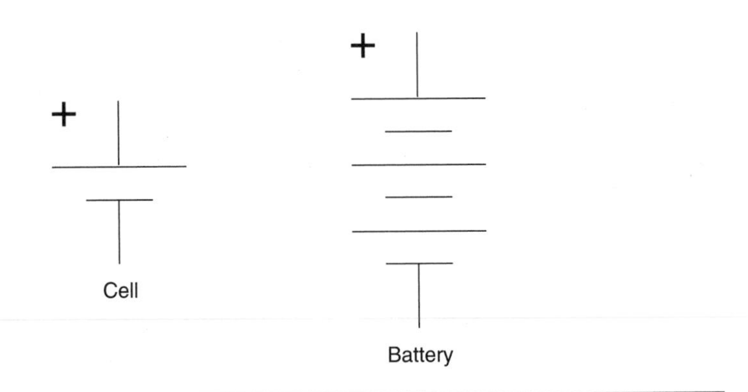

Figure 27–2 Schematic symbols used to represent an individual cell and a battery.

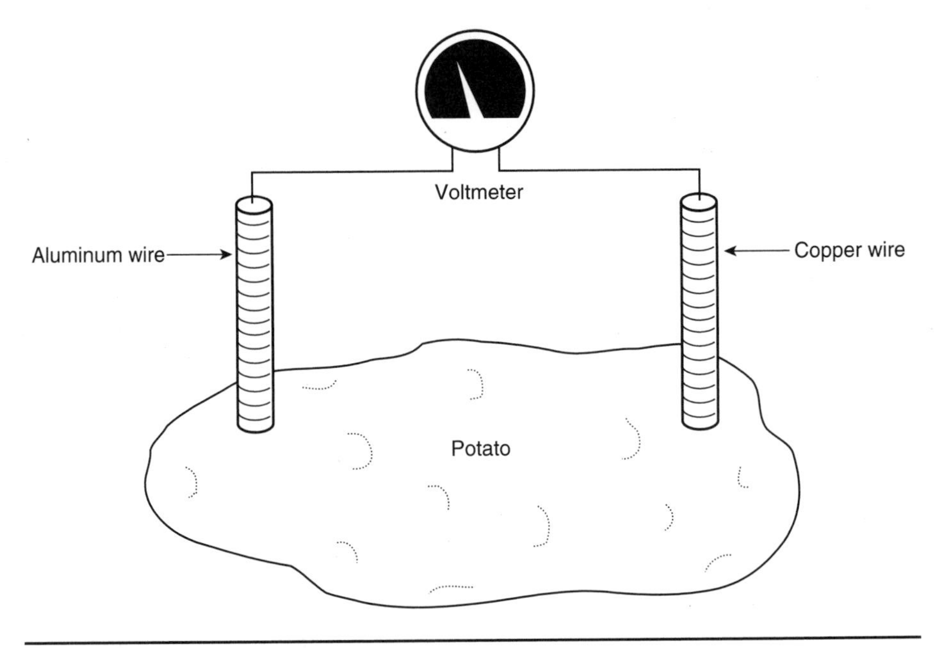

Figure 27–3 Simple voltaic cell constructed from a potato and two different metals.

Cells

A voltaic cell can be constructed using virtually any two unlike metals and an acid, alkaline, or salt solution. Figure 27–3 shows a very simple cell. In this example, a copper wire is inserted in one end of a potato and an aluminum wire is inserted in the other. The acid in the potato acts as the electrolyte. If a high-impedance voltmeter is connected to the wires, a small voltage can be measured.

Another example of a simple voltaic cell is shown in Figure 27–4. In this example, two coins of different metal, a nickel and penny, are separated by a piece of paper. The paper has been wet with saliva from a person's mouth. The saliva contains acid or alkaline, which acts as the electrolyte. When a high-impedance voltmeter is connected across the two coins, a small voltage is produced.

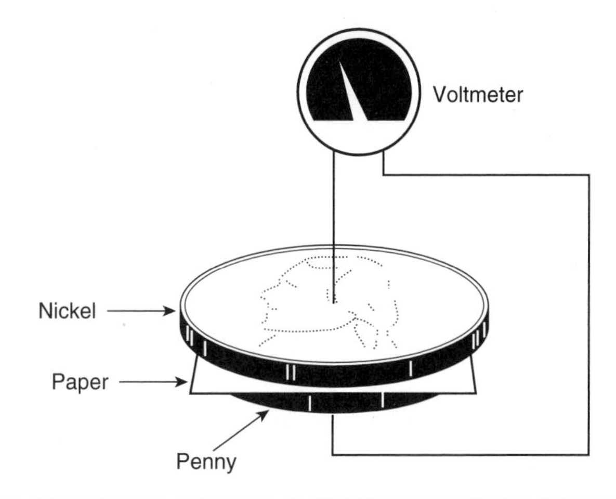

Figure 27–4 A voltaic cell constructed from two coins and a piece of paper.

Electromotive Series of Metals (Partial list)		
Lithium	Zinc	Lead
Potassium	Chromium	Antimony
Sodium	Iron	Copper
Calcium	Cadium	Mercury
Magnesium	Cobalt	Silver
Aluminum	Nickel	Platinum
Manganese	Tin	Gold

Figure 27–5 A partial list of the electromotive series of metals.

Cell Voltage

The amount of voltage produced by an individual cell is determined by the materials it is made of. When a voltaic cell is constructed, the plate metals are chosen on the basis of how easily one metal will give up electrons as compared to the other. A special list of metals, called the *electromotive series of metals,* is shown in Figure 27–5. This table lists metals in the order of their ability to accept or receive electrons. The metals at the top accept electrons more easily than those at the bottom. The farther apart the metals are on the list, the higher the voltage developed by the cell. One of the first practical cells to be constructed was the zinc–copper cell. This cell uses zinc and copper as the active metals and a solution of water and hydrochloric acid as the electrolyte. Notice that zinc is located closer to the top of the list than copper. Since zinc will accept electrons more readily than copper, zinc will be the negative electrode and copper will be the positive.

Although it is possible to construct a cell from virtually any two unlike metals and an electrolyte solution, not all combinations are practical. Some metals corrode rapidly when placed in an electrolyte solution, and some produce chemical reactions that cause a buildup of resistance. In actual practice, there are relatively few metals that can be used to produce a practical cell.

The table in Figure 27–6 lists common cells. The table is divided into two sections. One section lists primary cells and the other lists secondary cells. *A primary cell is a cell that cannot be recharged.* The chemical reaction of a primary cell causes one of the electrodes to be eaten away as power is produced. When a primary cell becomes discharged, it should be replaced with a new cell. *A secondary cell can be recharged.* The recharging process will be covered later in this section.

Primary Cells

An example of a primary cell is the zinc–copper cell shown in Figure 27–7. This cell consists of two electrodes, one zinc and one copper, suspended in a solution of water and hydrochloric acid. This cell will produce approximately 1.08 volts. The electrolyte contains positive H^+ ions that attract electrons from the zinc atoms that have two valence electrons. When an electron bonds with the H^+ ion, it becomes a neutral H atom. The neutral H atoms combine to form H_2 atoms of hydrogen gas.

These hydrogen atoms can be seen bubbling away in the electrolyte solution. The zinc ions, Zn^{++}, are attracted to negative Cl^- ions in the electrolyte solution.

The copper electrode provides another source of electrons that can be attracted by the H^+ ions. When a circuit is completed between the zinc and copper electrodes, as shown in Figure 27–8, electrons are attracted from the zinc electrode to replace the electrons in the copper electrode that are attracted into the solution. After some period of time, the zinc electrode dissolves as a result of the zinc ions being distributed in the solution.

The Carbon–Zinc Cell (Leclanché Cell)

One of the first practical primary cells was invented by Leclanché. Leclanché used a carbon rod as an electrode instead of copper, and a mixture of ammonium chloride, manganese dioxide, and granulated carbon as the electrolyte. The mixture was packed inside a zinc

Cell	*Negative Plate*	*Positive Plate*	*Electrolyte*	*Volts per Cell*
		Primary Cells		
Carbon-zinc (Leclanche)	Zinc	Carbon, Manganese dioxide	Ammonium chloride	1.5
Alkaline	Zinc	Manganese dioxide	Potassium hydroxide	1.5
Mercury	Zinc	Mercuric oxide	Potassium hydroxide	1.35
Silver-zinc	Zinc	Silver oxide	Potassium hydroxide	1.6
Zinc-air	Zinc	Oxygen	Potassium hydroxide	1.4
Edison-lalande	Zinc	Copper oxide	Sodium hydroxide	0.8
		Secondary Cells		
Lead-acid	Lead	Lead dioxide	Dilute sulfuric acid	2.2
Nickel-iron (Edison)	Iron	Nickel oxide	Potassium hydroxide	1.4
Nickel-cadmium	Cadmium	Nickel hydroxide	Potassium hydroxide	1.2
Silver-zinc	Zinc	Silver oxide	Potassium hydroxide	1.5
Silver-cadmium	Cadmium	Silver oxide	Potassium hydroxide	1.1

Figure 27–6 Voltaic cells.

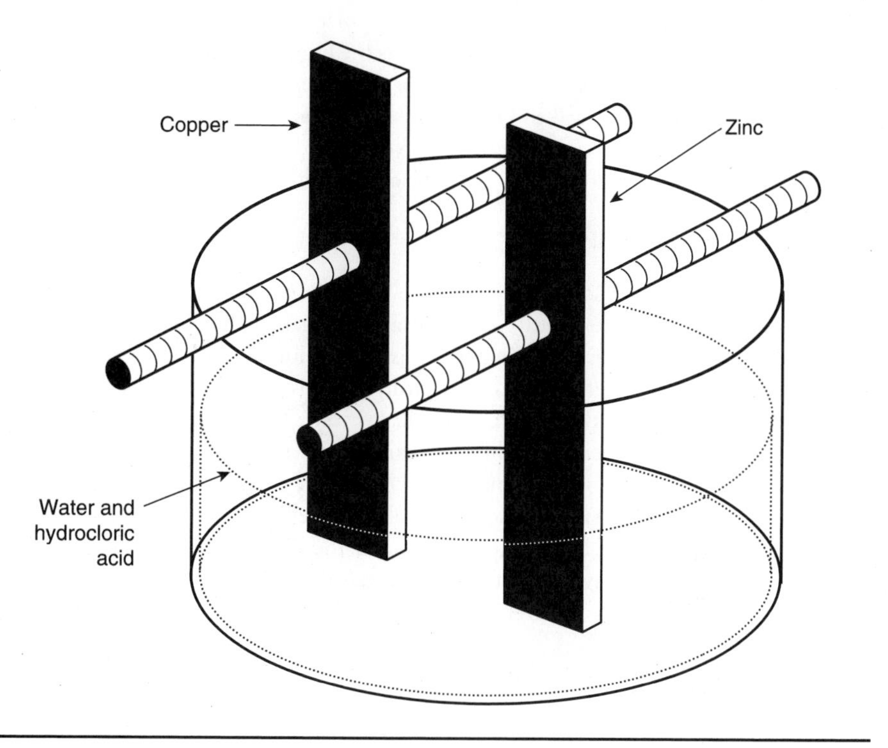

Figure 27–7 Zinc–copper cell.

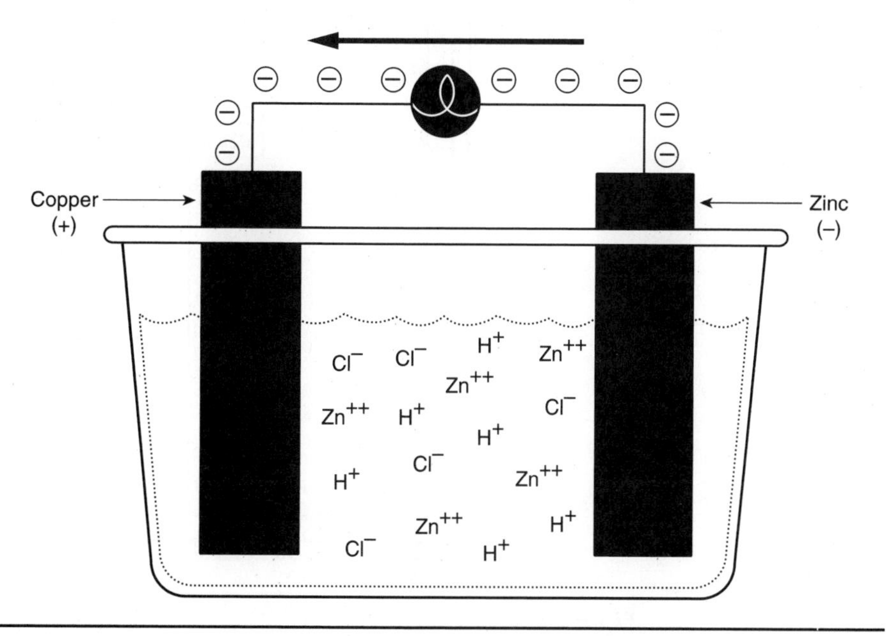

Figure 27–8 Electrons flow from the zinc to the copper electrode.

container, which acted as the negative electrode (see Figure 27–9). This cell is often referred to as a dry cell because the electrolyte is actually a paste instead of a liquid, which permits the cell to be used in any position without spilling the electrolyte. As the cell discharges, the zinc container is eventually dissolved. Once this happens, the cell should be discarded immediately and replaced. Failure to replace the cell can result in damage to equipment.

Alkaline Cells

Another cell similar to the carbon–zinc cell is the alkaline cell. This cell uses a zinc can as the negative electrode and manganese dioxide, MnO_2, as the positive electrode. The major electrolyte ingredient is a potassium hydroxide. Like the carbon–zinc cell, the alkaline cell contains a paste electrolyte and is considered a dry cell. The voltage developed is the same as the carbon–zinc cell, 1.5 volts. The major advantage of the alkaline cell is longer life. The average alkaline cell can supply from three to five times the power of a carbon–zinc cell, depending on the discharge rate when the cell is being used. The major disadvantage is cost. Although some alkaline cells are rechargeable, the number of charge and discharge cycles is limited. In general, the alkaline cell is considered to be nonrechargable. A cut-away view of an alkaline–manganese cell is shown in Figure 27–9.

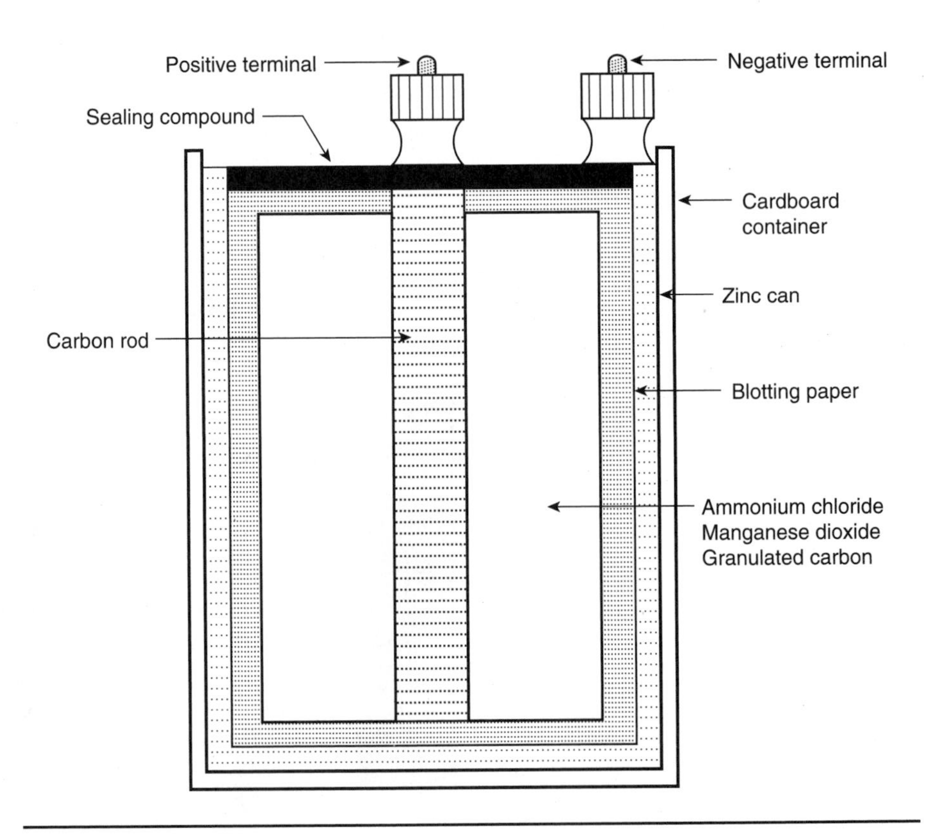

Figure 27–9 Carbon–zinc cell.

Button Cells

Another common type of primary cell is the button cell. The button cell is so named because its size and shape is similar to a button. Button cells are commonly used in cameras, watches, hearing aids, and hand-held calculators. Most button cells are constructed using mercuric oxide as the cathode, zinc as the anode, and potassium hydroxide as the electrolyte (see Figure 27–10). Although these cells are expensive as compared to other cells, they have a high-energy density and a long life. The mercury–zinc cell produces a voltage of 1.35 volts.

Another type of button cell that is less common because of its higher cost is the silver–zinc cell. This cell is the same as the mercury–zinc cell, except that it uses silver oxide as the cathode material instead of mercuric oxide. The silver–zinc cell does have one distinct advantage over the mercury–zinc cell. The silver–zinc cell develops a voltage of 1.6 volts, as compared to the 1.35 volts developed by the mercury–zinc cell. This increased voltage can be of major importance in some electronic circuits.

A variation of the silver–zinc cell uses divalent silver oxide as the cathode material instead of silver oxide. The chemical formula for silver oxide is Ag_2O. The chemical formula for divalent silver oxide is Ag_2O_2. One of the factors that determines the energy density of a cell is the amount of oxygen in the cathode material. Although divalent silver oxide contains twice as much oxygen as silver oxide, it does not produce twice the energy density. The increased energy density is actually about 10 to 15 percent. Using divalent silver oxide does have one disadvantage, however. The compound is less stable, which can cause the cell to have a shorter shelf life.

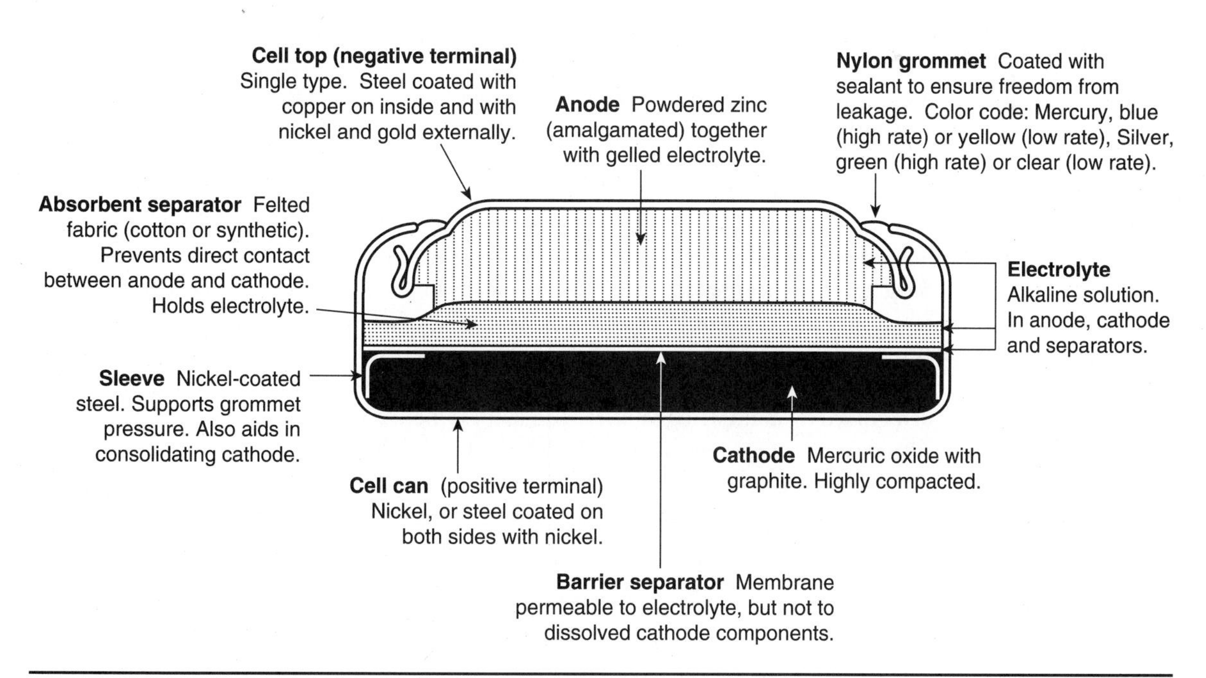

Figure 27–10 Mercury button cell.

Lithium Cells

Lithium cells should probably be referred to as the lithium *system,* because there are several different types of lithium cells. Lithium cells can have voltages that range from 1.9 to 3.6 volts, depending on the material used to construct the cell. Lithium is used as the anode material because of its high affinity for oxygen. For many years lithium had to be handled in an airless and moisture-free environment because of its extreme reactivity with oxygen. Several different cathode materials can be used, depending on the desired application. One type of lithium cell uses a solid electrolyte of lithium hydroxide. The use of a solid electrolyte produces a highly stable compound, which results in a shelf life that is measured in decades. This cell produces a voltage of 1.9 volts but has an extremely low current capacity. The output current of this cell is measured in microamps (millionths of an amp) and is used to power watches with liquid crystal displays and to maintain memory circuits in computers.

Other lithium cells use liquid electrolytes and can provide current outputs comparable to alkaline–manganese cells. Another lithium system uses a combination electrolyte–cathode material. One of these is sulfur dioxide, SO_2. This particular combination produces a voltage of 2.9 volts. In another electrolyte–cathode system sodium chloride, $SOCl_2$, is used. This combination provides a terminal voltage of 3.6 volts.

Although lithium cells are generally considered to be primary cells, some types are rechargeable. It should be noted, however, that the amount of charging current is critical for these cells. The incorrect amount of charging current can cause the cell to explode.

Current Capacity

The amount of current a particular type of cell can deliver is determined by the surface area of its plates. A D cell can deliver more current than a C cell, and a C cell can deliver more current than a AA cell. The amount of power a cell can deliver is called its *capacity.* To determine a cell's capacity, several factors must be included, such as the type of cell, the rate of current flow, the voltage, and the length of time involved. Primary cells are generally limited by size and weight, and therefore, do not contain a large amount of power.

One of the common ratings for primary cells is the milliampere–hour. A milliampere is 1/1000 of an amp. Therefore, if a cell can provide a current of one milliampere for one hour, it will have a rating of one milliampere–hour. An average D-size alkaline cell has a capacity of approximately 10,000 milliampere–hours. Some simple calculations would reveal that this cell should be able to supply 100 milliamperes of current for a period of 100 hours, or 200 milliamperes of current for 50 hours.

Another common measure of a primary cell's capacity is watt–hours. Watt–hours are determined by multiplying the cell's milliampere–hour rating by its terminal voltage. If the alkaline–manganese cell just discussed has a voltage of 1.5 volts, its watt–hour capacity would be 15 watt–hours (10,000 milliampere–hours × 1.5 volts = 15,000 milliwatt–hours or 15 watt–hours). The chart in Figure 27–11 shows the watt–hour capacity for several sizes of alkaline–manganese cells. The chart list the cell size, the volume of the cell, and the watt–hours per cubic inch.

The amount of power a cell contains not only depends on the volume of the cell, but also on the type of cell being used. The chart in Figure 27–12 compares the watt–hours per cubic inch for different types of cells.

Alkaline–Manganese Cell Size	Volume in Cubic Inches	Watt–Hours per Cubic Inch
D Cell	3.17	4.0
C Cell	1.52	4.2
AA Cell	0.44	4.9
AAA Cell	0.20	5.1

Figure 27–11 Watt–hours per cubic inch for different sized cells.

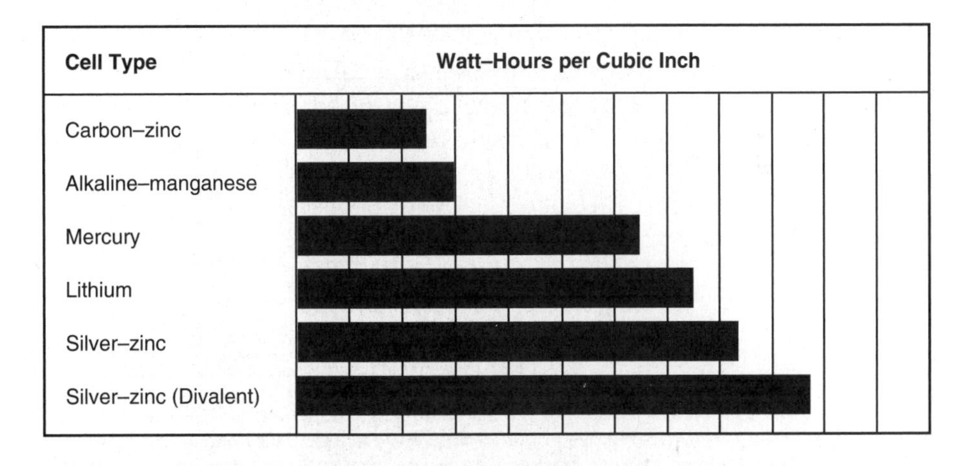

Figure 27–12 Watt–hours per cubic inch for different types of cells.

Internal Resistance

Batteries actually have two voltage ratings, one at no load and the other at normal load. The cell's rated voltage at normal load is the one used. The no-load voltage of a cell will be greater due to internal resistance of the cell. All cells have some amount of internal resistance. For example, the alkaline cell discussed previously had a rating of 10,000 milliampere–hours, or 10 amp–hours (AH). Theoretically, the cell should be able to deliver 10 amps of current for one hour or 20 amps of current for one-half hour. In actual practice, it would be found that the cell could not deliver 10 amperes of current, even under a short circuit condition. Figure 27–13 illustrates what happens when a DC ammeter is connected directly across the terminals of a D-size alkaline cell. It is assumed that the ammeter indicates a current flow of 4.5 amperes and the terminal voltage of the cell has dropped to 0.5 volts. By applying Ohm's law, it can be determined that the cell has an internal resistance of 0.111 ohms (0.5 volts / 4.5 amps = 0.111 Ω).

As the cell ages and power is used, the electrodes and electrolyte begin to deteriorate. This causes them to become less conductive, which results in an increase of internal resistance. As the internal resistance increases, the terminal voltage decreases.

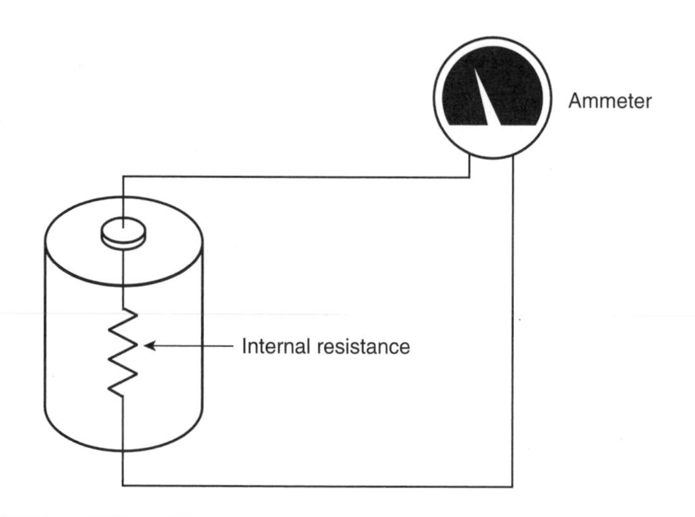

Figure 27–13 Short circuit current is limited by internal resistance.

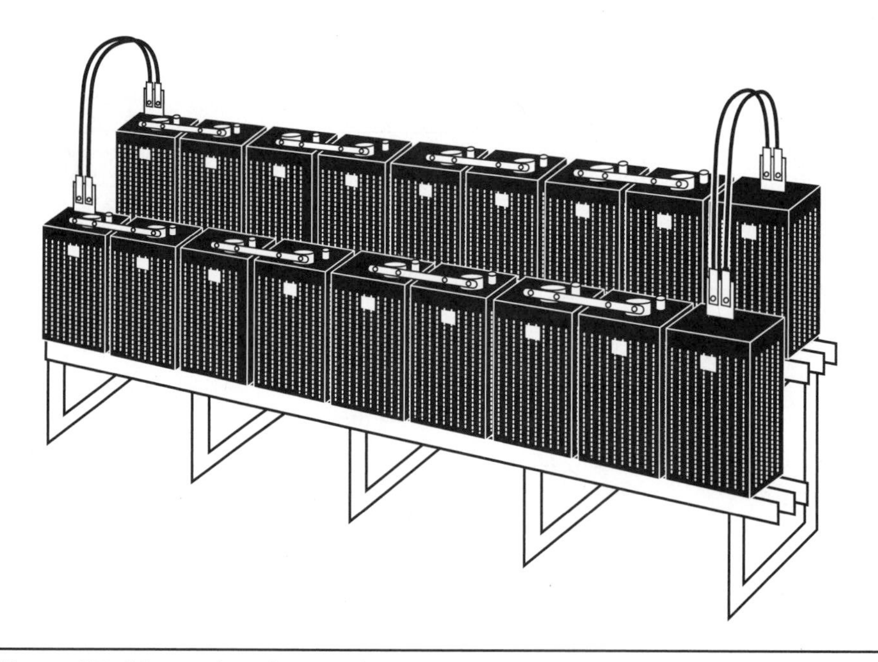

Figure 27–14 Lead–acid storage batteries.

Secondary Cells

The secondary cell is characterized by the fact that once its stored energy has been depleted, it can be recharged.

Lead–Acid Batteries

One of the most common types of secondary cells is the lead–acid cell. A string of thirty individual cells connected to form one battery, as shown in Figure 27–14. A single lead–acid cell consists of one plate made of pure lead, Pb, a second plate of lead dioxide,

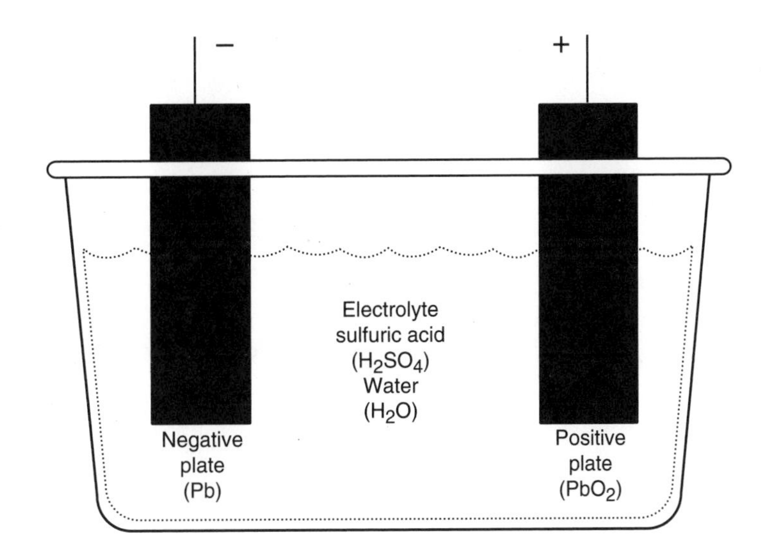

Figure 27–15 Basic lead–acid cell.

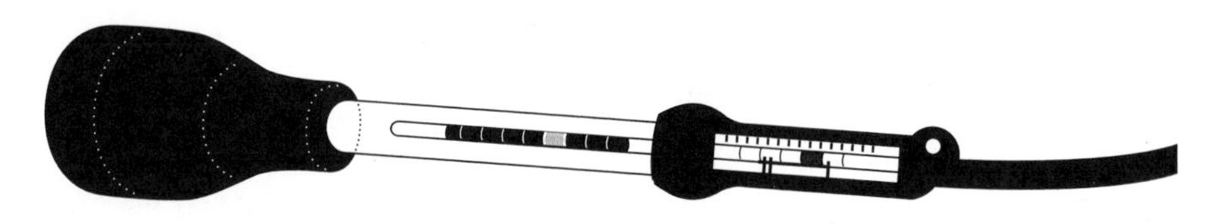

Figure 27–16 Hydrometer.

PbO_2, and an electrolyte of dilute sulfuric acid, H_2SO_4, with a specific gravity that can range from 1.215 to 1.28, depending on the application and the manufacturer (see Figure 27–15). Specific gravity is a measure of the amount of acid contained in the water. Water has a specific gravity of 1.000. A device used for measuring the specific gravity of a cell is called a *hydrometer,* shown in Figure 27–16.

Sealed Lead–Acid Batteries Sealed lead–acid batteries have become increasingly popular in the past several years. They come in different sizes, voltage ratings, amp–hour ratings, and case styles. These batteries are often referred to as *gel cells* because the sulfuric acid electrolyte is suspended in an immobilized gelatin state. This prevents spillage and permits the battery to be used in any position. Gel cells utilize a cast grid constructed of lead–calcium that is free of antimony. The calcium adds strength to the grid. The negative plate is actually a lead paste material and the positive plate is made of lead dioxide paste. A one-way pressure relief valve set to open at 2 to 6 P.S.I. (pounds per square inch) is used to vent any gas build-up during charging.

Ratings for Lead–Acid Batteries One of the most common ratings for lead–acid batteries is the amp–hour rating. The amp–hour rating for lead–acid batteries is determined by measuring the battery's ability to produce current for a 20-hour period at 80° F. A battery with the ability to produce a current of 4 amperes for 20 hours would have a rating of 80 amp–hours.

Another common battery rating, especially for automotive batteries, is cold-cranking amps. This rating has nothing to do with the amp–hour rating of the battery. Cold-cranking amps is the maximum amount of initial current the battery can supply at 20° C (68° F).

Testing Lead–Acid Batteries It is sometimes necessary to test the state or charge or condition of a lead–acid battery. The state of charge can often be tested with a hydrometer, as previously described. As batteries age, however, it will be seen that the specific gravity remains low even after the battery has been charged. When this happens, it is an indication that the battery has lost part of materials due to lead sulfate flaking off the plates and falling to the bottom of the battery, as described earlier. When this happens, there is no way to recover the material.

Another standard test for lead–acid batteries is the load test. This test probably reveals more information concerning the condition of the battery than any other test. To perform a load test, the amount of test current should be three times the amp–hour capacity. The voltage should not drop below 80 percent of the terminal voltage for a period of three minutes.

For example, an 80 AH, 12-volt battery is to be load tested. The test current will be 240 amps (80 × 3) and the voltage should not drop below 9.6 volts (12 × 0.80) for a period of three minutes.

Recharging Secondary Cells

Discharge Cycle When a load is connected between the positive and negative terminals, the battery begins to discharge its stored energy through the load, shown in Figure 27–17. The lead atoms on the surface of the negative plate each lose two electrons to become Pb^{++} ions. These positive ions attract SO_4^- ions from the electrolyte. As a result, a layer of lead sulfate, $PbSO_4$, forms on the negative lead plate.

The positive plate is composed of lead dioxide, PbO_2. Each of these molecules lacks four electrons which were given to the oxygen atom when the compound was formed to become Pb^{++++} ions. Each of the Pb^{++++} ions take two electrons from the load circuit to become Pb^{++} ions. The Pb^{++} ions can not hold the oxygen atoms which are released into the electrolyte solution and combine with hydrogen atoms to form molecules of water, H_2O. The remaining Pb^{++} ions combine with SO_4^- ions which forms a layer of lead sulfate around the positive plate also.

Notice that as the cell is discharged, two H^+ ions contained in the electrolyte combine with an oxygen atom liberated from the lead dioxide to form water. For this reason, the hydrometer can be used to test the specific gravity of the cell which is an indication of the state of charge. The more discharged the cell becomes, the more water that is formed in the electrolyte, and the lower the specific gravity reading becomes.

The Charging Cycle The secondary cell can be recharged by reversing the chemical action that occurred during the discharge cycle. This is accomplished by connecting a direct current power supply or generator to the cell. The positive output of the power supply connects to the positive terminal of the cell and the negative output of the power supply connects to the negative cell terminal, as in Figure 27–18.

The power supply must have a terminal voltage which is greater than that of the cell or battery. As current flows through the cell, hydrogen is produced at the negative plate and oxygen is produced at the positive plate. In a lead–acid cell, the cell is in a state of discharge,

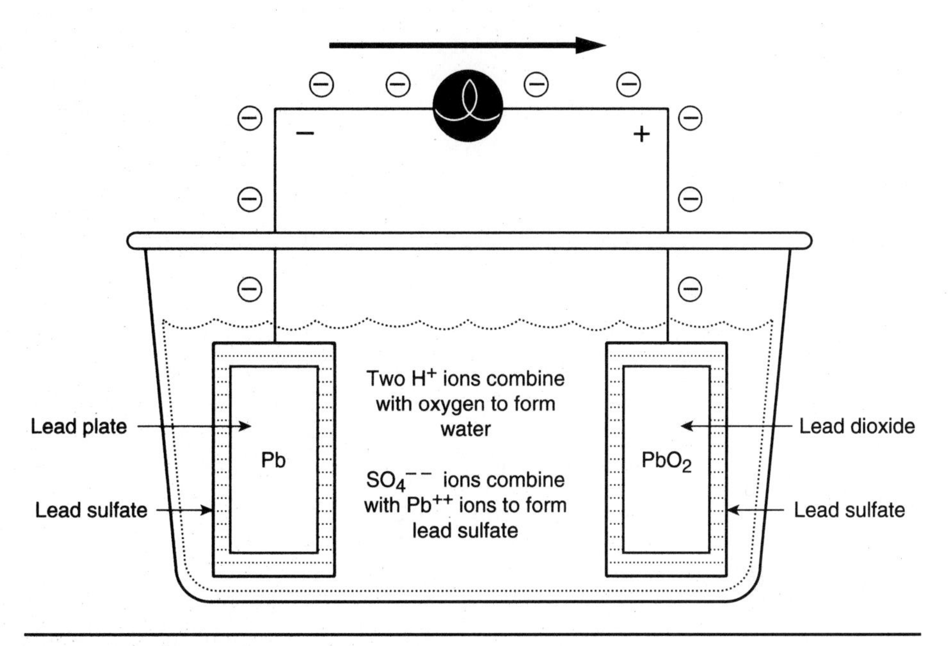

Figure 27–17 Discharge cycle of a lead–acid cell.

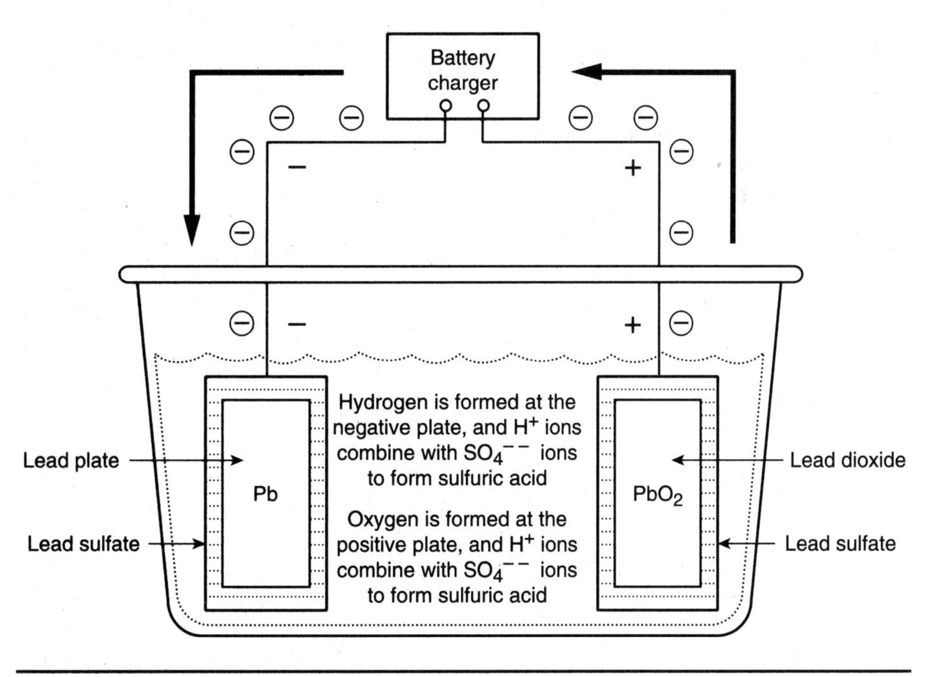

Figure 27–18 A lead–acid cell during the charging cycle.

both plates are covered with a layer of lead sulfate. The H^+ ions move toward the negative plate and combine with SO_4^- ions to form new molecules of sulfuric acid. As a result, Pb^{++} ions are left at the plate. These ions combine with electrons being supplied by the power supply and again become neutral lead atoms.

At the same time, water molecules break down at the positive plate. The hydrogen atoms combine with SO_4^- ions in the electrolyte solution to form sulfuric acid. The oxygen atom recombines with the lead dioxide to form a Pb^{++} ion. As electrons are removed from the lead dioxide by the power supply, the Pb^{++} ions become Pb^{++++} ions.

Notice that as the cell is charged, sulfuric acid is again formed in the electrolyte. The hydrometer can again be used to determine the state of charge. When the cell is fully charged, the electrolyte should be back to its original strength.

Most lead–acid cells contain multiple plates, as shown in Figure 27–19. One section of plates is connected together to form a single positive plate, while the other section forms a single negative plate. This increases the surface area which increases the capacity of the cell.

Cautions Concerning Charging Theoretically, a secondary cell should be able to be discharged and recharged indefinitely. In practice this is not the case, however. If the amount of charge current is too great, the lead sulfate does not have a chance to dissolve back into the electrolyte and become acid. High charging current or mechanical shock can cause large flakes of lead sulfate to break away from the plates and fall to the bottom of the cell. These flakes can no longer be recombined with the H^+ ions to become sulfuric acid. Therefore, the electrolyte is permanently weakened. If these flakes built up to a point that they touch the plates, they cause a short circuit and the cell can no longer operate.

Overcharging can also cause the formation of hydrogen gas. Since hydrogen is the most explosive element know, keep sparks or open flames away from batteries or cells, especially during the charging process. Overcharging also causes excess heat, which can permanently damage the cell. The accepted temperature limit for most lead–acid cells is 110° F.

The proper amount of charging current can vary from one type of battery to another, and manufacturers' specifications should be followed when possible. A general rule concerning charging current is that the current should not be greater than one-tenth the amp–hour capacity. An 80-amp–hour battery for instance should not be charged with a current greater than 8 amps.

Figure 27–19 Multiple plates increase the surface area and the amount of current the cell can produce.

Nickel–Iron Battery (Edison Battery)

The nickel–iron cell is often referred to as the Edison cell or Edison battery. The nickel–iron battery was developed in 1899 for use in electric cars being built by the Edison Company. The negative plate is a nickeled steel grid containing powdered iron. The positive plates are nickel tubes containing nickel oxides and nickel hydroxides. The electrolyte is a solution of 21 percent potassium hydroxide.

The nickel–iron cell is lighter in weight than lead–acid cells, but has a lower energy density. The greatest advantage of the nickel–iron cell is its ability to withstand deep discharges and recover without harm to the cell. The nickel–iron cell can also be left in a state of discharge for long periods of time without harm. Due to the fact that these batteries need little maintenance, they are sometimes found in portable and emergency lighting equipment. They are also used to power electric mine locomotives and electric fork lifts.

The nickel–iron battery does have two major disadvantages. One is high cost. These batteries cost several times more than comparable lead–acid batteries. The second disadvantage is high internal resistance. Nickel–iron batteries do not have the ability to supply the large initial currents needed to start gasoline or diesel engines.

Nickel–Cadmium (Ni–Cad) Battery

The nickel–cadmium cell was invented in 1898 in Sweden by Junger and Berg. The positive plate is constructed of nickel hydroxide mixed with graphite. The graphite is used to increase conductivity. The negative plate is constructed of cadmium oxide, and the electrolyte is potassium hydroxide with a specific gravity of approximately 1.2. Ni–cad batteries have extremely long life spans. On average, they can be charged and discharged about 2,000 times.

Nickel–cadmium batteries can produce large amounts of current, similar to the lead–acid battery, but do not experience the voltage drop associated with the lead–acid battery, shown in Figure 27–20.

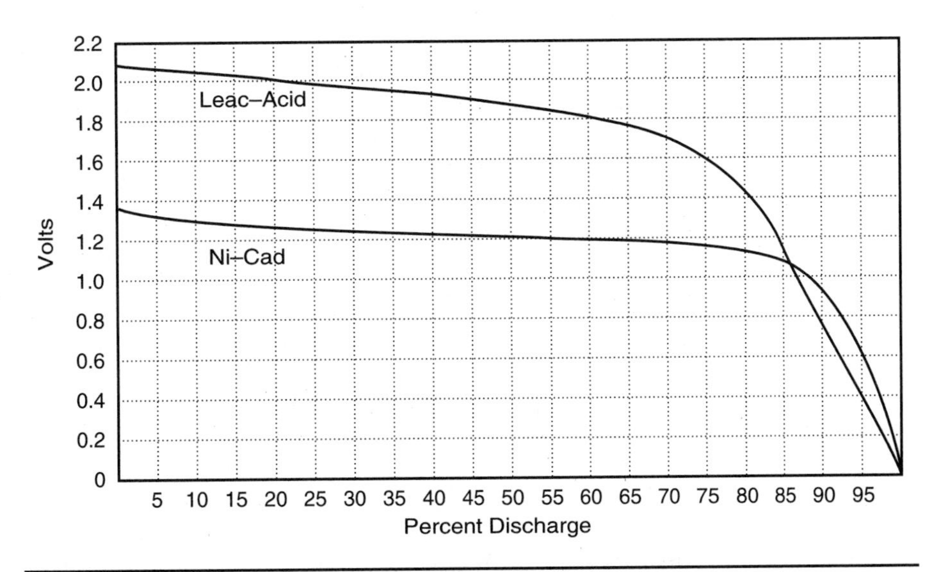

Figure 27–20 Typical discharge curves for ni–cad and lead–acid batteries.

Nickel–cadmium batteries do have some disadvantages. Some of these are as follows:

1. The ni–cad battery develops only 1.2 volts per cell, as compared to 1.5 volts for carbon–zinc and alkaline primary cells, or 2 volts for lead–acid cells.
2. Ni–cad batteries have higher initial cost than lead–acid batteries.
3. Ni–cad batteries remember their charge–discharge cycles. If they are used at only low currents and are permitted to discharge through only part of a cycle and then recharged, over a long period of time they will develop a characteristic curve to match this cycle.

Series and Parallel Battery Connections

When batteries or cells are connected in series, their voltages will add and their current capacity will remain the same. In Figure 27–21, four batteries, each having a voltage of 12 volts and 60 amp–hours, are connected in series. This has the effect of maintaining the surface area of the plates and increasing the number of cells. The connection shown in Figure 27–21 will have an output voltage of 48 volts and an amp–hour capacity of 60.

When batteries or cells are connected in parallel, as in Figure 27–22, it has the effect increasing the area of the plates. In this example, the same four batteries are connected in parallel. The output voltage will remain 12 volts, but the amp–hour capacity has increased

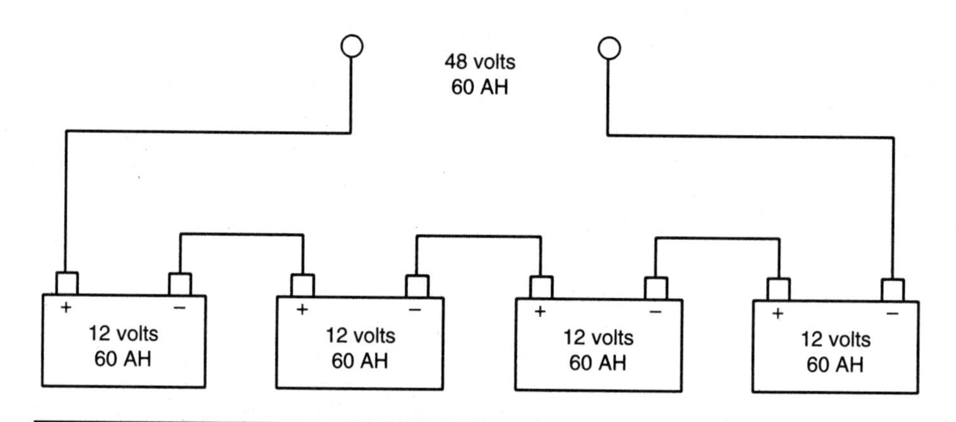

Figure 27–21 When batteries are connected in series, their voltages add and the amp–hour capacities remain the same.

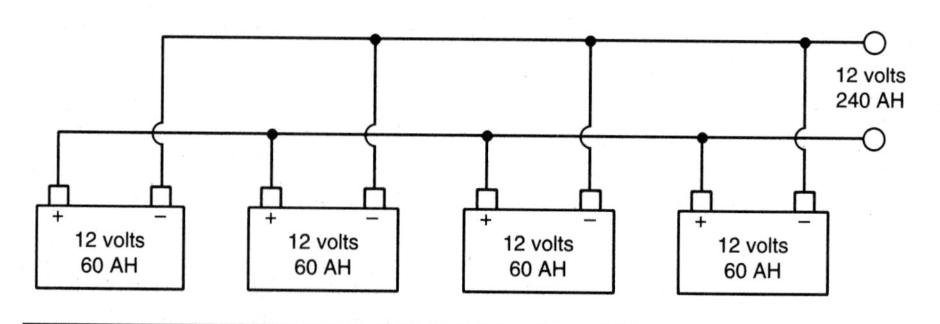

Figure 27–22 When batteries are connected in parallel, their voltages remain the same and their amp–hour capacities add.

to 240 amp–hours. It should be noted that batteries of different voltages should *never* be connected in parallel. This can cause damage to the battery and in some cases can cause one of the batteries to explode.

It is also possible to connect batteries in a series–parallel combination. In Figure 27–23, the four batteries have been connected in such a manner that the output will have a value of 24 volts and 120 amp–hours. To make this connection, the four batteries were divided into two groups of two batteries each. The batteries of each group were connected in series to produce an output of 24 volts at 60 amp–hours. These two groups were then connected in parallel to provide an output of 24 volts at 120 amp–hours.

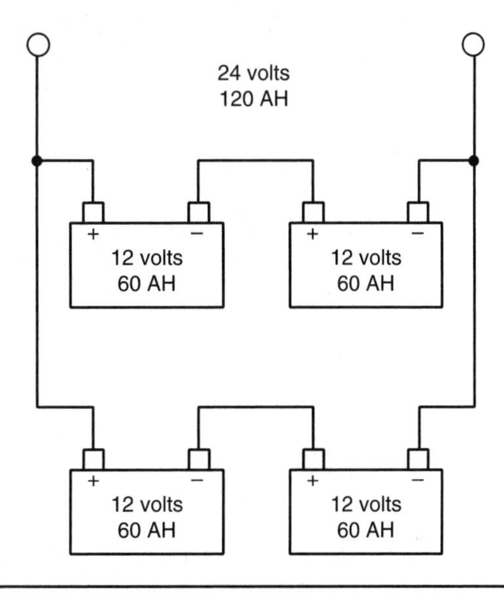

Figure 27–23 Series–parallel connection.

CHAPTER 28

Small Sources of Electricity

Solar Cells

Although batteries are the largest source of electricity after alternators and generators, they are not the only source. One source of electricity is the photovoltaic cell or solar cell. Solar cells are constructed from a combination of P- and N-type semiconductor material. Semiconductors are made from materials that contain four valence electrons. The most common semiconductor materials are silicon and germanium. Impurities must be added to the pure semiconductor material to form P- and N-type materials. If a material containing three valence electrons is added to a pure semiconductor material, P-type material is formed. P-type material has a lack of electrons. N-type material is formed when a material containing five valence electrons is added to a pure semiconductor material. N-type material has an excess of electrons.

Light is composed of particles called *photons.* A photon is a small package of pure energy that contains no mass. When photons strike the surface of the photocell, the energy contained in the photon is given to a free electron. This causes the electron to cross the junction between the two types of semiconductor material producing a voltage (see Figure 28–1).

The amount of voltage produced by a solar cell is determined by the material it is made of. Silicon solar cells produce an open circuit voltage of 0.5 volt per cell in direct sunlight. The amount of current a cell can deliver is determined by the surface area of the cell. Since the solar cell produces a voltage in the presence of light, the schematic symbol for a solar cell is the same as that used to represent a single voltaic cell, with the addition of an arrow to indicate it is receiving light, as shown in Figure 28–2.

It is often necessary to connect solar cells in series and/or parallel to obtain desired amounts of voltage and current. For example, assume that an array of photovoltaic cells is to be used to charge a 12-volt lead–acid battery. The charging voltage is to be 14 volts and the charging current should be 0.5 amp. Now assume that each cell produces 0.5 volts with a short circuit current of 0.25 amp. In order to produce 14 volts, it will be necessary to connect 28 solar cells in series. This will produce an output of 14 volts with a current capacity of 0.25 amp. To produce an output of 14 volts with a current capacity of 0.5 amp, it will be necessary to connect a second set of 28 cells in series and parallel them with the first set, as shown in Figure 28–3.

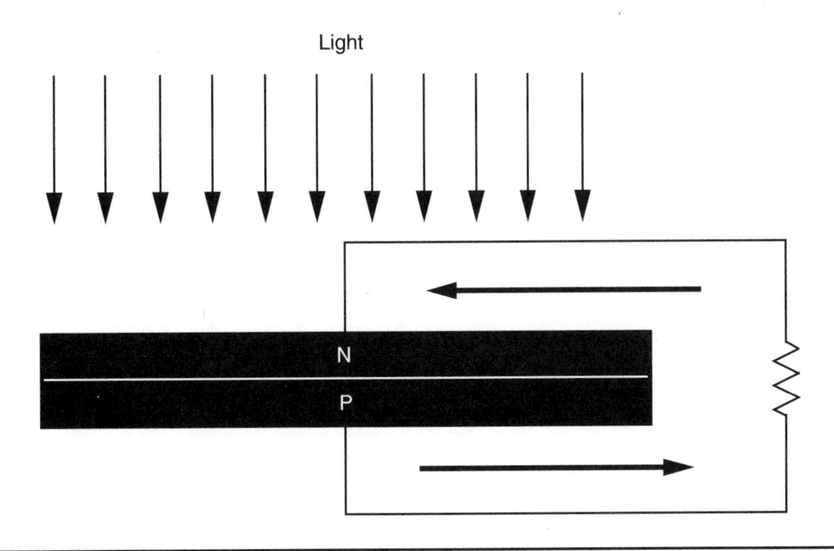

Figure 28–1 Solar cells are formed by bonding P- and N-type semiconductor materials together.

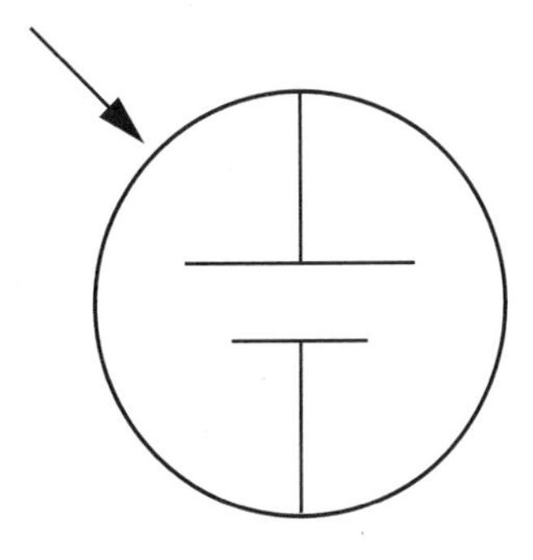

Figure 28–2 Schematic symbol for a solar cell.

Thermocouples

In 1822, a German scientist named Seebeck discovered that when two dissimilar metals are joined at one end, and that junction is heated, a voltage is produced (see Figure 28–4). This is known as the Seebeck effect. The device produced by the joining of two dissimilar metals for the purpose of producing electricity with heat is called a *thermocouple.* The amount of voltage produced by a thermocouple is determined by the type materials used to produce the thermocouple and the temperature difference of the two junctions. The chart in Figure 28–5 shows common types of thermocouples. The different metals used in the construction of thermocouples is shown as well as their normal temperature range.

The amount of voltage produced by a thermocouple is small, generally in the order of millivolts (1 millivolt = 0.001 volt). The polarity of the voltage of some thermocouples is determined by the temperature. For example, a type J thermocouple produces zero volt at about 32° F. At temperatures above 32° F, the iron wire is positive and the constantan wire is negative. At temperatures below 32° F, the iron wire becomes negative and the constantan wire becomes positive. At a temperature of +300° F, a type J thermocouple will produce a voltage of about +7.9 millivolts. At a temperature of –300° F, it will produce a voltage of about –7.9 millivolts.

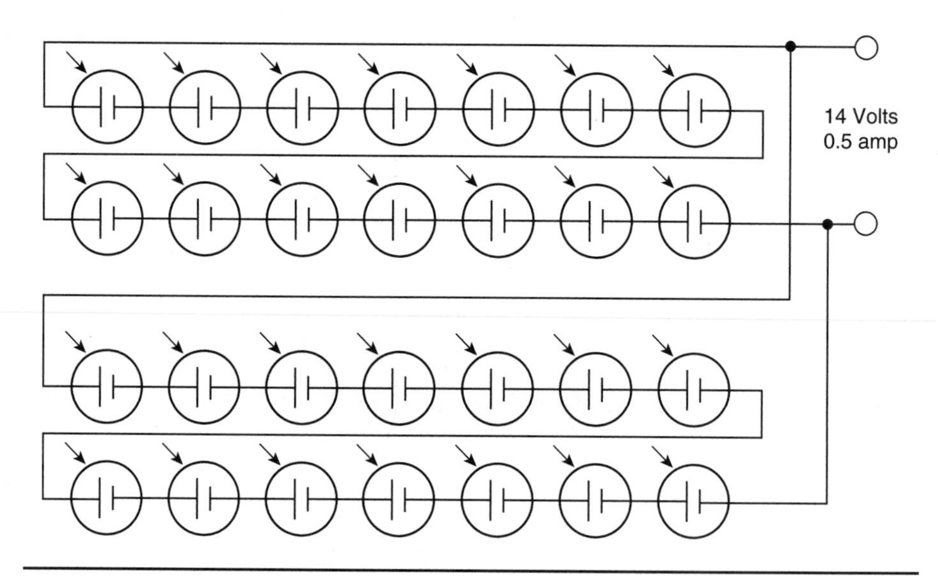

Figure 28–3 Series–parallel connection of solar cells.

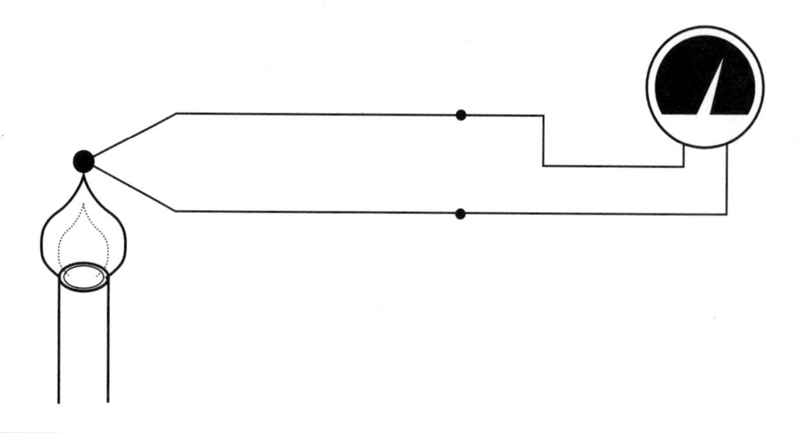

Figure 28–4 A thermocouple is made by forming a junction of two different types of metal.

Since thermocouples produce such low voltages, they are often connected in series as shown in Figure 28–6. This connection is referred to as a *thermopile.* Thermocouples and thermopiles are generally used for making temperature measurements and are sometimes used to detect the presence of a pilot light in appliances that operate with natural gas. The thermocouple is heated by the pilot light. The current produced by the thermocouple is used to produce a magnetic field that holds a gas valve open to permit gas to flow to the main burner. If the pilot light should go out, the thermocouple ceases to produce current and the valve closes, as shown in Figure 28–7.

Type	Material		Degrees F	Degrees C
J	Iron	Constantan	–328 to +32 +32 to +1432	–200 to 0 0 to 778
K	Chromel	Alumel	–328 to +32 +32 to 2472	–200 to 0 0 to 1356
T	Copper	Constantan	–328 to +32 +32 to 752	–200 to 0 0 to 400
E	Chromal	Constantan	–328 to +32 +32 to 1832	–200 to 0 0 to 1000
R	Platinum 13% Rhodium	Platinum	+32 to +3232	0 to 1778
S	Platinum 10% Rhodium	Platinum	+32 to +3232	0 to 1778
B	Platinum 30% Rhodium	Platinum 6% Rhodium	+992 to 3352	533 to 1800

Figure 28–5 Thermocouple chart.

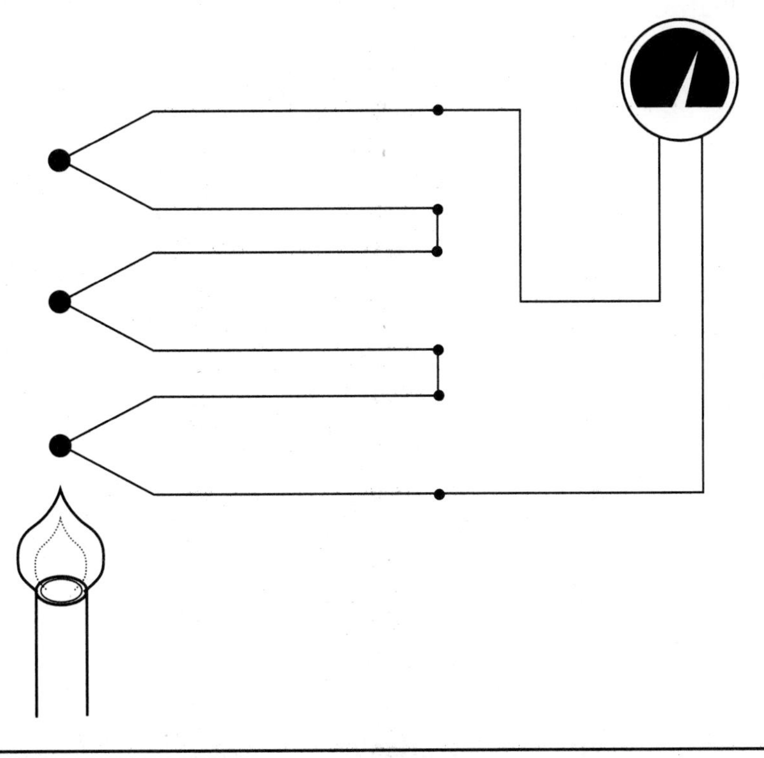

Figure 28–6 A thermopile is a series connection of thermocouples.

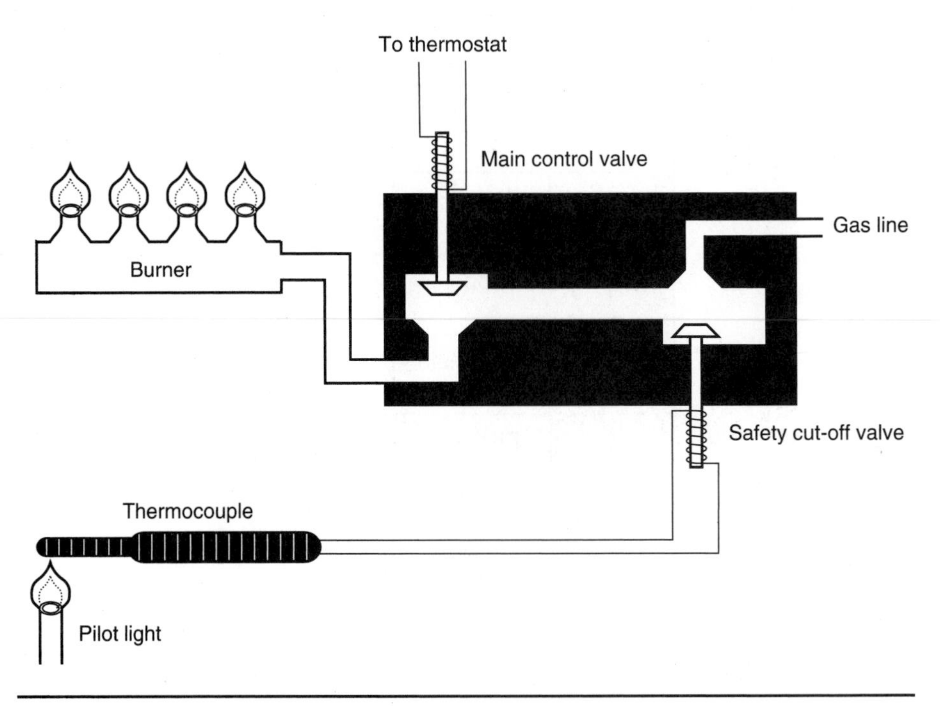

Figure 28–7 A thermocouple provides power to the safety cut-off valve.

Piezoelectricity

The word *piezo,* pronounced pee-ay′-zo, is derived from the Greek word for pressure. Piezoelectricity is produced by some materials when they are placed under pressure. The pressure can be caused by compression, twisting, bending, or stretching. Rochelle salt (sodium potassium tartrate) is often used as the needle or stylus of a phonograph. Groves in the record cause the crystal to be twisted back and forth, producing an alternating voltage. This voltage is amplified and is heard as music or speech. Rochelle salt crystals are also used as the pick-up for microphones. The vibrations caused by sound waves produce stress on the crystals. This stress causes the crystals to produce an alternating voltage, which can be amplified.

Another crystal used to produce the piezoelectric effect is barium titanate. Barium titanate can actually produce enough voltage and current to flash a small neon lamp when struck by a heavy object (see Figure 28–8). Industry uses the crystal's ability to produce voltage in transducers for sensing pressure and mechanical vibration of machine parts.

Quartz crystal has been used for many years as the basis for crystal oscillators. In this application, an AC voltage close to the natural mechanical vibration frequency of a slice of quartz is applied to opposite surfaces of the crystal. This will cause the quartz to vibrate at its natural frequency. The frequency is extremely constant for a particular slice of quartz. Quartz crystals have been used to change the frequency range of two-way radios for many years.

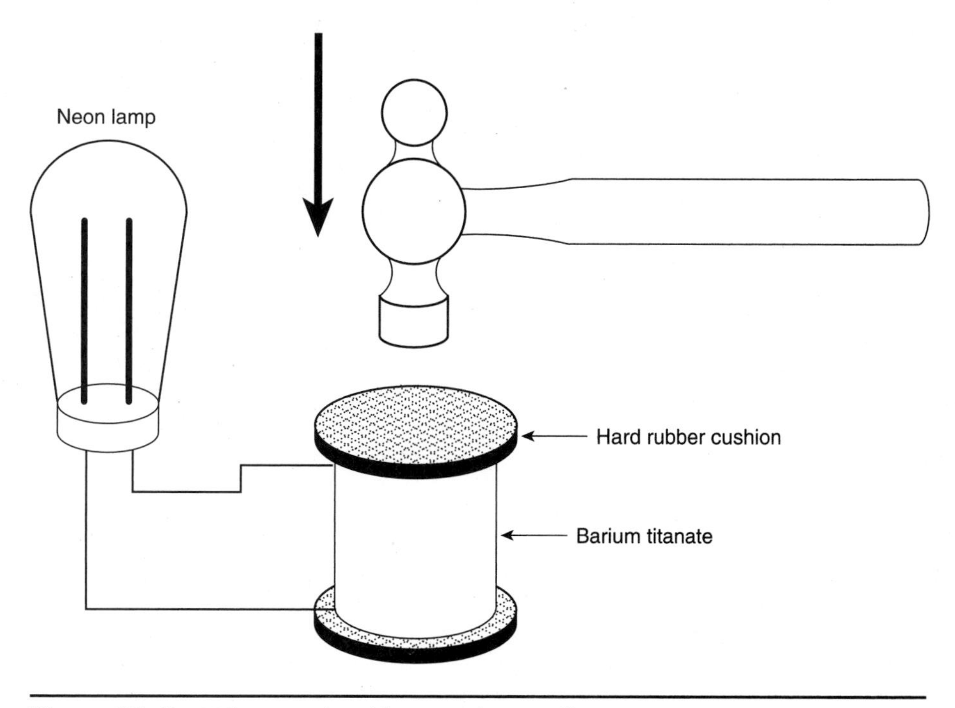

Figure 28–8 Voltage produced by piezoelectric effect.

INDEX